Lecture Notes in Mathematics

W9-ABS-017

Edited by A. Dold and B. Eckmann

968

Numerical Integration of Differential Equations and Large Linear Systems

Proceedings of two Workshops
Held at the University of Bielefeld
Spring 1980

Edited by Juergen Hinze

Springer-Verlag
Berlin Heidelberg New York 1982

Editor

Juergen Hinze
Fakultät für Chemie, Universität Bielefeld
4800 Bielefeld, Federal Republic of Germany

AMS Subject Classifications (1980): 65, 81, 34, 35, 39

ISBN 3-540-11970-1 Springer-Verlag Berlin Heidelberg New York
ISBN 0-387-11970-1 Springer-Verlag New York Heidelberg Berlin

This work is subject to copyright. All rights are reserved, whether the whole or part of the material is concerned, specifically those of translation, reprinting, re-use of illustrations, broadcasting, reproduction by photocopying machine or similar means, and storage in data banks. Under § 54 of the German Copyright Law where copies are made for other than private use, a fee is payable to "Verwertungsgesellschaft Wort", Munich.

© by Springer-Verlag Berlin Heidelberg 1982
Printed in Germany

Printing and binding: Beltz Offsetdruck, Hemsbach/Bergstr.
2146/3140-543210

Math
Sef

Introduction

Differential equations of all types play a central role in all of physical sciences. With the advent of electronic computers, methods for the numerical solution of differential equations have become highly developed and sophisticated. These developments are carried out not only by applied mathematicians, numerical analysts in general, but also by chemists, physicists and engineers. The latter, faced with specific problems, look for, adapt and develop selected methods, which appear to be most suitable for the specific physical problem at hand.

It was the purpose of the two consecutive workshops on the "Numerical Integration of Differential Equations" and "Large Linear Systems; Eigenvalue and Linear Equations" held at the Centre for Interdisciplinary Studies (ZiF, Zentrum für interdisziplinäre Forschung) of the University of Bielefeld in Spring 198o, to bring together numerical analysts and chemical physicists in order to further the progress of the numerical methods used in chemical physics by promoting the contact between these two groups so each may profit from the advances of the other. The same purpose is to be served by this volume, a proceedings of these workshops; a volume which could appear as well in a series on mathematics, chemistry or physics.

As this volume appears in a series on mathematics it was deemed appropriate to emphasize its interdisciplinary character, by beginning with those contributions, papers (1) through (1o), which focus on specific applied problems in chemical physics. Valuable additional information on the numerical methods used in scattering theory, the topic of the first 4 contributions, can be found in the NRCC Proceedings No. 5 "Algorithms and Computer Codes for Atomic and Molecular Quantum Scattering Theory", Volume I and II of the Lawrence Berkeley Laboratory (LBL-95ol, UC-4, CONF-79o696). Also the topic on chemical kinetics, approached from a mathematical point of view in the articles (9) and (1o), could be supplemented by a recent review by a chemist, D. Edelson, Science 214, 981 (1981).

The following ten contributions of this volume, articles (11) through (2o), focus on specific improvements in the methodology of integrating various types of differential equations and error estimates or bounds for such methods. The major emphasis in the procedures is on finite difference methods.

As all discretization algorithms of differential equations lead to large linear equations or eigenvalue problems, the volume concludes with eight contributions which focus on the efficient solution of such large linear systems, where the coefficient matrices are of special structure or sparse.

It is my pleasure to express my gratitude to the directorship and staff of the Centre for Interdisciplinary Studies, who made the workshops possible through generous financial and administrative assistence. My special thanks go to Mrs. K. Mehandru for her help in assembling and completing the manuscripts. But most important to the success of the workshops have been the active participants, contributing through their intellectually stimulating discussions, creating a productive dialog between chemists and mathematicians, enlivened by many limericks, thus I will close with one:

Mathematicians and Chemists at ZiF
Were prepared to discuss without tiff
Each contribution
To the solution
Of equations non linear and stiff.

Bielefeld, August 1982

The editor
Juergen Hinze

Table of Contents

AN OVERVIEW OF THE TECHNIQUES IN USE FOR SOLVING THE COUPLED EQUATIONS OF SCATTERING THEORY

by

Don Secrest
School of Chemical Sciences
University of Illinois
Urbana, Illinois 618o1

I will give a very general outline of the mathematical methods in use at present for solving scattering problems and leave the detailed exposition of a few of them to other participants. There are two major problems in applied mathematics which are at present the center of focus in scattering theory. As physical theorists we are more interested in the physical results of our investigations than in the mathematical problems. Thus the fact that so many of us are spending our time on these problems only incidentally related to our interests is an indication of their importance to us.

The first problem, the solution of the coupled scattering equations, is probably of most interest to the present audience. I shall discuss that problem first to provide the background for the second problem which I will discuss later.

Let me first write down in very formal terms the differential equation we want to solve.

$$[H_a + H_b + T + V(\eta R)]\psi(\eta,R) = E\psi(\eta,R) \qquad (1)$$

This equation describes the collision of two systems (atoms or molecules) a and b colliding with energy E. The differential operators H_a and H_b describe the isolated systems a and b. We will assume that we know all about these systems. That is we know the solutions to the eigenvalue problem

$$H_c\varphi_{cn} = E_n\varphi_{cn} \qquad (2)$$

where c is either a or b. The subscript n is a set of quantum numbers designating the state for which φ_{cn} is an eigenfunction and E_n is the eigenvalue or energy of system c in that state. The second order differential operator T describes the

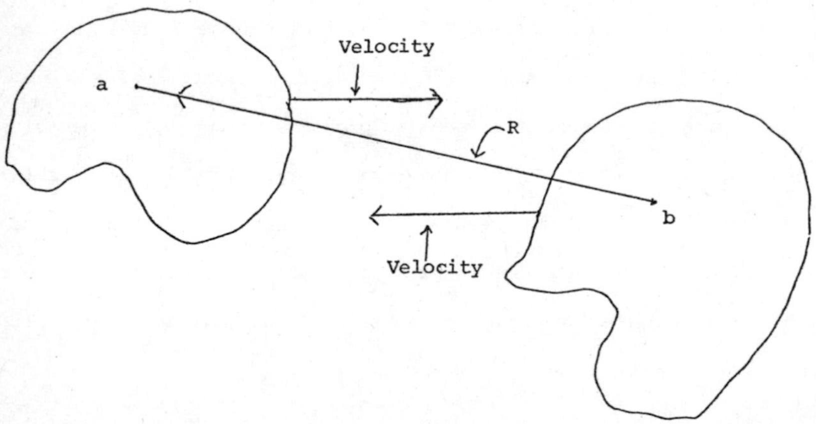

Fig. 1. Coordinates for the collision between system a and b.

kinetic energy of approach of these two systems and $V(\eta,R)$ is the energy of interaction between these two systems. The quantity R is the magnitude of the vector connecting the center of mass of the two systems as shown in fig. 1 and η is a symbol representing all of the internal coordinates of systems a and b as well as the 2 angles which describe the orientation of the vector R. When a and b are far apart, of course, they do not interact, and this implies that

$$V(\eta R) \underset{R \to \infty}{\to} 0 \tag{3}$$

Equation (1) is not an eigenvalue problem, or rather the solution sought is in the continuum and the eigenvalue E is known. The boundary condition is that ψ remain finite everywhere. Eq. (2) is a discrete eigenvalue problem, but it may also have a continuum. If the collision excites the continuum of system c it will dissociate. This is the subject of Dr. Diestlers contribution and I shall not discuss it further. In what follows I shall always assume that any continuum of the systems a and b lie at too high an energy to be excited by the collision.

By far the most common approach to solving this problem is to expand ψ in some complete set of functions of the internal coordinates,

$$\psi = \sum_n f_n(\eta) F_n(R). \tag{4}$$

Often f_n is taken to be products of φ_{ai_n}, φ_{bj_n} and $Y_{\ell_n}^{m_n}(\theta\varphi)$ where θ and φ are the angles describing the orientation of R and Y_ℓ^m is a spherical harmonic. It is often better to use some other set for f_n, but it doesn;t really matter what set is used if enough terms are carried in Eq. (4).

Substituting Eq. (4) in Eq. (1), multiplying by f_i and integrating over all of the bounded variables η we obtain the coupled equations for F_i

$$\frac{\partial^2}{\partial R^2} F_i(R) + \sum_n V_{in}(R) F_n(R) = 0 \tag{5}$$

Here I have included E and the internal eigenenergies of the systems in the coupling matrix V_{in}. This system is in general infinite, but if the f_n are chosen cleverly we can often carry enough terms to obtain the accuracy we want. Now since I have included a number of terms into the V matrix which were not in the original $V(R\eta)$ it no longer approaches zero as R approaches infinity, but it does approach a constant diagonal matrix

$$V(R)_{in} \to \delta_{in} k_n^2 \qquad (6)$$
$$R \to \infty$$

Some of the constants k_n^2 are positive and some are negative. Since ψ must remain finite the boundary condition is obvious and must be some linear combination of the following physical boundary condition

$$F_n \to \frac{1}{k_n^{1/2}} e^{-ik_I R} \delta_{In} - S_{nI} e^{ik_n R} \qquad (7)$$

for k_n^2 positive. If k_n^2 is negative the asymptotic solution must approach zero or F_n will not remain finite. The constants S_{nI} are unknown. Since I have added a subscript I to the boundary condition I will add it to F_n also and refer to it henceforth as F_{nI}. The first subscript labels the expansion function and the second labels the boundary condition.

The differential equation, Eq. (5) with boundary condition Eq. (7) is well defined but extremely difficult to solve. That is to say, it is difficult to find F. The function F is wiggly and remains finite as R approaches infinity. None of the methods I shall describe actually find F. We are in fact not interested in knowing F at all. The result we want as the solution to this problem is the matrix S_{nI} of Eq. (7). This is the so called S-matrix or scattering matrix. To explain the curious form of Eq. (7) it is necessary to discuss the physics of the problem in some detail. The S-matrix is finite since there are only a finite number of states n for which k_n^2 is positive. All of the other F_n vanish asymptotically. The factor $k_n^{-1/2}$ in Eq. (7) is placed there to force the S-matrix to be symmetric and unitary as may be easily shown to be the case using only Eq. (5) to (7). A physicist or chemist would write Eq. (7) somewhat differently to give a different phase to S than the expression given here, but the phase is unimportant and is used by scientists for historical reasons.

Before I go on I would like to give a short physical description of what is being described by these equations.

We are considering systems a and b colliding. Initially they are in states defined by the collective index I. After collision they will have a probability of being in any of a number of possible states n. What is possible is determined by the collision energy. Both a and b may be only in states such that their total energy is less than the total energy of the system initially. This is a time dependent process. It is convenient to get rid of time both theoretically and experimentally. This is done by a steady state process in which a beam of a systems is collided with a beam of b systems and the detector looks at the scattered systems and ideally determines what states the scattered systems are in. Thus we know the state of the system before collision and measure it after the collision. That is all we can know about the system; how it looked before and how it looks after the collision. The boundary condition Eq. (7) tells us this. The k_i^2 are the kinetic energies of the system in state i for the known total energy E. When k_i^2 is positive it is a physically possible kinetic energy, and there are only a finite number of these. When k_i^2 is negative there is not enough energy to excite the states i and F approaches zero asymptotically indicating that these states never reach the detector, but they must be carried in the quantum mechanical calculation. These are referred to by physicists as closed channels while the energetically accessible states are called open channels. In the boundary condition, Eq. (7) the $e^{-ik_I R}$ represents an incoming wave of unit amplitude with the systems in state I and $e^{ik_n R}$ is an outgoing wave of systems in state n. The absolute square of S_{nI} is the probability of finding the systems in states n after collision if they were in states I before. From this S-matrix we can compute cross sections, spectral line broadening and, in fact, any measurable quantity related to the collision. The s-matrix has all of the information one needs about the collision. I should mention that its phase is important. The magnitude tells us only the probability of the final state. The phase is necessary in the computation of other properties. The absolute phase of the s matrix is, of course, meaningless as Eq. (5) is a linear equation, but the relative phases of the individual elements is important.

That is probably enough physics for the present purposes, so we shall return to mathematics. The boundary condition, Eq. (7) is the physical boundary condition and is not very convenient mathematically as it would lead to complex arithmetic in solving a real differential equation. In practice the boundary condition usually used is

$$F_{nI} \sim \frac{1}{k_n^{1/2}} \sin(k_I R)\delta_{nI} + K_{nI} \cos(k_n R) \tag{8}$$

The K_{nI} matrix is usually referred to as the reactance matrix. If the K-matrix is known one may easily compute the S-matrix from it. Thus we need only real arithmetic for most methods, though I will mention some methods later which require complex arithmetic. Now the way I have developed these equations is called for historical reasons close coupling and this approach is essentially exact if one can carry enough terms. There are numerous approximations in use which are entirely different, but most of these also lead to equations of exactly the same form as Eq. (5) with boundary conditions which are cast in the form of Eq. (8). Thus the various approach I will discuss have a utility in scattering theory far beyond what I have described so far.

Solution of the Coupled Equations

There are two basically different approaches to solving Eq. (5). The first of these I shall call, for want of a better name, the approximate solution approach. This approach is the usual approach of numerical analysis in which one integrates numerically. The second approach I shall call the approximate potential approach. In this approach one approximates the potential matrix, $V(R)_{in}$, in Eq. (5) by some function for which one may integrate the equation analytically. This must usually be a simple function so one approximates the potential only over a finite range by this function, integrates a bit and then again approximates the potential over another range and continues. For each of these approaches. There are two common techniques which classify most of the methods in common use. In Table I have listed at least one method in each of these four classifications.

Table I[a]

	Approx. Soln. Approach small steps higher energy smaller little work/step must repeat for each energy	Approx. Pot large steps pot slow small steps pot fast lots of work per step stepsize little dep on E. can save work for new E	
solution follow technique	Sams & Kouri[2] (SAMS) DeVogelaere (DEVOG) Lester	Gordon[4] (GORDON) Light[5] et al	Must stabilize
Invarient embedding	Log Derivative (LOGD) Johnson[6]	R-Matrix (RMAT) Light[7] et al	No stabilization
	short range	long range	

[a]The acronyms in parentheses are programs available from NRCC.[1]

Let me start with the solution following technique in the approximate solution approach. At small R the elements of the potential matrix become large positive. The potential is trying to keep the two systems from coming too close together. The solution becomes very small in this region. Thus, most methods usually start integration here by choosing some small value for the F_{nI} and integrating stepwise into the asymptotic region. Of course since we don't know the proper values to start with, when we reach the asymptotic region we would find that we did not satisfy the boundary condition. We could iterate the solution, and I will say more about that later. The method usually used is to solve the system for a complete set of initial conditions. There are in general 2 N solutions to Eq. (5) where N is the number of states carried in the expansion Eq. (4). The condition that the solution must remain finite everywhere eliminates the N solutions which are singular at the origin. Thus it is only necessary to choose N independent solutions which grow as R increases at some small value of R. Then, when we approach the asymptotic region, we may take linear combinations of these solutions which will satisfy a complete set of boundary conditions of the form of Eq. (7). This algorithm also has difficulty. When we reach the asymptotic region we find that the solution matrix is singular. Of course, if we start with a linearly independent set of solutions they should be independent when we get to the asymptotic region. The problem is the finite arithmetic of the computer. Near the origin all of the solutions are growing rapidly. Some are growing more rapidly than others. Since we started with an arbitrary combination of solutions all contain some portion of the most rapidly growing solution. Thus as we proceed all solutions begin to look like the one which grows most rapidly at the origin and this solution is usually one of the close channels in which we have no interest at all. Thus in order to maintain linear independence it is necessary to resum the solutions from time to time to assure this property. This is referred to by scattering theorists as stabilizing. In the early days when this game was young people wrote their programs to stabilize every rth step. They found that r must be rather small and stabilization was time consuming. A number of efficient techniques for stabilizing were developed then. Later it was found that stabilization was necessary soon after starting and then a bit later and one could use increasingly larger intervals before stabilizing. Thus only four or five stabilizations are needed during a calculation if the stabilization point is well chosen and it is now a trivial part of the calculation.

Near the origin the potential is rapidly varying. As we move farther out the potential becomes more slowly varying and has a long weak tail in most practical problems. Thus near the origin the potential determines the stepsize. As the integration proceeds into the region where the potential is not so large the solution becomes wiggly and the solution determines the stepsize. That is, we must take several steps per oscillation of the most rapidly varying F_{nI}. It is in this region that the approximate potential approach becomes advantageous.

In the approximate potential approach as I mentioned earlier one approximates the potential matrix over a finite range by a simple function (a constant, a linear function or a parabola). This is still not analytically soluble unless the potential matrix happens to be diagonal. Thus, one chooses an interval, finds a transformation which will diagonalize V at some point cleverly chosen in the range and then approximates the potential by a diagonal one which is correct only at the point of diagonalization. Then he may integrate analytically through this range.

Then he chooses another range and diagonalizes again. Of course, the transformation will be different from that of the previous range and so the solution developed to this point must also be transformed the new space to match with the solution in the new range. Thus at each step there is a matrix diagonalization required and a transformation of the solution.

The approximate potential approach would thus appear to require a lot of work per step. There are advantages however. In regions in which the potential is slowly varying large steps may be taken, over several oscillations of the solutions. Another advantage of the approximate potential approach is evident when calculations must be performed at several energies E. As the energy is increased the solutions become more oscillatory requiring smaller stepsizes in the approximate solution approach. The stepsize in the approximate potential approach depends entirely on the potential and thus the same steps may be used for all energies. Furthermore the diagonalization may be done at the first energy and saved along with their corresponding transformations. These may then be used at all higher energies and the work per step at the other energies is almost the same as for the approximate solution approaches.

Another advantage becomes practical with the approximate potential approach. Since a matrix diagonalization is necessary at each step one may easily change the expansion functions, $f_n(\eta)$ of Eq. (4), at each step and use a set more appropriate to the potential for the interacting system. This has the advantage of reducing the size of the matrix V_{ni} required. I mentioned earlier that one must include several states in the calculation which lie at such a high energy they cannot be excited, the closed channels. These states die out asymptotically making no direct contribution to the S matrix, but they must be included in the calculation. Why is this so? The reason is purely mathematical. Asymptotically the $f_n(\eta)$ used in Eq. (4) are usually the states of the noninteracting system. These are an ideal representation for the system before and after collision, and their use is implied by the boundary conditions Eqs. (7) and (8). But when the two systems interact we need large numbers of these asymptotic functions to properly describe the system. John Light[5] and his collaborators were the first to take advantage of the matrix diagonalization to change basis set at every step. When this is done far fewer close channels are required and the size of the coupled system is correspondingly reduced. This in the jargon of scattering theory is referred to as an adiabatic basis set. The use of such a set is particularly important when the internal states of one of the systems supports a near continuum. In this case very likely continuum states would be needed to represent the interacting system if only the asymptotic system states are used in the expansion. The use of an adiabatic basis set works well and the continuum does not enter the problem.[8]

There are invariant embedding techniques which use both approaches. These techniques are applied by breaking the potential matrix into finite parts and solving the problem for each part, and then joining the solutions to form a solution to the whole problem. That is to say the potential is taken to be equal to the actual potential over some range of R and zero outside of that range. The scattering problem is then solved for that potential with a boundary condition like Eq. (8) at the right boundary. In this case however, it is necessary to include the closed channels also in Eq. (8). They are decaying exponentials but as the potential has just abruptly gone to zero they have not had time to decay away. The problem must be solved for a similar boundary condition on the left side of the interval. Then using these K matrices a solution for the entire problem may be pieced together exactly. The only approximation is that made in solving the individual pieces. In the approximate solution approach the potential pieces are taken so small they may be solved by a single integration step. The most commonly used method of this sort is the Log Derivative method of Johnson[6] (see Table 1) in which the integration formula used is the Simpson rule. The only approximate potential method using the invariant imbedding technique, I am aware of is the R-matrix method of John Light[7] which will be discussed later by John in some detail. In this approach the truncated potential problem may be solved analytically for the approximate potential and then the accumulation of the K matrix for the complete problem is constructed exactly, as will be described by John.

One of the great advantages of the invariant imbedding technique is that it is completely stable. There are scattering problems which are so ill conditioned that solution following techniques fail no matter how much stabilization is used. I have never known the log derivative method to fail on any problem.

The National Resource for Computers in Chemistry (NRCC) held a series of workshops last year in which various workers were brought together to compare the methods they were using on a reasonably representative set of problems. All tests were run on the same computer. The acronyms in parentheses in Table I are names of some of the computer programs used in these tests and all are available from NRCC.[1] On a typical problem no program stood out as clearly the best. On problems which were mainly short range problems with a rapidly varying potential the approximate solution approach was best. For problems where the long range potential dominated the approximate potential approach worked best. But typical problems have both kinds of regions. After the workshop a group of participants got together and produced a new program which is a hybrid of these two approaches. The log derivative method, an approximate solution approach, is used at small R where the potential is varying rapidly. Farther out an approximate potential approach is used. Both methods use invariant imbedding techniques. This program is the work of G. A. Parker, B. R. Johnson and J. C. Light.[9] It is called VIVAS and it is also available from NRCC.[1] This program should be a good general program for most scattering problems. Even this hybrid program is not the best program for all problems. One of the NRCC problems was an electron scattering problem with an exchange correction. For this problem a method using an approximate solution approach with a solution following technique was fastest. This hybrid program would have been about as fast for the second energy, but for this particular problem, the approximation was energy dependent and $V(R\eta)$ depended on E. Thus at the new energy the whole process would have to be repeated no matter what method was used.

Some Specific Problems Which Arise

One of our biggest problems at present does not concern the actual solving of the coupled equations but involves the size of the system. Before we can even begin to solve the problem we must reduce it to a reasonable size. There are a number of things we can do related to the symmetry of the problem which reduce the size of the coupled systems immensely. These procedures introduce no approximation. Even after all of these reductions are made we are often still left with a system too large to handle.

To illustrate these problems let us consider a specific problem. We will discuss an atom-diatomic molecule collision. The molecule can rotate and vibrate but we will assume that the collision energy is low enough that we can neglect electronic excitation completely. Most work in scattering theory has been done on such systems. There is a little work in the literature on polyatomic molecule collisions and diatom-diatom collisions. There is no basic difference between these systems and the one we will discuss, except that the matrices are a little larger.

We will assume that we know the solutions of the Schrodinger equation for the internal states of the rotating vibrating diatomic molecule.

$$H\varphi_{njm_j} = E_{nj}\,\varphi_{njm_j} \quad . \tag{9}$$

Here the φ_{njm_j} are the eigenfunction of the free molecule in vibration state n and rotation state j with a projection of j on the z axis given by m_j. The internal energy of this molecule only depends on n and j and is independent of the orientation of the rotation of the molecule which is described by the quantum number m_j. In fact φ_{njm_j} is given by

$$\varphi_{njm_j} = f_{nj}(r)\, Y_j^{mj}(\Omega) \tag{10}$$

where $f_{nj}(n)$ is called the radial wavefunction and depends only on r, the distance between the two atoms in the diatomic molecule and $Y_j^{m_j}$ is the spherical harmonic which depends depends on the orientation of the molecule in space. This is represented by Ω which in polar coordinates would be the angles θ_j and φ_j. Since we are neglecting electronic excitations in this problem the atom is considered to have no internal structure. This is an extremely good assumption for many atom-molecule systems at low enough energy. Thus no H_b occurs in Eq. (1), only H_a which is the H of Eq. (9). We may now expand the ψ of Eq. (1) according to Eq. (4) as

$$\psi = \sum_{njm_j\ell m} f_{nj}(r)\, Y_j^{m_j}(\Omega)\, Y_\ell^{m}(\theta\varphi)\, F_{njm_j\ell m}(r) \qquad (11)$$

Here θ and φ describe the orientation of the vector R between the center of mass of the molecule and the atom. Substituting ψ from Eq. (11) into Eq. (1) and multiplying by coefficients of F in Eq. (11) we obtain Eq. (5) were the V matrix is given by

$$V_{ii'}(R) = V_{njm_j\ell m,n'j'm'_j\ell'm'} + \left(\frac{\ell(\ell+1)}{R^2} - k_{jn}^2\right)\delta_{ii} \qquad (12)$$

where

$$k_{jn}^2 = E - E_{nj} \qquad (13)$$

The size of the $V_{ii'}$ matrix, Eq. (12) depends on how many terms we must carry in the expansion, Eq. (11) and this in turn depends on what system we choose to study. Let us consider a very simple case where the molecule is H_2 and the atom is He and the energy is just enough to excite the second vibrational state of H_2. For this case we have found that the first five vibrational states n is enough to give answers correct to 3 decimal places. Thus there are 5 values of n. For each of these we must carry up to J=8 rotation states. The H_2 molecule, being homonuclear has only even or odd rotation states. Thus for even states this means we need only carry j = 0, 2, 4, 6 and 8, for each vibrational state. The quantum number m_j' runs from -j to j. Thus there is one j=0 state 5 j=2 states and in general 2j+1 m_j states or each j, or 45j states for each of the 5 vibration states for a total of 225 rotation-vibration states. From calculation we find that we may truncate the sum over states at around ℓ = 100. Again for each ℓ/state there are $2\ell+1$ m states. Furthermore all ℓ states must be included giving a total of $\sum_{\ell=0}^{100} (2\ell+1) = 101^2$ ℓ m states for each of the 225 rotation-vibration states. This means that the V matrix for this system of 2295225 x 2295225 is accurate enough to give us a three figure answer. This is clearly a rather large problem, and numerical analysis alone is not going to solve it. We have not as yet used any of the physical properties of the system to simplify the problem.

We have stated the problem in the center of mass coordinate system, which is already a simplification. But we could have chosen a different orientation for the coordinate system. If we had the differential equation it would be entirely unchanged. It is clear that the results of an isolated collision in space should not depend on the orientation we chose for our coordinate system and neither should the equations which describe it. But the basis functions we used in Eq. (11) do depend on the coordinate system. If we use a basis set expansion we are stuck with this problem. However from group theory we know that if our basis functions transform like an irreducible representation of the full rotation group, there will be no mixing between representations. Thus instead of using a simple product of the spherical harmonics we should use linear combinations of these which transform like irreducible representations of the full rotation group. These linear combinations are well known from group theory and are given by

$$\mathcal{Y}^{JM}_{j\ell}(\Omega,\theta,\varphi) = \sum_{m_j m} (jm_j \ell m|JM)\, Y^{m_j}_{j}(\Omega)\, Y^{m}_{\ell}(\theta\varphi) \tag{14}$$

The coefficients $(jm_j \ell m|JM)$ are the well known Clebsch-Gordan coefficients of group theory. Using these functions instead of simple products of the spherical harmonics in Eq. (11), Eq. (5) becomes

$$\left[-\frac{\partial^2}{\partial R^2} + \frac{\ell(\ell+1)}{R^2} - k^2_{jn}\right] G^J_{nj\ell} + \sum_{n'j'\ell'} V^J_{nj\ell n'j'\ell'}\, G^J_{n'j'\ell'} = 0 \tag{15}$$

The $G^J_{nj\ell}$ are, of course, just linear combinations of the $F_{njm_j\ell m}$. We have simply performed a linear transformation on Eq. (11) which made the potential matrix block diagonal. It is now diagonal in the quantum numbers J and M. These quantum numbers have a physical meaning. J is the total angular momentum of the whole system. That the equation is block diagonal in J tells us that the total angular momentum of the system is conserved. The M quantum number tells us the projection of the total angular momentum on the z axis. The matrix is still 2295225 x 2295225 but now it consists of 10000 uncoupled blocks labelled by J and M. The J runs from 0 to 100 and M runs from -J to J. Thus there are 2J+1 different Ms for the blocks with quantum number J. Now we notice that the differential equation Eq. (15) does not depend on M. This is reasonable since the result of the collision should not depend on the orientation of the angular momentum vector. Thus the 2J+1 different M blocks are all identical for each J and furthermore their solutions are identical. There are then only 100 different blocks each of which may be solved independently. The ℓ subscript n a particular J block now runs from $|J-j|$ to J+j. Thus for J large enough there are only 2j+1 different ℓs associated with each j in the block. A further simplification comes from the cylindrical symmetry of a diatomic molecule from which we may show that states with ℓ even do not interact with states of odd ℓ. Thus each J block breaks into two diagonal blocks one of even ℓ and one of odd ℓ. The largest of these blocks has $\ell = J-j, J-j+2, \ldots J+j$. For the problem we described then $j=0$ $\ell=J$ is the first state $j=2$ $\ell=J-2$, $j=2$ $\ell=J$, $j=2$ $\ell=J+2$ are the next three states and in all there are 25 $j\ell$ states for each of the 5 vibration states for a 125 x 125 matrix. The matrix of the other parity in ℓ is slightly smaller and when J is less than j the matrices become even smaller. Using the physical symmetries of the problem and a little elementary mathematics we have reduced the coupled system from a 2295225 x 2295225 to 200 systems of 125 x 125 or less. This problem is soluble, but this pushes the limits of what we can solve. If we substitute N_2 for H_2 the problem gets much worse. At about the same energy the vibration remains much the same and 5 vibration states is enough. Since N_2 is heavier however, the rotation levels are closer together. Whereas with H_2, up to j=8 is enough for N_2 we need to go to j=30. This leads to matrices of 1000 x 1000 after all simplifications are made. I feel that at this point in the development of scattering theory, it is the size of the problem rather than numerical techniques which are the stumbling block. There are still a few things we can do. The methods I have discussed require us to find a complete set of solutions to the problem. In the solution following technique we need the complete set in order to stabilize, and in the invariant imbedding technique a complete set of solutions is intimately involved in the method. In practice we are usually interested in only one column of the S matrix or at most a few. Recently an attempt has been made to find only one column of the S matrix by iteration.[10] Clearly if the integration is started near the origin the violently growing solutions which introduced the need for stabilization in the solution following technique will defeat any iteration starting at the origin. Thomas[10] starts his iteration in the asymptotic region and integrates toward the origin where all physical solutions are decaying. There are still problems. For one the complex boundary conditions are necessary if only one column of S is to be computed. Thomas has found a way to overcome the problem of solutions which become singular at the origin and has been able to solve for one of the blocks labelled by J of size of the order of 210 x 210. This is the largest problem ever to be solved. A complete problem has not been solved by this technique as yet (a complete set of J states) and it remains to be seen how valuable this approach will be, and whether it will be able to handle 1000 x 1000 problems or not.

So much for exact methods of solving the problem. At present the most viable approach seems to be approximation. And the most promising approach to approximation for these problems seems now to be the dimensionality reducing approximations pioneered by Herschel Rabitz.[11] Since Rabitz first work a number of workers have developed improved techniques which work well in many, but not all, cases.

The philosophy of this approximation goes as follows. In vibration problems the difficulty comes from the large number of rotation states. In particular the large amount of ℓ coupling, the 2j+1 ℓ's associated with each j state, which comes from the transitions between the 2j+1 orientations of the j state. In most beam experiments these orientations are ignored in any case and only the vibrational state n and the rotational state j are measured. Thus if the orientation coupling is ignored the problem is diagonalizable into still smaller blocks. This works as follows. There is a linear transformation over ℓ in Eq. (15) which makes the V matrix diagonal in ℓ and leads to the equation,

$$[- \frac{\partial^2}{\partial R^2} + \frac{J(\Lambda,Jj)}{R^2} - k_{jn}^2] g_{jn}^{\Lambda} + \sum_{n'j'} V_{jnj'n'}^{\Lambda} g_{j'n'}^{\Lambda} + \frac{J^+(\Lambda+1,Jj)}{R^2} g_{jn}^{\Lambda+1} \tag{16}$$

$$+ \frac{J^-(\Lambda-1Jj)}{R^2} g_{jn}^{\Lambda-1} = 0$$

The ℓ quantum number is now gone and is replaced by the diagonal quantum number Λ. The g_{jn}^{Λ} also should have a J index but I draped it for simplicity. Now for H_2-He the V and G matrices in Eq. (15) were 225 x 225 for each J. In Eq. (16) for $\Lambda = 0$

and J the V and g are 25 x 25 matrices. There are 9 blocks all told. The sizes for $\Lambda = 0, \ldots, 8$ are 25, 20, 20, 15, 15, 10, 10, 5 and 5. Eq. (16) is exact and now the coupling has moved from V to the last two terms. The approximation consists of dropping these terms.[12] Then the equations are decoupled and since the effort in solving a system goes like the cube of the matrix size we may solve the new system in 1/30 of the time required for the original system. For the N_2 problem where Eq. (15) gave approximately 1000 x 1000 the largest matrix one would have to solve using the approximation of Eq. (16) is 75 x 75. This brings the problem into the realm of possibility. For H_2-He which can be solved exactly this approximation works extremely well. The elastic transitions are given almost exactly. The transitions for large rotation state become worse as the change in rotation state becomes greater, but at its worst the error is only about 6% and this large error is for the least important (ie. the smallest cross section) transitions. Thus, this appears to be a truly excellent approximation. As the energy of the system increases k_{jn}^2 gets larger and it is argued that the neglect of the $J\pm/R^2$ terms become even more negligible. This isn't a very convincing argument, but it is verified that the approximation does indeed improve as the energy increases up to the highest energy for which exact calculations can be made. If one looks only at the vibrational transitions, summed over all final rotation states the approximation is even better. In the N_2-He problem the rotation states are so close together that many beam experiments can only measure the vibrational transition summed over rotation states. Thus this is expected to be a good approximation for this problem, but we don't really know because the N_2-He or any other problem involving heavy molecules like N_2 cannot be solved exactly for vibrational transitions.

This approximation is usually called the CS approximation which means coupled states or centrifugal sudden depending on whom you are talking to.[12] It does not solve all of our problems however. When I say it gives good answers for H_2-He, I mean for calculations in which one does not look at the orientation of the rotation states. If the individual oriented states are studied, and it is possible to do this experimentally, the CS approximation gives much worse results, as it must since it is the coupling between these states which is neglected. Even for these transitions the approximation is not terrible, but that is merely an indication that for this case the coupling is not important and that its neglect gives good results is not surprising. For a similar problem, the scattering of Li^+ from H_2 the CS approximation is not good. This has been explained as follows. It is a fact that the lithium ion, Li^+, interaction potential with H_2 is much longer range than that of He and thus we must carry many more J states. That is we must solve many more blocks at much higher J. The term we neglect in the CS approximation get large as J gets large. Thus the approximation is felt to be bad for long range potentials.

Recently calculations[13] were made both exactly and in the CS approximation for Ne-HD. The interaction potential of He with H_2 and Ne with HD are similar. Since deuterium is twice as heavy as hydrogen the center of mass of HD does not lie at the geometrical center of the molecule. Thus the reflection symmetry of the molecule is lost and both odd and even transitions become allowed. In this case the coupled states approximation is very bad. We do not know why. Another calculation[14] for which the CS approximation was not good was Ar-CH_4. Methane is a heavy polyatomic molecule, but only the hydrogens contribute to its moment of inertia. As a result its rotation levels are widely separated and exact calculations are possible. As in the H_2 case the interaction potential is relatively short ranged and from experience with diatomics the CS approximation was expected to be excellent. We do not know why it fails here. There are a number of other cases where this approximation fails also. What is the point of my discussion of the coupled states approximation? This is just one of a number of these dimensionality reducing approximations.[11b] There are more that I haven't discussed. The point I want to make is this, the scattering problem for any reasonable system leads to a large set of coupled equations. Before we can even hope to solve the system we must use every analytical tool we can to reduce the set to a reasonable size. We have developed methods which we can afford to use to solve up to the order of 100 coupled equations. With these numerical methods we may solve a few simple systems exactly, but most of the problems we are interested in are too large to handle. There are numerical methods in the works which might double or triple the size of matrix we can handle. But this is not enough. Evan the simple N_2 problem I discussed is an order of magnitude larger and completely beyond our capability of handling exactly. Some more interesting problems which we would like to solve are two to three orders of magnitude larger. The point is then that I think we have just about reached the limit of what we can do with numerical methods. I feel that the next step is the use of applied mathematics to reduce the size of the problem to the point where our methods can handle them. These dimensionality reducing approximations appear to be the proper approach. They lead to equations of the form we know how to solve. They are often good approximations. But we do not yet know how to predict under what set of circumstances they are good. And often we do know when the CS approximation is not good, but we do not know what we can do in that case. What we need are approximations we can trust for each situation. We need to know how accurate the approximation is and also what unknown features of the problem it is likely to miss.

I have tried in this overview to give you an idea of what our problem is, how large it is and some of the problems encountered. I have discussed the types of numerical approaches we are using to solve the coupled equations. I have stressed the point that the systems we want to solve are too big to handle by our present methods and I have indicated my belief that in addition to the development of numerical methods capable of handling bigger systems there is a need for reliable approximations to reduce the size of the systems we need to solve.

I would like to end by saying a few words about my own work. What I will say now doesn't fall into the category of an overview, but is an indication of a possible approach to developing better approximations and to evaluating the ones we now have.

The coupled states and many other of the dimensionality reducing approximations which I have not discussed consist of replacing our known problem with a slightly modified problem-neglecting the $J^{\pm}$ terms in Eq. (16) is the case of the CS approximation. Then we proceed to solve this new reduced problem accurately numerically. We may rewrite Eq. (16) by moving the $J^{\pm}$ terms to the right hand side and write it in matrix notation as

$$\underset{\sim}{H}_o \underset{\sim}{g} = \underset{\sim}{J} \underset{\sim}{g} \tag{17}$$

This is an exact equation. The CS approximation consists of solving

$$H_o g_o = 0 \qquad (18)$$

where g_o is the coupled states solution. A next approximation we can make is to solve the equation,

$$H_o g_1 = J g_o \qquad (19)$$

Equation (19) is an inhomogeneous equation for which we know a complete set of homogeneous solutions. It can in theory be solved though it is not as easy as it looks. My research group has recently developed some efficient approaches to solving this equation.[15] We do not have much experience with this approach yet, but in the cases we have tried the improvement in the results are dramatic. This allows us not only to arrive at a better approximation, but there are many problems for which g_o may be good, but the system is too large to solve accurately. From the solution to Eq. (19), g_1, we can get an idea as to how accurate g_o is. You will remember that the solution to Eq. (17), the exact problem consisted of solving on the order of 100 uncoupled blocks of 100 x 100 equations. The solutions to Eq. (19) could be used to test only a few of these to see if g_o is accurate enough. If not the whole set could be computed much more efficiently than solving the entire system. This approach gives us not only a better approximation, but allows us to study approximations we know.

Acknowledgment

This work was supported in part by a grant from the National Science Foundation.

References

1. National Resource for Computation in Chemistry software library, Lawrence Berkeley Laboratory, University of California, Berkeley.

DEVOG	cat. no. KQ11	by L. D. Thomas
GORDON	KQ05	by M. H. Alexander
LOGD	KQ06	by B. R. Johnson
RMAT	KQ09	by T. G. Schmalz
SAMS	KQ07	by K. D. McLenithan and D. Secrest
VIVAS	KQ04	by G. A. Parker, J. V. Lill, and J. C. Light

2. W. N. Sams and D. J. Kouri, J. Chem. Phys. 51, 4809 (1969), see also D. Secrest in Methods of Computational Physics Vol. 10, p. 243, Ed. B. Alder, S. Fernbach, and M. Rotenberg, Academic Press, New York (1971).

3. W. A. Lester, J., J. Comp. Phys. 3, 322 (1968), Methods of Computational Phys., loc. cit., p. 211.

4. R. G. Gordon, J. Chem. Phys. 51, 14 (1969), Methods of Computational Phys., loc. cit., p. 81.

5. C. C. Rankin and J. C. Light, J. Chem. Phys. 51, 1701 (1969), G. Miller and J. C. Light, J. Chem. Phys. 54, 1635 (1971), J. C. Light, Methods of Computational Physics, loc. cit., p. 111.

6. B. R. Johnson, J. Comp. Phys. 13, 445 (1973).

7. J. C. Light and R. B. Walker, J. Chem. Phys. 65, 4272 (1976).

8. A. L. Scherzinger and D. Secrest, J. Chem. Phys. 73, 1706 (1980).

9. G. A. Parker, J. C. Light, and B. R. Johnson, submitted for publication.

10. L. D. Thomas, J. Chem. Phys. 70, 2979 (1979).

11. (a) H. Rabitz, J. Chem. Phys. 57, 1718 (1972). (b) See also Modern Theoretical Chemistry, Vol. I, p. 33. Edited by W. H. Miller, Plenum Press, New York (1976).

12. For a review of these methods see D. J. Kouri in Atom Molecule Collision Theory, Chap. 9, p. 301. Edited by R. B. Bernstein, Plenum Press, New York (1979).

13. U. Buck, F. Huisken, J. Schleusener, and J. Schäfer, J. Chem. Phys. 72, 1512 (1980).

14. T. G. Heil and D. Secrest, J. Chem. Phys. 69, 219 (1978).

15. K. McLenithan and D. Secrest, J. Chem. Phys. 73, 2513 (1980).

WEYL'S THEORY FOR SECOND ORDER DIFFERENTIAL EQUATIONS AND ITS APPLICATION TO SOME PROBLEMS IN QUANTUM CHEMISTRY

by

ERKKI BRÄNDAS

QUANTUM CHEMISTRY GROUP
FOR RESEARCH IN
ATOMIC, MOLECULAR AND
SOLID STATE THEORY
UPPSALA UNIVERSITY
S-751 20 UPPSALA
SWEDEN

ABSTRACT

Weyl's complex eigenvalue theory is examined with respect to analyticity properties of solutions and associated Green's functions. Numerical aspects are discussed and some applications in quantum chemistry reviewed.

I. INTRODUCTION

In the present workshop on "Numerical Integration of Differential Equations" both initial value problems as well as boundary value problems have been extensively discussed. In the field of quantum chemistry both types occur. In a scattering theory formulation of chemical reactions for instance the problem is to find an asymptotic oscillatory solution of a differential equation with a given initial value (and eigenvalue parameter), while the traditional bound state formulation of quantum chemistry and solid state physics is the problem of finding an eigenvalue for which the solution fulfils prescribed values at the boundaries. In most textbooks of quantum mechanics these aspects are usually presented in terms of spectral decompositions of abstract operators in hilbert space. Although this theory is well-known the general many-body problem is of such a complicated nature that approximations are necessary. Although numerical difficulties may be formidable there exist methods of various degrees of accuracy that may be employed. However, if one wants to treat the dynamics of the system, it may nevertheless not be sufficient to look at the eigenvalue equation only, but rather one needs to investigate the whole differential equation with respect to analyticity properties etc. A careful investigation of the latter will determine the appropriate time and/ or temperature behaviour of our system.

In order to stay within such a general framework, we will study the simple second order linear differential equation from a slightly more general view-point than usual. This lecture will therefore be concerned with a careful analysis of analytic properties of the solutions, associated Green's functions, properties of the potential, various types of limiting procedures, etc. We will employ a theory, due to Weyl [1], which preceeded both the development of the hilbert space as well as Schrödinger's [2] original paper "Quantization as an eigenvalue problem".

We have previously applied Weyl's theory to the Stark effect in the hydrogen atom [3], to predissociation in mercury hydride [4,5] and analysed the tunnelling problem in general.[6] In the first application, i.e. to the hydrogen atom in an electric field, it was found that the spectral function directly obtained from the Weyl-Titchmarsh formula [7,8], see below, could be analytically continued into a higher order Riemann sheet. The uniqueness of the spectral function and its continuation yields a complex pole corresponding to a metastable state defined by its position and lifetime. It is interesting to point out that eventhough the Stark hamiltonian does not fall in the category of dilatation analytic operators[9,11], i.e. essentially containing two body potentials that are $(\Delta-\varepsilon)$- bounded. Our analyticity studies of the Green's function, extensions of the Hellman-Feynman theorems etc. could be rigorously justified from Weyl's limit-point theory [12]. For the definition of limit-point-limit-circle classifications (the Stark hamiltonian belongs to the former) see below.

In the HgH study tunnelling probabilities through the centrifugal barrier arising from the angular separation of the Schrödinger equation for the center of nuclear masses for large enough rotational quantum numbers K were calculated using a potential due to Stwalley[13]. See also fig. 1. Experimentally, the spectral lines appear to be broadened by an amount corresponding to the probability of tunnelling through the potential barrier. See fig. 2. Following the classical terminology the width Γ is related to the complex pole $(E_{res} - i\Gamma/2)$ and to the lifetime of the metastable state by the uncertainity relation $\tau = \hbar/\Gamma$. In particular it was found that the resonances with K=9 and vibrational quantum numbers v=3 and v=4 were so close, see table IV, separated only by about 20 cm^{-1}, that careful analytic continuation procedures of the full complex Green's function were needed [5], see also figures 3 and 4.

A recent application to non-rotational potential barrier problems [14] like for instance the energy-curve of the B $^2\Sigma^-$ state of the CH-radical see e.g. fig. 5 showed good consistency between calculated linewidths and experimental life-time measurements. This consistency includes enchancement by the centrifugal term giving rise to predissociation by rotation and orbiting resonance phenomena.

After a review of the various steps in the Weyl-Titchmarsh development, we will discuss some of the numerical aspects, i.e. treatment of origin and infinity, choice of numerical techniques, accuracy, timing and general strategy of a resonance calculation, as well as some comments on the future development.

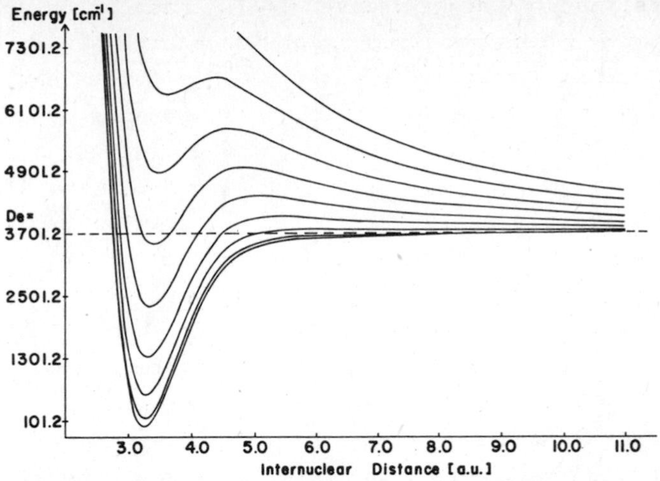

FIG. 1. Effective potential for $K = 0, 5, 10, 15, 20, 25, 30, 35, 40$, between $R = 2.0$ and $R = 11.0$ a.u.

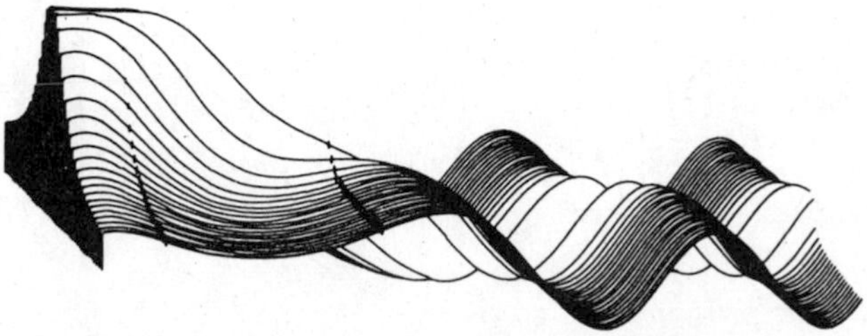

FIG. 2. Wavefunctions $\psi(E, R)$, normalized to unit asymptotic amplitude according to Eqs. (65)–(69) for $3.5 \leq R \leq 10$ a.u. and $5311 \leq E \leq 5315.5$ cm^{-1}. The energy is varied from the front to the back. The classical turning points are marked. Note the oscillating behaviour of $\psi(E, R)$ outside the right turning point and the sudden change of phase in the resonance region.

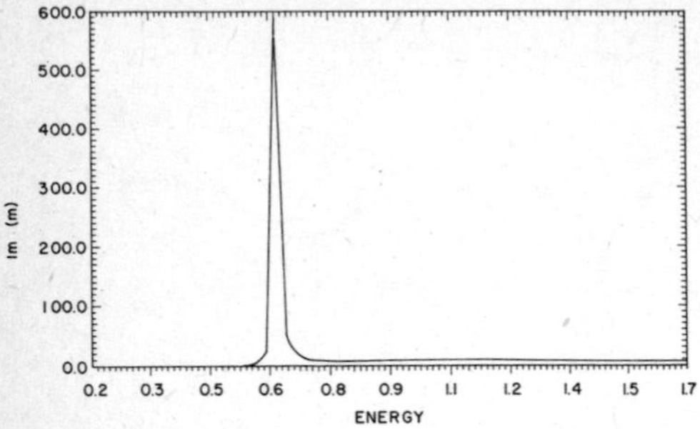

FIG. 3. Imaginary part of m function for $K=9$. Note sharp peak at $E\sim 0.66$ m.u. associated with $v=3$ and the very shallow maximum to the right for $v=4$.

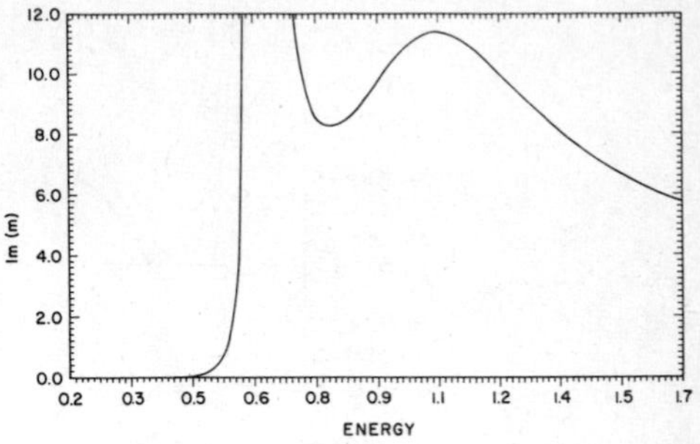

FIG. 4. Magnification of the maximum in Im$m(\lambda)$ associated with $v=4$, same data as Fig. 3.

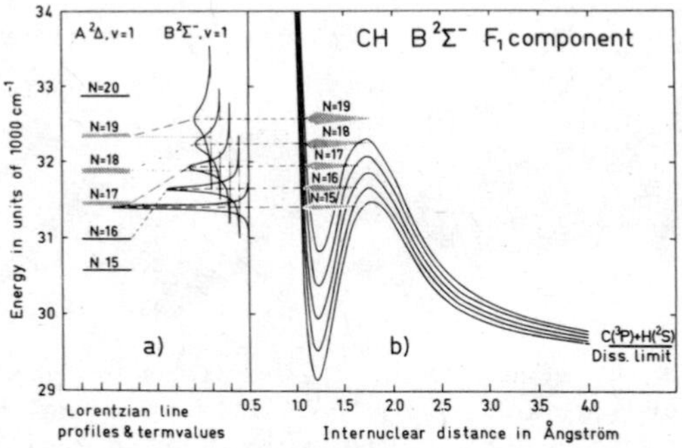

Fig. 5. The $A^2\Delta$–$B^2\Sigma$ accidental predissociation. (a) CH $A^2\Delta$ term values from [5] and the calculated spectral densities for the (b) F_1 spin component potential energy curves of N = 15–19 of the $B^2\Sigma$ state.

Figures 1 - 4 are taken from references 4 and 5, and figure 5 from reference 14.

II. FORMAL DEVELOPMENT OF WEYL'S THEORY

a) Introductory remarks

The theoretical development to be discussed here is essentially taken from references 1, 3 - 8. The problem that Weyl studied[1] in his thesis, given to him by Hilbert, was to extend the Sturm-Liouville theory to an infinite interval. Given the differential equation (p, q real and $p>0$ on $[0,b]$)

$$H u = (p u')' + q u = E u \quad ; \qquad [o,b] \tag{1}$$

with the real boundary conditions (to make the differential operator H formally self-adjoint)

$$\begin{aligned} u \sin\alpha - p u' \cos\alpha &= 0 \qquad x=0 \\ u \cos\beta + p u' \sin\beta &= 0 \qquad x=b \end{aligned} \tag{2}$$

one can show that there exists a complete orthonormal set of eigensolutions u_{bk} with eigenvalues (discrete) E_{bk}. The completeness theorem states that for an arbitrary function f, which is square integrable over the interval $[0,b]$ one has

$$\int_0^b |f|^2 dx = \sum_{k=0}^{\infty} |c_{bk}|^2 \tag{3}$$

$$c_{bk} = \int_0^b u_{bk}(x) f(x)\, dx \tag{4}$$

The problem is now to study what happens when b extends to infinity. The trick involved in this analysis is to make the eigenvalue complex, i.e. to study

$$H\chi = \lambda\chi ; \quad \lambda = E + i\epsilon ; \quad \epsilon \neq 0 \tag{5}$$

on the finite interval $[o,b]$. We are then faced with two types of limiting procedures, i.e. $b\to\infty$ and $\epsilon\to 0$. The only theorem necessary for our discussion is the classical Green's formula, which relates solutions φ and ψ of (1) or (5) at the boundary points o and b with a certain integral over the whole interval $[0,b]$, i.e.

$$\int_0^b \{\varphi^* H\psi - (H\varphi)^*\psi\}\,dx = \int_0^b \frac{d}{dx}\left([\psi\varphi]\right)dx = [\psi\varphi](b) - [\psi\varphi](0) \tag{6}$$

$$[\psi\varphi](x) = p(x)\left(\psi(x)\,\varphi'(x)^* - \psi'(x)\,\varphi(x)^*\right) = W(\psi\varphi^*). \tag{7}$$

We have here introduced the complex bracket notation of Weyl, which is essentially a Wronskian between two functions of which the last one is taken complex conjugate. In the following we will show how one obtains Weyl's limit-point limit-circle theory and how one arrives at a meromorphic function $m(\lambda)$, the so called Weyl-Titchmarsh m-function, which exhibits poles and cuts identical with those of the Green's function. To simplify the development we have partitioned the formulation into seven distinct steps.

b) Development.

Step 1.

Consider the initial value solutions $\varphi(\lambda;x)$ and $\psi(\lambda;x)$

$$\varphi(\lambda;0) = \sin\alpha \;; \qquad \psi(\lambda;0) = \cos\alpha \tag{8}$$

$$p(0)\varphi'(\lambda;0) = -\cos\alpha \;; \qquad p(0)\psi'(\lambda;0) = \sin\alpha$$

Before proceeding we note that the initial conditions are chosen so that $[\varphi\psi^*] = W(\varphi\psi) = 1$ for all x in the interval $[0,b]$. We also note that for $\alpha = \frac{\pi}{2}$; $\psi(\lambda;0) = 0$, i.e. when $\epsilon \to 0$ ψ would be the regular solution at the origin, provided the potential has no singularity there. In most casesthe potential q(x) will be singular at x=0. As we will see, however, we should then study the interval $[a,b]$;$0<a<b$ and then determine the angle α at a so that ψ becomesregular at origin. For the time being we assume that the only singularity in our problem arises when $b\to\infty$ and this takes us onto step 2.

Step 2

Using φ and ψ from (8) we construct the general solution

$$\chi_b(\lambda;x) = \varphi(\lambda;x) + m_b(\lambda)\,\psi(\lambda;x) \tag{9}$$

where the coefficient $m_b(\lambda)$; $\lambda = E+i\epsilon$; $\epsilon \neq 0$ is determined so that

$$\chi_b(\lambda;b)\cos\beta + p(b)\,\chi_b'(\lambda;b)\sin\beta = 0. \tag{10}$$

The boundary condition (10) implies that the logarithmic derivative of χ is real, which for an arbitrary β can be expressed as

$$[\chi_b\,\chi_b] = 0 \tag{11}$$

Step 3

Insertion of (9) into (11) yields Weyl's circle, which after some manipulations may be written as

$$|(m_b - c_b)|^2 = r_b^2 \tag{12}$$

$$c_b = -\frac{[\varphi\psi](b)}{[\psi\psi](b)} \; ; \qquad r_b = \frac{1}{|[\psi\psi](b)|} \; ; \tag{13}$$

i.e. the center of the circle and the radius are given by c_b and r_b respectively.

Step 4

Using Green's formula (6) we express r_b as

$$r_b = \frac{1}{2\epsilon\int_0^b |\psi|^2 dx} \; ; \qquad \epsilon \neq 0 \tag{14}$$

from which one deduces that r_b shrinks as b increases.

Step 5

We distinguish between two cases. Either $\lim_{b\to\infty} r_b = 0$, i.e. the circle shrinks to a point, the limit point case (l.p.) or $\lim_{b\to\infty} r_b =$ const >0, the limit circle case (l.c.). Denoting the space of square integrable functions $\mathcal{L}^2[0,\infty)$ we obtain that l.p. implies $\psi \notin \mathcal{L}^2[0,\infty)$ and l.c. implies $\psi \in \mathcal{L}^2[0,\infty)$.

Step 6

In the l.p. case we have the unique limit

$$\lim_{b\to\infty} m_b(\lambda) = m_\infty(\lambda); \quad \lambda = E + i\epsilon; \quad \epsilon \neq 0 \tag{15}$$

while the l.c. case yields an m-coefficient that depends on the angle β . However, since (using Green's formula (6))

$$\lim_{b\to\infty} \int_0^b |\chi_b|^2 dx = \lim_{b\to\infty} \frac{m_b(\lambda) - m_b^*(\lambda)}{\lambda - \lambda^*}; \quad \epsilon \neq 0 \tag{16}$$

we always find that $\lim_{b\to\infty} \chi_b \in \mathcal{L}^2[0,\infty)$,albeit β dependent in the l.c. case. Noting that $\lim_{b\to\infty} W(\psi\chi_b) = -1$ we observe that in the l.c. case all solutions of (5) in the limit $b\to\infty$ $\in \mathcal{L}^2[0,\infty)$, while the l.p. case only gives one unique solution, i.e.

$$\lim_{b\ \infty} \chi_b(\lambda;x) = \chi(\lambda;x); \quad \chi \in \mathcal{L}^2[0,\infty) \tag{17}$$

Step 7

In order to study what happens when $\epsilon\to 0$, showing the significance of the Weyl-Titchmarsh m-function as defined in (9 - 15), we return to the Sturm-Liouville problem. Considering the initial value solutions in (8) at an eigenvalue E_{bk} ; $b<\infty$ we realize from (2) that u_{bk} and $\psi(E_{bk};x)$ satisfy the same initial condition and hence

$$u_{bk} = r_{bk}\psi(E_{bk}). \tag{18}$$

In (18) r_{bk} is a proportionality factor that relates $\psi(E_{bk})$ (with a given initial condition) with u_{bk} , which is normalized to unity i.e. $\int_0^b |u_{bk}|^2 dx = 1$ as prescribed by Sturm-Liouville theory. Using the completeness theorem (3) on $\chi_b(\lambda;x)$, $\epsilon \neq 0$ obtained in (9-14) one gets

$$\int_0^b |\chi_b|^2 dx = \sum_{k=0}^{\infty} |c_{bk}|^2 \tag{19}$$

with

$$c_{bk} = \int_0^b u_{bk}(x)\ \chi_b(\lambda; x)\, dx \tag{20}$$

Using (18) we write (19) as a Stieltjes integral

$$\int_0^b |\chi_b(\lambda;x)|^2 dx = \int_{-\infty}^{+\infty} |g_b(\omega)|^2 d\rho_b(\omega) \tag{21}$$

with

$$g_b(E_{bk}) = \int_0^b \psi(E_{bk}; x)\ \chi_b(\lambda; x)\, dx \tag{22}$$

and

$$\Delta \rho_b(E_{bk}) = | r_{bk} |^2 \tag{23}$$

Again employing Green's formula (6) on both sides of (21) one finds

$$\frac{m_b(\lambda) - m_b^*(\lambda)}{\lambda - \lambda^*} = \sum_{k=0}^{\infty} \frac{| r_{bk} |^2}{|\lambda - E_{bk}|^2} \tag{24}$$

Taking the limit $b \to \infty$ see steps 5 and 6 it is obtained (im means imaginary part of)

$$im\{m_\infty(\lambda)\} = \int_{-\infty}^{+\infty} \frac{\epsilon\, d\rho_\infty(\omega)}{(E-\omega)^2 + \epsilon^2} \tag{25}$$

As can be seen from (23) the spectral function $\rho_b(\omega)$ is a step function that increases with the value $|r_{bk}|^2$ at E_{bk}. In the limit $b \to \infty$ $\rho_\infty(\omega)$ may still be discontinuous at ω and we will say that ω is an element of the discrete spectrum σ_D of the differential operator H defined in (1 - 2) over the interval $[0,\infty)$. If $\rho_\infty(\omega)$ is continuous but not constant at ω then ω is an element of the continuous spectrum σ_C and if ρ is constant at ω then ω is not in the spectrum of H. It may happen that $\rho_\infty(\omega)$ has discontinuities in the continuous part of the spectrum. Such points ω are said to belong to the point-continous spectrum σ_{PC}. A general relation between m_∞ and ρ_∞ is given by Stieltjes inversion formula, which amounts to taking the desired limit $\epsilon \to 0$ i.e.

$$\lim_{\epsilon \to 0} \int_E^{E+\Delta E} im\{m_\infty(E+i\epsilon)\}\, dE = \pi\{\rho_\infty(E+\Delta E) - \rho_\infty(E)\}\ . \tag{26}$$

When $E \in \sigma_c$, i.e. $\rho_\infty(E)$ is continuous and analytic in a region around E (true if the potential satisfies appropriate conditions, see below) then letting $\epsilon \to 0$ in (25) gives the dispersion relation

$$\lim_{\epsilon \to 0+0} \operatorname{im}\{m_\infty(E+i\epsilon)\} = \pi \left(\frac{d\rho}{d\omega}\right)_{\omega=E} . \tag{27}$$

c) Some comments

Considering the resolvent or Green's function in $\mathcal{L}^2[0,\infty)$

$$(H-\lambda I)^{-1} = \psi(\lambda; x_<)\, \chi(\lambda; x_>); \tag{28}$$

where $x_<$ ($x_>$) denotes the smaller (larger) of the two independent variables occurring in ψ and χ we obtain for an arbitrary $f \in \mathcal{L}^2[0,\infty)$ that $(E \in \sigma_c)$

$$\lim_{\epsilon \to 0\pm 0} (H-\lambda I)^{-1} f = \pm\, \pi \left(\frac{d\rho}{d\omega}\right)_{\omega=E} \psi(E;x) \int_0^\infty \psi(E;\xi) f(\xi)\, d\xi . \tag{29}$$

From (29) we see that the imaginary part of m_∞ via (27) has the character of a spectral density, whose variation with $E \in \sigma_c$ reveals important information about the structure of the continuum. We will come back to this point below, but first we collect some general characteristics of $m_\infty(\lambda)$. In the l.p. case $m_\infty(\lambda)$ must have poles on the real axis corresponding to discrete eigenvalues $E_k \in \sigma_D$. Since $\psi(\lambda;x)$ satisfy the boundary condition (2) at $x=0$ and $\chi_b(\lambda;x)$ the right boundary condition (2) at $x=b$, which at the limit $b \to \infty$ yields a $\chi(\lambda) \in \mathcal{L}^2[0,\infty)$, a pole in $m_\infty(\lambda)$ implies quantization i.e. the existence of a unique solution satisfying boundary conditions at both endpoints. It also follows from (28) that the analytic properties of m_∞ are identical to those of the resolvent. Provided that the potential fulfil certain criteria, we may deduce that m_∞ may have a meromorphic continuation passing the real energy axis either from above or below. A complex pole of m_∞ at $\mathcal{E} = E_{res} - i\Gamma/2$ obtained in such a way corresponds to a metastable state with a resonance energy given by E_{res} and a life time given by $\tau = \hbar/\Gamma$.

There are several criteria for potentials that belong to the l.p. case. We will here give the condition that in particular assures that the Stark effect is associated with a differential operator containing a l.p. type potential. If p=1 and $q(x) \geq -k^2x^2$ for some real k then (1) is l.p. at $b \to \infty$.

In the Stark problem as in many problems in physics and chemistry, the asymptotic form is known. In particular we are interested in these asymptotic solutions of exponential type that can be analytically continued into the complex plane. In ordinary scattering theory they are named Jost solutions, to include the Stark effect as well, we will refer to them as Weyl solutions. Knowing the unique asymptotic form $\chi(E+i\epsilon,x)$; $\epsilon > 0$, $m_\infty(\lambda)$ is directly obtained from a formula due to Titchmarsh[7], see also reference 8,

$$m_\infty(\lambda) = \frac{W(\varphi\chi)}{W(\chi\psi)} \tag{30}$$

where the Wronskian may be evaluated in the asymptotic region where χ is known. Furthermore if the real part of $\lambda \notin \sigma_D \cup \sigma_{PC}$ then $W(\chi\psi) \neq 0$ and $\lim_{\epsilon\to 0\pm 0} m_\infty(E+i\epsilon)$ exists. Moreover, if we define $k = i\, W(\chi^+\chi^-)$, where $\chi^\pm = \lim_{\epsilon\to 0\pm 0} \chi(E+i\epsilon)$ and $f_J^\pm(E) = W(\chi^\pm, \psi(E))$ then an extended theorem originally due to Kodaira[15] reads, see also reference 4,

$$\mathrm{im}\left\{ m_\infty(E)\right\} = \frac{k}{|f_J(E)|^2} \tag{31}$$

If it happens that the origin is a singular point then one studies first the interval $[x_0, a]$, $0<x_0<a<b$. Keeping a fixed, one investigates the limiting procedure $x_0 \to 0$, $\lambda = E + i\epsilon$; $\epsilon \neq 0$. Giving initial conditions at a (step1) and constructing a general solution $\chi_0(\lambda; x)$ in $[x_0, a]$ (step 2), then carrying out steps 1 - 7 a unique square integrable solution on the interval $(0, a]$ is found. The logarithmic derivative of that (regular) solution defines the angle α, i.e. $\frac{p(a)\, \chi_0'(\lambda, a)}{\chi_0(\lambda; a)} = \mathrm{tg}\,\alpha$, to be used on the interval $[a, b]$ with initial solutions defined as

$$\varphi(\lambda; a) = \sin\alpha \qquad \psi(\lambda; a) = \cos\alpha$$
$$p(a)\, \varphi'(\lambda; a) = -\cos\alpha \qquad p(a)\, \psi'(\lambda; a) = \sin\alpha \tag{8'}$$

The angle α could also be obtained from a power series expansion, from minimizing the length $\{\chi_0(\lambda; x_0)\}^2 + \{p(x_0)\, \chi_0'(\lambda; x_0)\}^2$ as $x_0 \to 0$ or by simply assuming a hard core at x_0, i.e. putting $\chi_0(\lambda; x_0) = 0$ at some x_0 close to origin. We will come back to those procedures in the next section.

The use of (15) or (30) combined with the boundary conditions (8') therefore allows the study of the differential operator (5) on the interval $(0, \infty)$. From the unique meromorphic function $m_\infty(\lambda)$ the various types of spectra can be found. In the l.p. case the total spectrum σ has the disjoint decomposition

$$\sigma = \sigma_D \cup \sigma_q \cup \sigma_{pq} \tag{32}$$

see e.g. reference 16 for a detailed study.

III. NUMERICAL APPLICATIONS OF WEYL'S THEORY

It is evident that a resonance determination requires highly accurate solutions of the differential equation. It is therefore necessary to carefully test the various limiting procedures employed, the appropriate division of the interval $(0, \infty)$, best use of matching formulas, efficient use of analytical properties of m as well as accurate and reliable analytic continuation procedures. Let us again divide the calculation into 6 stages. We list them first and then proceed to a more detailed discussion. We have mainly used the following computational strategy[4,5] $(p=1)$.

1. Divide the interval $(0, \infty)$ into $[R_1, R_2]$ and $[R_2, \infty)$ $\{$and $(0, R_1]\}$

2. Use the hard core approximation at R_1 to obtain $\psi'/\psi = tg\alpha$ at R_2

3. Determine $iz = \chi'/\chi$ at R_2 where χ is Weyl's solution.

4. Evaluate $m_\infty = \dfrac{iz \sin\alpha + \cos\alpha}{\sin\alpha - iz\cos\alpha}$; α and z evaluated at R_2

5. Look for sign change in $re\{m_\infty(E)\}$ (re means real part of)

6. Evaluate $\mathcal{E} = E - i\epsilon$ by analytic continuation.

In the first stage we divide the interval as prescribed. The simplest procedure is then to put the regular solution ψ equal to zero at a small distance R_1 at which one starts the outward integration. At R_2 usually chosen to be around the minimum of the potential curve the angle α is determined. To test the reliability of this method we used two alternative ways based on inward integration from R_2 towards the origin automatically checking the accuracy of α to decide where to stop. Although this determines R_1 provided some convergence condition is given, it means that both independent solutions have to be evaluated from R_2 to R_1, hence involving extra labour compared to the hard core case. For instance using (8') at $a=R_2$ and initially given α_I we can directly calculate

$$m_{R_1}(\lambda) = -\frac{\varphi(\lambda;R_1)}{\psi(\lambda;R_1)} \tag{33}$$

As the convergence criteria for α we use the radius r_{R_1}

$$r_{R_1} = \frac{1}{2\epsilon\int_{R_1}^{R_2} |\psi(\lambda;x)|^2 dx} \tag{34}$$

This procedure has the additional disadvantage of requiring $\mathrm{im}\lambda = \epsilon \neq 0$. The other method due to McIntosh[17] is based on the idea that in the l.p. case d(x) given by

$$d(x) = \left\{ \chi_0^2(E;x) + (p(x)\,\chi_0'(E;x))^2 \right\} \tag{35}$$

approaches zero when x gets smaller. The trick is to evaluate χ_0'/χ_0 at R_2 expressed in $\varphi(E,R_1)$ and $\psi(E,R_1)$; φ and ψ initially given at R_2, minimizing $d(R_1)$. For details see reference 4.

In table 1 we have tested the hard core approximation against the more rigorous methods. We see that at convergence all three methods lead to identical results proving the reliability of the Runge-Hutta methods that were used. Although the hard core results are obtained with optimal efficiency, it might still be a good thing to use the minimization procedure for checking purposes and to choose the best R_1. In stage 3 we employ Riccati's equation[4,18]

$$-iz'(x) = E - q(x) - z^2(x) \tag{36}$$

$$iz(x) = \frac{\chi'}{\chi} \tag{37}$$

to determine $iz(R_2)$. Several alternatives are here possible. Integrating Riccati's differential equation directly inwards from the asymptotic region to R_2, or integrating the second order equation via invariant imbedding, one may find $iz(R_2)$ to be used in stage 4.

TABLE I. Test of distance R_1 for hard core approximation [$K=0$, R_2 $(=a)=3.289$ a.u., classical t.p. at 2.847 a.u.].

Distance R_1 (a.u.)	α (hard core) outward, real	α (minimiz.) inward, real	α (limit p.) inw., compl.	Radius Weyl c.
2.7587	1.567 124 949	1.478 067 468	1.567 864 354	9.8×10^{5}
2.5542	1.537 950 389	1.536 445 232	1.537 960 025	6.6×10^{5}
2.3570	1.537 286 276	1.537 284 829	1.537 286 287	5.2×10^{3}
2.1696	1.537 285 593	1.537 285 593	1.537 285 593	5.3×10^{-1}
1.9950	1.537 285 593	1.537 285 593	1.537 285 593	1.8×10^{-6}

TABLE II. Timings of Runge–Kutta routines (potential assumed to be stored in numerical table).

Type of R.–K. method	Steps/second (IBM 370/155) [a]
4th order, 1 real solution	700
4th order, 2 real solutions	565
4th order, 2 complex solut.	185
6th order, 1 real solution	333
6th order, 2 real solutions	245
6th order, 2 complex solut.	60

[a]Double precision.

TABLE III. Accuracies of 4th- and 6th-order Runge-Kutta methods, interval $[R_2, R_3]$ $R_2=2.0$ a.u., $R_3=3.5$ a.u.

Number of steps	Number of significant figures	
$[R_2, R_3]$	4th o. R.–K.	6th o. R.–K.
3	0–1	2
6	2–3	3–4
15	4	6
30	5	8
75	6	10
150	8	11

TABLE IV Resonance energies and widths for the broadest levels in the rotation–vibration spectrum of HgH, calculated with Stwalley's IC1-potential. All energies given in cm^{-1}. Reference energy $E(0,0) = 677.03$ cm^{-1} [or $E(0,0) = -3024.7$ if dissociation energy taken as zero point].

(v, K)	$\{V_{max} - E(0,0)\}$	$\{E(v,K) - E(0,0)\}$ Weyl(r)	Weyl(c)	Phase shift method	$\Gamma(v,K)$ Weyl(r)	Weyl(c)	Phase shift method
(0,39)	7025.85	7216.93	7211.39	···	338.	343.	···
(0,38)	6754.52	6909.64	6907.96	···	253.	265.	···
(0,37)	6500.71	6612.00	6613.89	···	178.	184.	···
(0,36)	6267.29	6320.62	6320.58	···	115.	115.	···
(0,35)	6054.24	6034.43	6034.46	6041.9	67.5	67.4	61.4
(0,34)	5848.10	5751.71	5751.71	5751.8	34.3	34.3	32.2
(1,29)	4915.81	4988.62	4993.72	4984.9	171.	154.	118.
(1,28)	4749.14	4777.60	4778.84	4775.5	95.3	93.8	77.1
(1,27)	4589.49	4569.24	4569.40	4569.7	48.3	48.3	42.8
(2,22)	3956.07	4056.47	4048.33	~3960	212.	197.	···
(2,21)	3851.71	3922.15	3923.70	~3860	133.	128.	···
(2.20)	3753.84	3787.91	3788.35	3784.0	81.3	80.1	62.3
(3.12)	3196.99	3234.17	3234.39	~3200	92.6	85.2	···
(3,11)	3153.04	3176.81	3177.41	~3150	65.0	64.8	···
(3,10)	3088.04	3124.95	3125.12	3113.3	53.8	48.3	26.5
(3,9)	3078.41	3064.19	3064.18	3077.9	0.765	0.763	97.2[a]
(4,9)	3078.41	~3080	3085.5	···	~53	36.1	···
(4,8)	3058.61	3062.8	3062.8	···	19.2	17.0	···
(4,7)	3048.02	3048.3	3048.3	3048.3	5.01	5.01	3.79

Tables I - IV are taken from references 4 and 5.

Alternatively one may take φ and ψ at R_2 and integrate outwards towards the asymptotic region or outwards until the Riccati expansion converges. The m-coefficient then could be obtained from (30). If only im$\{m\}$ is needed it will be enough to integrate ψ outwards into the asymptotic region and then applying formula (31).Since the detection of very sharp resonances are simplified by looking for a sign change in re$\{m\}$ it is usually very valuable to obtain the full m-function. Also when the resonance is very broad or when several resonances may be interacting the full analytic information of the total m-coefficient should be used. For the analytic continuation procedure, based on rational fractions we refer to 5. A different, more rigorous procedure will be published elsewhere.

In tables II and III we have further tested our numerical procedures. In table III the accuracies obtained in the 4th and 6th order Runge-Kutta procedures are shown. It has recently been pointed out to us that 6th order method used should be viewed as a 5th order method. 19) Nevertheless we found the higher order method to be more economic to use. The timings shown in table II together with the data in table III support this view.

Although the present lecture has only delt with a simple linear second order differential equation, extensions to coupled equations are straight forward. Multiple channel analysis with one or more channels closed should then preferably be discussed within the framework of Weyl's theory. This as well as improved analytic continuation techniques are currently being investigated and applications in time resolved spectroscopy of small molecules [20] are under way.

ACKNOWLEDGEMENTS

I am indebted to my collegues and coworkers in this project, particularly I like to mention docent Harold V McIntosh, docent Michael Hehenberger and docent Nils Elander.

IV REFERENCES

1. H. Weyl, Math. Ann. 68, 220 (1910).
2. E. Schrödinger, Ann. Phys. (Leipz.) (4) 80, 437 (1926).
3. M. Hehenberger, H.V. McIntosh, and E. Brändas, Phys. Rev. A 10, 1494 (1974).
4. M. Hehenberger, B. Laskowski, and E. Brändas, J. Chem. Phys. 65. 4559 (1976).
5. M. Hehenberger, P. Froelich, and E. Brändas, J. Chem. Phys. 65, 4571 (1976).
6. Z.S. Herman and E. Brändas, Mol. Phys. 29, 1545 (1975).
7. E.C. Titchmarsh, Eigenfunction Expansions Associated with Second Order Differential Equations. (Clarendon, Oxford, 1946; 1962), Vol. 1, (1958), Vol. II.
8. E. Brändas, M. Hehenberger and H.V. McIntosh, Int. J. Quant. Chem. 9, 103 (1975).
9. J. Aguilar and J.M. Combes, Commun. Math. Phys. 22, 269 (1971).
10. E. Balslev and J.M. Combes, Commun. Math. Phys. 22, 280 (1971).
11. B. Simon, Ann. Math. 97, 247 (1973).
12. E. Brändas and P. Froelich, Phys. Rev. A16, 2207 (1977).
13. W.G. Stwalley, J. Chem. Phys. 63, 3062 (1975).
14. N. Elander, M. Hehenberger and P.R. Bunker, Phys. Scripta 20, 631 (1979).
15. K. Kodaira, Amer. J. Math. 71, 921 (1949).
16. I. Chaudhuri and W.N. Everitt, Proc. Roy. Soc. Edinburgh (A) 68, 95 (1968).
17. H.V. McIntosh, Semniar Notes (unpublished) Uppsala Quantum Chemistry Group, (1971, 72).
18. E. Brändas and M. Hehenberger, in Lecture Notes in Mathematics. A. Dold and B. Eckmann, Eds. (Springer-Verlag, Berlin, 1974), Vol. 415, 316.
19. H.J. Stetter, private communication.
20. P. Erman, Chem. Rev. (in press) (1980).

THE DISCRETIZATION OF CONTINUOUS INFINITE SETS OF COUPLED ORDINARY LINEAR DIFFERENTIAL EQUATIONS: APPLICATION TO THE COLLISION-INDUCED DISSOCIATION OF A DIATOMIC MOLECULE BY AN ATOM

Dennis J. Diestler
Department of Chemistry
Purdue University
West Lafayette, Indiana 47907/USA

1. Nature of the Physical Problem

Before addressing the mathematical problem, I shall, primarily for the benefit of the mathematicians, describe the physical process and its measurement. Let us consider from a classical viewpoint the collision of an atom A with a diatomic molecule BC. Initially, (i.e. before the encounter) the atoms B and C of the diatomic are rotating and vibrating with respect to each other as the atom A translates at uniform speed toward them. Depending upon the specific nature of the intermolecular forces between A and BC and upon the initial conditions of the collision, especially the speed (or kinetic energy) of A relative to BC, the encounter can result in the following outcomes:

(i) elastic scattering, in which A simply bounces off of BC with no energy being transferred between the internal motions of BC (i.e. rotation and vibration) and the translation of A relative to BC;

(ii) inelastic scattering, in which there is a transfer of energy between translation and rotation-vibration;

(iii) reactive scattering, in which a new molecule AC (or AB) along with a new atom B (or C) is produced;

(iv) collision-induced dissociation (CID), in which all three atoms A, B and C fly freely away from each other after the encounter.

It is the last process with which we are particularly concerned in this lecture.

All of the scattering processes described can be investigated experimentally by crossing a beam of atoms A with a beam of diatoms BC (usually at right angles) in a vacuum and detecting the scattered (product) molecules.[1] The current of scattered particles (i.e. the number of particles scattered per unit time) corresponding to a given process relative to the initial intensity (i.e. current per unit area) of the unscattered beams provides a measure (the *cross section*) of the probability of that process, which of course depends upon the intermolecular interactions and initial conditions. Indeed, by measuring the dependence of the cross sections upon the collision parameters (e.g. initial relative kinetic energy), one may be able to "invert" the data to obtain the intermolecular forces.

Although molecular beams provide the most direct way of studying microscopic scattering processes, one is often concerned with chemical processes occurring in bulk, i.e. in gas or in liquid solution. In the bulk experiment, one measures simply the concentrations of reactants or products as a function of time. Then, from the dependence of the *rate of change* of these concentrations upon their initial values one calculates a thermal rate constant k(T). For sufficiently *dilute gases* the overall reaction occurs via sequences of the microscopic scattering processes described above. In this case k(T) can be written as a convolution of the cross section with the relative kinetic energy distribution function.[2] Thus, one can see that the rate constant is also related to the nature of the intermolecular forces, although much less directly than is the cross section itself. Incidentally, thermal rate constants are essentially the "data" that go into the nonlinear first-order ordinary differential equations of "chemical kinetics," which are dealt with by Dahlquist and Ames in this set of lectures.

It should now be clear why a sizeable fraction of the chemists are concerned with the calculation of cross sections. I shall next deal with the quantitative description of microscopic scattering processes, that is with the calculation of cross sections.

2. Quantitative Description of Collision Processes

To simplify the consideration, while retaining all of the relevant features of the mathematical problem, I shall restrict my attention from now on to *collinear* (one-dimensional) collisions, i.e. encounters in which A approaches BC along its line of centers, as depicted in Fig. 1. In order to describe the collision quantitatively

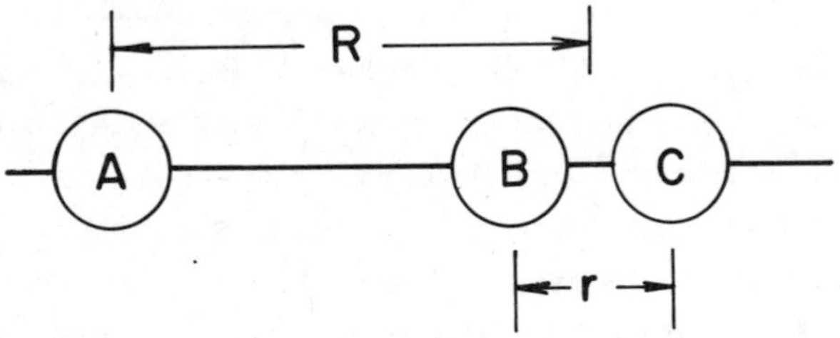

Fig. 1. Collinear collision of atom A with diatomic molecule BC.

we need two spatial coordinates, which can be conveniently taken to be R, the distance between A and the center of mass of BC, and r, the distance between B and C. The momenta conjugate to R and r are P and p, respectively. Then the classical expression for the total energy, that is the total Hamiltonian, can be written as

$$H = H^o_{BC} + P^2/2\mu + \overline{V}(R,r) \ . \qquad (2.1)$$

In (2.1) H^o_{BC} is the Hamiltonian for the isolated diatomic, given by

$$H^o_{BC} = p^2/2\mu_{BC} + V_{BC}(r) \ , \qquad (2.2)$$

where the first term is the vibrational kinetic energy and the second is the potential energy binding B and C; $P^2/2\mu$ is kinetic energy of A relative to BC; $\mu[=m_A(m_B + m_C)/(m_A + m_B + m_C)]$ and $\mu_{BC}[=m_B m_C/(m_B + m_C)]$ are, respectively, the reduced masses of A and BC and of B and C, where m_X is, of course, the mass of atom X; $\overline{V}(R,r)$ is the intermolecular interaction. Chemists refer to the sum $V_{BC} + \overline{V}$ as the potential-energy surface (PES), since it is commonly represented as a (mathematical) surface above the R-r plane, or more commonly as equipotential contours in the R-r plane. For present purposes I shall assume that the PES is known, although it should be noted that another good-sized fraction of chemists devote their efforts to calculating such surfaces.

With this background I can now proceed to outline the calculation of the probability of any given scattering process within the framework of classical mechanics. Note that in the case of collinear collisions one calculates scattering probabilities rather than the cross sections of the corresponding three-dimensional process.

2.1 Classical treatment[3]

According to the basic principles of classical mechanics, given the positions and momenta of all the atoms at any instant (i.e. given the initial conditions), we can determine the positions and momenta of these atoms for all future (and past) time. A record of the positions and momenta as a function of time constitutes the *trajectory* of the system. Of course, the trajectory depends upon the PES and the initial conditions {R(0), r(0), P(0), p(0)}, where 0 specifies an arbitrary origin of time, which we shall take to be long before the encounter.

Skipping ahead a little bit, I wish to take into account that (according to the principles of quantum mechanics) the energy of the initially *bound* isolated BC can take only certain values ε_j. Hence, in the usual classical calculation, one fixes the total energy H^o_{BC} of the vibration to correspond to one of these allowed values, say ε_i. One also fixes the *total* energy of collision H. Thus, since $\overline{V} = 0$ initially (i.e. long before the encounter), it follows from (2.1) that P(0) is fixed. Further, since the total vibrational energy is fixed, r and p are not independent [see (2.2)]. In fact, the trajectory of the isolated molecule is represented by a closed curve in the r-p plane. The remaining initial conditions can be chosen as follows: R(0) is taken so large that $\overline{V}$ is less than some very small fraction of the total energy;

r(0) and p(0) are determined by picking a point off the vibrational trajectory with equal weight given to all portions of the curve. Now for each set of initial conditions, one integrates Hamilton's equations

$$\dot{R} = \partial H/\partial P \quad ; \quad \dot{P} = -\partial H/\partial R$$
$$\dot{r} = \partial H/\partial p \quad ; \quad \dot{p} = -\partial H/\partial r \ , \qquad (2.3)$$

following the trajectory sufficiently long to ascertain which of the scattering processes occurs. Equations (2.3) constitute a set of first-order nonlinear differential equations, whose efficient numerical integration is also of great concern to chemists, but beyond the scope of this lecture. The probability of a given scattering process is equal to the number of trajectories resulting in that process divided by the total number of trajectories computed.

2.2 Quantal treatment[4]

In quantum mechanics we cannot determine, nor should we even think of (lest the gods of quantum theory become angry), trajectories. We must be content with a less precise description, all knowable information concerning the system being carried by a state vector $|\Psi(t)\rangle$. The state vector is represented in configuration space (R,r) at time t by a wavefunction $\Psi(R,r;t)$ which evolves in accordance with Schrödinger's equation

$$\hat{H}\ \Psi = i\hbar\ \partial\Psi/\partial t \ , \qquad (2.4)$$

where h is Planck's constant and the *operator* $\hat{H}$ is the total Hamiltonian (2.1) with the following operator substitutions:

$$P \longrightarrow -\ i\hbar\ \partial/\partial R \quad , \quad p = -\ i\hbar\ \partial/\partial r \ . \qquad (2.5)$$

A physical interpretation of Ψ is the following: the probability that a measurement at time t finds the system lying in a spatial element dRdr is given by $|\Psi(R,r;t)|^2 dRdr$, provided that Ψ is normalized, i.e.

$$\int dR \int dr\ |\Psi(R,r;t)|^2 = 1 \ . \qquad (2.6)$$

Of course, the precise way in which the system evolves depends upon the initial wavefunction $\Psi(R,r;0)$, which corresponds to the initial conditions characterizing a classical trajectory.

Before proceeding with the quantal treatment of scattering processes, I need to discuss in some detail the energy states of the isolated diatomic BC. The binding interaction $V_{BC}(r)$ has the form pictured in Fig. 2 for a typical diatomic.

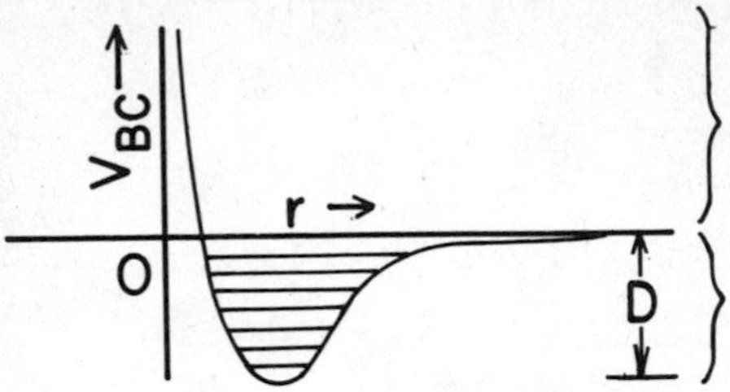

Fig. 2. Schematic of the potential energy binding a typical diatomic molecule.

We take the zero of energy to correspond to the atoms B and C at rest at $r = \infty$; D is the classical dissociation energy. If the total classical vibration energy is less than zero, then the diatomic undergoes periodic motion, its trajectory being a closed curve in the p-r plane, as pointed out in Section 2.1. Within the framework of quantum mechanics, the total vibrational energy can take only certain discrete values $\varepsilon_j < 0$. These allowed energy levels have associated eigenfunctions ϕ_j that satisfy the relation

$$\hat{H}^o_{BC}\, \phi_j\,(r) = \varepsilon_j\, \phi_j\,(r) \quad , \quad j = 1,2, \ldots N_b \; , \tag{2.7}$$

where $\hat{H}^o_{BC}$ is the vibrational Hamiltonian operator, given explicitly by

$$\hat{H}^o_{BC} = -\,(\hbar^2/2\mu_{BC})\partial^2/\partial r^2 + V_{BC}(r) \tag{2.8}$$

and N_b is the total number of bound states. The vibrational eigenfunctions can be taken to be orthonormal, i.e.

$$\int dr\, \phi_i^{\,*}\,(r)\, \phi_j\,(r) = \delta_{ij} \; , \tag{2.9}$$

where δ_{ij} is the Kronecker delta.

Now if the total classical vibrational energy is greater than zero, the atoms B and C fly apart. From a quantum-mechanical viewpoint the diatomic is now in a *continuum* state having an eigenfunction $\phi(k;r)$ that also satisfies the equation

$$\hat{H}^o_{BC}\, \phi(k;r) = \varepsilon(k)\, \phi(k;r) \; , \tag{2.10}$$

where ε can take *any* value greater than zero. The continuous label k denotes the asymptotic wavenumber, which is related to the energy by

$$\varepsilon = \hbar^2 k^2 / 2\mu_{BC} \tag{2.11}$$

The continuum eigenfunctions are also orthonormal, i.e.

$$\int dr\, \phi^*(k';r)\, \phi(k;r) = \delta(k'-k) \ , \tag{2.12}$$

except that now δ is the Dirac delta function. Our central mathematical problem is due precisely to the existence of this continuum of states, as will become apparent shortly.

2.2.1 Time-dependent method

Since $\hat{H}$ is explicitly time-independent, (2.4) can be integrated formally to yield

$$\Psi(\underset{\sim}{r},t) = \exp(-i\hat{H}t/\hbar)\, \Psi(\underset{\sim}{r};0) \tag{2.13a}$$

$$= (1 - i\hat{H}t/\hbar + \ldots)\, \Psi(\underset{\sim}{r};0) \ , \tag{2.13b}$$

where $\underset{\sim}{r}$ denotes the point (R,r). One might suppose that, if the time interval of interest were divided into sufficiently small increments Δt, then the wavefunction could be propagated from one instant (t) to the next $(t + \Delta t)$ simply by operating upon the wavefunction at that instant, $\Psi(\underset{\sim}{r};t)$, with $(1 - i\hat{H}\Delta t/\hbar)$. However, such an explicit numerical procedure turns out to be unstable. If, instead, one left multiplies both sides of (2.13a) by $\exp(i\hat{H}t/2\hbar)$ and expands the resulting exponentials, retaining terms only through those linear in $\hat{H}$, one obtains the Crank-Nicholsen approximation:

$$[1 + i\Delta t\hat{H}/2\hbar]\Psi(\underset{\sim}{r};\Delta t) \simeq [1 - i\Delta t\hat{H}/2\hbar]\Psi(\underset{\sim}{r};0). \tag{2.14}$$

Using a suitable numerical method, one can solve (2.14) for $\Psi(\underset{\sim}{r};\Delta t)$, given $\Psi(\underset{\sim}{r};0)$. Upon replacing $\Psi(\underset{\sim}{r};0)$ by $\Psi(\underset{\sim}{r};\Delta t)$ in (2.14) an equation for $\Psi(\underset{\sim}{r};2\Delta t)$ is obtained which can be solved as before. Repetition of this procedure allows the wavefunction to be propagated in time in a stable fashion.

The Crank-Nicholson method has been applied to both reactive scattering[5] and CID[6,7] in collinear atom-diatom collisions. In all cases the initial wavefunction is taken as a product of the vibrational wavefunction $\phi_i(r)$ describing the initial (i) bound state of BC times a localized translational wavepacket ψ_{tr}, i.e.

$$\Psi(R,r;0) = \psi_{tr}(R)\, \phi_i(r) \tag{2.15}$$

with ψ_{tr} chosen to correspond to A approaching BC with someititial *average* momentum, say P_o. The wavefunction is propagated by the procedure outlined above until sufficiently long after the encounter has taken place. The probabilities for various outcomes are computed as expectation values of appropriate operators. For example, the probability of an inelastic transition from the initial state i to final state f is given by

$$P_{i\to f} = \lim_{t\to\infty} \int dR \, \left| \int dr \phi_f^*(r) \;\; \Psi(R,r;t) \right|^2 . \tag{2.16}$$

The CID probability is most conveniently obtained by difference from

$$P_{i\to D} = 1 - \sum_f P_{i\to f} - \sum_{f'} P^{(R)}_{i\to f'} , \tag{2.17}$$

where the superscript R denotes a reactive-scattering probability, given by an expression analogous to (2.16). The relation (2.17) depends upon the fact that probability is conserved, i.e. the sum over the probabilities of all possible processes is unity.

2.2.2 Stationary-wave method

An alternate way to calculate quantal scattering probabilities, which is equivalent to the time-dependent method, is by solving the time-independent Schrödinger equation

$$\hat{H} \psi_i (\underset{\sim}{r}) = E \psi_i (\underset{\sim}{r}) \tag{2.18}$$

for the *stationary* scattering wavefunction ψ_i. Here $\hat{H}$ is still the total Hamiltonian operator and E is the *total* energy of the collision. The subscript i on ψ_i denotes the initial vibrational state of the diatom.

If for the moment we assume that only elastic and inelastic collisions are possible, then ψ_i has the following asymptotic ($R \to \infty$) form:

$$\psi_i(\underset{\sim}{r}) = \sum_j F_j(R) \, \phi_j(r) , \tag{2.19a}$$

where

$$\lim_{R\to\infty} F_j(R) = (2\pi\hbar)^{-1/2} \, [\delta_{ij} \exp(-iP_i R/\hbar)$$

$$-(2\pi i\mu/P_j) \, T_{ji} \exp(iP_j R/\hbar)] , \; j \leq N_\varepsilon \tag{2.19b}$$

In (2.19b) P_j is the momentum of A relative to BC, with BC in bound state j. Since the total energy is fixed, we have

$$E = \varepsilon_i + P_i^2/2\mu = \varepsilon_j + P_j^2/2\mu , \; j \leq N_\varepsilon , \tag{2.20}$$

where the ε_j's are the discrete allowed vibrational energy levels of BC. T_{ji} is the transition matrix, or simply T-matrix, in terms of which the scattering probabilities are given by

$$P_{i\to j} = (2\pi)^2\mu^2|P_i|^{-1}|P_j|^{-1}|T_{ji}|^2 \;,\; i,j \le N_\varepsilon \;. \tag{2.21}$$

In (2.19)-(2.21) N_ε is the total number of levels accessible at the given total energy. Clearly, $N_\varepsilon \le N_b$.

A natural way to attempt to solve (2.18) is simply to substitute the expansion (2.19a) for ψ_i, multiply both sides of the resulting equation by $\phi_j^*(r)$, and integrate over r invoking relation (2.9), to obtain in matrix form

$$[\underset{\sim}{1}d^2/dR^2 + \underset{\sim}{\overline{E}} - \underset{\sim}{\overline{V}}(R)]\cdot\underset{\sim}{F}(R) = 0 \;, \tag{2.22}$$

where $\underset{\sim}{1}$ is the $N_\varepsilon \times N_\varepsilon$ unit matrix,

$$(\underset{\sim}{\overline{E}})_{ij} = 2\mu\hbar^{-2}(E - \varepsilon_j)\delta_{ij} \tag{2.23a}$$

$$[\underset{\sim}{\overline{V}}(R)]_{ij} = 2\mu\hbar^{-2} \int dr\, \phi_i^*(r)\, \overline{V}(R,r)\, \phi_j(r) \;,\; i,j \le N_\varepsilon \tag{2.23b}$$

and $\underset{\sim}{F}$ is the column vector of functions $\{F_1, F_2, \ldots F_{N_\varepsilon}\}$. Equation (2.22) represents a set of N_ε second-order ordinary linear differential equations, the so-called *close-coupled* equations. The asymptotic condition (2.19b) along with the physically reasonable condition

$$\underset{\sim}{F}(0) = \underset{\sim}{0} \tag{2.24}$$

is sufficient to ensure a unique solution, which can be determined numerically by any of the various techniques discussed by Secrest, Light and Parker elsewhere in this volume.

3. Close-Coupled Equations for CID

The close-coupling method generally works well for inelastic and reactive scattering. However, if the energy is sufficiently high to permit CID, then the expansion (2.19a) breaks down, simply because it does not take into account transitions into the continuum of BC. An obvious way to rectify this deficiency is to supplement the expansion with a continuum contribution as

$$\psi_i(\underset{\sim}{r}) = \sum_{j=1}^{N_b} F_j(R)\phi_j(r) + \int dk\, F(k;R)\phi(k;r) \;. \tag{3.1}$$

Strictly, (3.1) ignores the possibility of reactive scattering, although the relevant mathematical features of the problem remain unaltered. Substituting (3.1) into (2.18), left-multiplying both sides of the resulting equation successively by the bound ϕ_j^* and continuum $\phi^*(k;r)$ eigenfunctions, integrating over r, and finally making use of the orthonormality relations (2.9) and (2.12), we obtain the analogue of (2.22), namely

$$\{\underset{\sim}{1} d^2/dR^2 + \underset{\sim}{\overline{E}} - \underset{\sim}{\overline{V}}(R)\}\cdot \underset{\sim}{F}(R) = 0 \tag{3.2}$$

where now

$$(\underset{\sim}{1})_{ij} = \delta_{ij},\ i,\ j \leq N_b\ ;\ (\underset{\sim}{1})_{k;k'} = \delta(k-k'),\ k,k' \geq 0 \tag{3.3a}$$

$$(\underset{\sim}{\overline{E}})_{ij} = 2\mu\hbar^{-2}(E-\varepsilon_i)\delta_{ij},\ i,\ j \leq N_b;\ (\underset{\sim}{\overline{E}})_{k;k'} = 2\mu\hbar^{-2}[E-\varepsilon(k)] \times \delta(k-k'),\ k,k' \geq 0 \tag{3.3b}$$

$$[\underset{\sim}{\overline{V}}(R)]_{ij} = 2\mu\hbar^{-2} \int dr\ \phi_i^*(r)\ \overline{V}(R,r)\ \phi_j(r),\ i,\ j \leq N_b \tag{3.3c}$$

$$[\underset{\sim}{\overline{V}}(R)]_{k;j} = 2\mu\hbar^{-2} \int dr\ \phi^*(k;r)\ \overline{V}(R,r)\ \phi_j(r),\ j \leq N_b,\ k \geq 0 \tag{3.3d}$$

$$[\underset{\sim}{\overline{V}}(R)]_{i;k'} = 2\mu\hbar^{-2} \int dr\ \phi_i^*(r)\ \overline{V}(R,r)\ \phi(k';r), i \leq N_b,\ k' \geq 0 \tag{3.3e}$$

$$[\underset{\sim}{\overline{V}}(R)]_{k;k'} = 2\mu\hbar^{-2} \int dr\ \phi^*(k;r)\ \overline{V}(R,r)\ \phi(k';r),\ k,k' \geq 0 \tag{3.3f}$$

The column vector $\underset{\sim}{F}$ comprises the finite number N_b of $F_j(R)$ corresponding to the bound states of BC plus the uncountably infinite number of F(k;R) corresponding to the continuum states. The asymptotic form of the bound-state F's is given, as before, by (2.19b). The continuum-state F's behave asymptotically as

$$\lim_{R\to\infty} F(k;R) = -\ 2\pi i\mu[P(k)]^{-1}\ (2\pi\hbar)^{-1/2}\ T_i(k) \times \exp(i\ P(k)R/\hbar)\ , \tag{3.4}$$

where

$$[P(k)]^2/2\mu + \varepsilon(k) = E \tag{3.5}$$

and $T_i(k)$ is the T-matrix element describing the transition from initial bound vibrational state i to final continuum state k.

The matrices defined by (3.3) are all continuous and infinite, so that (3.2) represents an infinite, continuous set of coupled, second-order ordinary linear differential equations. In principle, (3.2) could be solved in the same fashion as (2.22). The (mathematical) problem, our central problem, is: how do we handle an uncountably infinite number of d.e.'s simultaneously? The naive answer is to discretize the continuous portion of (3.2). The obvious approaches to discretization fall into two categories:

(i) transformations from the continuous (k) representation to a discrete one;

(ii) alternate methods of solving (2.12), i.e. methods other than close-coupling.

I shall outline in the next Section two techniques of type (i).

4. Discretization of the Continuum

4.1 Transformation method

One way that discretization of the close-coupled equations can be achieved is simply by expanding $F(k;R)$ as

$$F(k;R) = \sum_n L_n(k) G_n(R) \;, \tag{4.1}$$

where L_n constitute a complete orthonormal set of functions (to be chosen). Combining (3.1) and (4.1), we can recast the stationary collision wavefunction in the form

$$\psi_i = \sum_{j=1}^{N_b} F_j(R)\, \phi_j(r) + \sum_{n>N_b} G_n(R)\, \chi_n(r) \tag{4.2}$$

in terms of discrete *pseudocontinuum* states χ_n, which are related to the L_n by a Fourier-type transform

$$\chi_n(r) = \int dk\, L_{n-N_b}(k)\, \phi(k;r),\; n = N_b + 1, \ldots N \;. \tag{4.3}$$

We can choose the set $\{L_n\}$ such that $\hat{H}^0_{BC}$ is diagonal in the basis $\{\chi_n\}$. Then in the new totally discrete representation the close-coupled equations (3.2) become

$$\{\underset{\sim}{1}\, d^2/dR^2 + \underset{\sim}{\overline{E}}' - \underset{\sim}{\overline{V}}'(R)\}\cdot \underset{\sim}{G}(R) = \underset{\sim}{0} \;, \tag{4.4}$$

where $\underset{\sim}{1}$, $\underset{\sim}{\overline{E}}'$, $\underset{\sim}{\overline{V}}'$, and $\underset{\sim}{G}$ are the totally discrete analogs of $\underset{\sim}{1}$, $\underset{\sim}{\overline{E}}$, $\underset{\sim}{\overline{V}}$, and $\underset{\sim}{F}$, respectively [see (2.22)]. In particular, we have

$$(\underset{\sim}{\overline{E}}')_{ij} = 2\mu\hbar^{-2}(E - \varepsilon_j)\delta_{ij} \;, \tag{4.5}$$

where now ε_j are eigenvalues of the true bond states for $j \le N_b$ and of the pseudocontinuum states for $j > N_b$.

The discrete set of close-coupled equations (4.4) can now be solved by more-or-less standard methods (mentioned at the end of Section 2) to yield "non-physical" T-matrix elements for transitions into the pseudostates. These "non-physical" elements can then be combined linearly to construct the required (i.e. "physical") T-matrix, from which the CID probabilities can finally be computed.

The "transformation" technique has been applied[8] to the following model: V_{BC} is a semi-infinite square well; $\overline{V}$ is a repulsive exponential. The parameters are chosen so that BC has only one bound state. The functions $L_n(k)$ are taken to be

$$L_n(k) = [2^{2n-1}(2n-1)!]^{-1/2}(2\beta/\sqrt{\pi})^{1/2} \times \exp(-\beta^2k^2/2)\, H_{2n-1}(\beta k)\ ,\ n = 1, 2, \ldots, \tag{4.6}$$

where $H_{2n-1}(\beta k)$ is the Hermite polynomial of degree 2n-1. The parameter β controls the spread of the $L_n(k)$ in k space or, equivalently, the spread of $\chi_n(r)$ in coordinate space. For a given set of parameters determining the PES (i.e. $V_{BC} + \overline{V}$) the close-coupled equations (4.4) were solved for ranges of values of β, N (the total number of pseudostates), and E (the total energy). We first fixed E and β and solved the close-coupled equations repeatedly, increasing N until the dissociation probability $P_{i\to D}$ had converged. We then fixed E and solved the equations for a range of values of β, ensuring convergence (as a function of N) of $P_{i\to D}$ for each value of β. We found that for a fixed E, the final (converged) value of $P_{i\to D}$ is independent of β and N, as it must be if the technique is valid.

4.2 Diagonalization of $\hat{H}^o_{BC}$

An alternate, and more direct, way to discretize (3.2) is simply to diagonalize the vibrational Hamiltonian in an arbitrary, discrete (square-integrable) basis $\{\chi_j\}$. The lowest N_b eigenvalues

$$\int dr\, \tilde{\chi}_i^*(r)\, \hat{H}^o_{BC}\, \tilde{\chi}_j(r) = \lambda_i\, \delta_{ij}\ ,\ i,\ j \le N_b \tag{4.7}$$

are (by the variational principle of quantum mechanics) upper bounds on the true bound-state energies ε_j; the corresponding eigenfunctions $\tilde{\chi}_j$ approximate the true bound-state functions ϕ_j, of course. The remaining set of $(N-N_b)$ eigenstates are (localized) pseudocontinuum states and are precisely analogous to the χ_n defined by (4.3). In the basis $\{\tilde{\chi}_j\}$ the close-coupled equations assume a discrete form identical to (4.4) and may be solved in the same fashion.

The "diagonalization" technique has been utilized[9] to compute CID probabilities for the following model: V_{BC} is a Morse potential with parameters chosen to give five (5) bound states; $\overline{V}$ is again taken to be a repulsive exponential. For reasons that may seem "natural" to the chemists, we chose to diagonalize $\hat{H}^o_{BC}$ in the basis

$$\psi_n(y) = [(n-1)!\alpha/\Gamma(b+n)]^{1/2} z^{(b+1)/2}$$
$$\text{X} \exp(-z/2)\, L_{n-1}^{(b)}(z) \ , \tag{4.8}$$

where $z \equiv \exp(-\alpha y)$, $y[\equiv(r-r_{eq})/r_o]$ is a reduced internuclear distance, and $L_n^{(b)}$ is the generalized Laguerre polynomial. In this basis $\hat{H}_{BC}^o$ is tridiagonal and the diagonalization is computationally straightforward. In testing the validity of the method we have to distinguish carefully between two parameters: N_d, the size of basis used to diagonalize $\hat{H}_{BC}^o$ and N, the number of pseudostates included in the close-coupled expansion. Since the $\{\chi_n\}$ are essentially determined by specification of the PES, we must check the convergence of the CID probabilities as a function of N_d, N and E. Fixing E such that a large portion of the continuum is (energetically) accessible, we have solved the close-coupled equations as a function of N_d and N. In actuality, we fixed N_d and then solved the equations repeatedly, increasing N until the results converged. We then repeated this process for a succession of increasing values of N_d. Unfortunately, N required for convergence at a given N_b increases with N_b, so we eventually reached the limit of our computer's capability. We were able, however, to achieve partial convergence (to about 1% in the best cases).

4.3 Discretization by the finite-difference method

As yet a third alternative, the finite-difference method (FDM) could be used to generate a discrete set of pseudostates. Thus, approximating the second derivative in (2.7) with the three-point finite-difference formula, one obtains

$$(\chi)_{n+1} - [2 + (V_{BC})_n - \varepsilon](\chi)_n - (\chi)_{n-1} = 0 \ , \tag{4.9}$$

where $(\chi)_n$ is the approximate value of the pseudo-eigenfunction at grid point n, $(V_{BC})_n$ is the value of the binding potential at that point, and ε is the eigenvalue. (Note that energy is given here in units of $\hbar^2/2\mu_{BC}\Delta r^2$, where Δr is the distance between grid points.) Specifying the physical boundary conditions

$$\chi(0) = \chi(L) = 0 \ , \tag{4.10}$$

where L is taken sufficiently large that $V_{BC}(L) \simeq 0$, converts (4.9) into a standard eigenvalue problem of dimensionality N_p, the number of grid points. Since the relevant FDM matrix is tridiagonal, the diagonalization procedure is computationally very simple, as noted in Section 4.2. As far as I am aware, this approach has not been applied, although it certainly appears worthy of serious investigation.

5. Alternate Methods of Solving the Schrodinger Equation for CID

Instead of the close-coupling Ansatz (2.19a), one could employ alternative numerical techniques to solve (2.18). An example, which has been applied[10] to CID recently, is the finite-element method (FEM). The application of the FEM to scattering problems has been described in great detail by Askar, et al.[11] Unfortunately, the model treated in Ref. 10 is not "realistic" in that the motion of atoms B and C is along a line perpendicular to the direction of motion of A. Even so, it does retain the essential mathematical feature of the problem, namely the continuum of BC. The numerical results seem to have converged and thus the FEM appears to offer a viable alternative. I am informed[12] that additional FEM computations on the model described in Section 4.2 are underway and am eagerly awaiting the results.

References

1. J. Ross, ed. "Molecular Beams," in Adv. Chem. Phys. (Interscience, New York, 1966), Vol. 10.
2. M. A. Eliason and J. O. Hirschfelder, J. Chem. Phys. 30 (1959) 1426.
3. M. Karplus, R. N. Porter and R. D. Sharma, J. Chem. Phys. 43 (1965) 3259.
4. J. R. Taylor, *Scattering Theory* (Wiley, New York, 1972).
5. E. A. McCullough and R. E. Wyatt, J. Chem. Phys. 54 (1971) 3578.
6. L. W. Ford, D. J. Diestler and A. F. Wagner, J. Chem. Phys. 63 (1975) 2019.
7. K. C. Kulander, J. Chem. Phys. 69 (1978) 5064.
8. E.-W. Knapp, Y.-W. Lin and D. J. Diestler, Chem. Phys. Letters 49 (1977) 379.
9. E.-W. Knapp and D. J. Diestler, J. Chem. Phys. 67 (1977) 4969.
10. G.-D. Barg and A. Askar, preprint.
11. A. Askar, A. Cakmak and H. Rabitz, Chem. Phys. 33 (1978) 267.
12. D. Secrest and D. Malik, private communication.

Extraction of Continuum Properties from L^2 Basis Set Matrix Representations of the Schrödinger Equation: the Sturm Sequence Polynomials and Gauss Quadrature

John T. Broad
Fakultät für Chemie
Universität Bielefeld
Postfach 8640
D-48 Bielefeld
Germany

I. Introduction.

We are concerned in these workshops with the solution of large linear systems and of differential equations. I want to discuss here a way of reformulating the Schrödinger differential equation for scattering problems with the help of a square integrable (L^2) basis into a system of linear equations. Since this is the standard approach used in calculating the (negative energy) bound states of many electron atoms and molecules, one goal is to extend the approach to scattering (positive energy) states in such a way as to exploit as much as possible of the considerable technology which quantum chemists have developed for bound states. While this is the basis of several new L^2 scattering calculation methods[1] which show promise of making accurate electron polyatomic molecule scattering calculations feasible in the near future, the theoretical justification of how L^2 functions can represent a continuum is not yet complete. What follows here provides this justification for the one dimensional radial Schrödinger equation in certain special L^2 basis sets.[2]

In Part II, I review the properties of the solutions of the radial Schrödinger differential equation for a Hamilton operator H^o containing the radial kinetic energy and some short range potential V^o and examine further how these solutions are perturbed by adding another, shorter range potential ΔV. Then in Part III, the regular solution of the differential equation is expanded formally in a complete orthogonal L^2 basis with the special property that the resulting matrix representation of H^o is tridiagonal. The completeness of the basis set and the solutions of the differential equation **combined** with this tridiagonal structure are shown to imply the orthogonality of the Sturm sequence polynomials with respect to an explicitly constructed weight function. This in turn yields a Gauss quadrature scheme ideally suited for approximating matrix elements of operator functions of H^o. Furthermore, when a potential of finite rank in the L^2 basis is added to H^o, the same structure prevails, with only an appropriate modification of the weight function. The relations obtained between the spectral density and phase shift of the solutions of the differential equation to the quadrature weight functions and the spacing of the abscissas are then particularly illuminating of how the L^2 basis represents the continuum. Finally, in Part IV, I touch on the extension of these results to general L^2 bases and to many particle scattering systems. The Appendix, which follows part IV, arose as a comment to Dennis Diestler's talk[3] and applies this formalism to the one dimensional Morse oscillator in the hope of illuminating how the three particle break-up problem can be approximated by a close coupling approach with Morse oscillator pseudostates.

II. Solutions of the Radial Schrödinger Equation for H^o.[4]

Consider the radial Schrödinger equation, written symbolically as

$$(H^o - E)\psi^o(r;E) = 0 \qquad 1)$$

where

$$H^o = -1/2\ d^2/dr^2 + V^o(r) \qquad 2)$$

and $E=k^2/2$* on the semi-infinite interval $r\varepsilon[0,\infty)$. Here it is assumed that $V^o(r)$ vanishes faster than $1/r$ at infinity, but may well be infinite at the origin, for it will usually contain the centrifugal term $\ell(\ell+1)/2r^2$. The solutions of physical interest, $\psi^o(r;E)$, obey boundary

---

*Dimensionless coordinates generated by choosing m=1 are used.

conditions at $r=0$ and $r\to\infty$ whose form depends on E. While ψ^o must vanish at the origin at all energies and die off at large r fast enough to keep it square integrable for $\mathrm{im}(k) > 0$ (the physical sheet), in the limit $\mathrm{Im}(k) \to 0^+$ it becomes oscillatory at large r.[5]

One straightforward way of determining ψ^o, is to express it in terms of two linearly independent solutions $\varepsilon^o_\pm(r;k)$ which fulfill the initial conditions for starting at large r and integrating inward

$$\varepsilon^o_\pm(r;k) \sim e^{\pm ikr} \quad \text{as} \quad r \to \infty, \text{ and} \qquad 3)$$

which have the constant Wronskian

$$\mathrm{Wr}(\varepsilon^o_+,\varepsilon^o_-) \equiv \varepsilon^o_+ d\varepsilon^o_-/dr - \varepsilon^o_- d\varepsilon^o_+/dr = -2ik \qquad 4)$$

as expected from the form of the differential equation (1). ε^o_+ and ε^o_- are called outgoing and incoming waves, respectively, from the prescription that the solutions to the time dependent Schrödinger equation be written as $\Psi(r;t) = \psi(r;E)\exp(-iEt)$.

Under certain commonly satisfied conditions on the potential $V^o(r)$[4], there is a single solution of the differential equation which is regular at the origin, i.e. which satisfies an initial condition at the origin to vanish as some arbitrary constant times a functional form determined by $V^o(r)$. Using $\varepsilon^o_+(r;k^*) = \varepsilon^{o*}_+(r;k) = \varepsilon^o_-(r;k)$, this regular solution can be written as the real linear combination

$$R^o(r;k) = [W^o_-(k)\varepsilon^o_+(r;k) - W^o_+(k)\varepsilon^o_-(r;k)]\,/2ik \qquad 5)$$

where the Wronskians

$$W^o_\pm \equiv \mathrm{Wr}(\varepsilon^o_\pm,R^o) = W^o e^{\pm i\phi^o(k)} \qquad 6)$$

are known as the Jost functions. From the large r behavior of $\varepsilon^o_\pm$ in Eqn. (3), the regular solution behaves as

$$R^o(r;k) \sim W^o(k)\sin(kr+\phi^o)/k \quad \text{as} \quad r \to \infty. \qquad 7)$$

Since the regular solution is unique and the physical solution must also be regular, they can only differ by an energy dependent normalization. If the physical solution is normalized at real positive energies as

$$\int_0^\infty dr\psi^{o*}(r;E)\psi^o(r;E) = \delta(E-E), \text{ then} \quad 8)$$

$$\psi^o(r;E) = \sqrt{2k}/\pi \; R^o(r;k)/W_+^o(k) \quad 9)$$

$$= \sqrt{2/\pi k} \; i/2(\varepsilon_-^o(r;k) - W_-^o(k)/W_+^o(k)\varepsilon_+^o(r;k))$$

which behaves asymptotically as $r \to \infty$ as

$$\sim \sqrt{2/\pi k} \; \sin(kr + \phi^o)e^{i\phi^o} \quad 10)$$

On the other hand, the physical solutions $\psi_{bd}(r)$ at negative energies (positive imaginary k), the bound states, must be regular at the origin and square integrable. A look at the large r behavior of the regular solution in Eqns. (5) and (7) indicates that such bound states can only exist at those energies, E_{bd}, where $W_+^o(k_{bd}) = 0$, and there $\psi_{bd}(r)$ is proportional to $R^o(r;k_{bd})$ and $\varepsilon_+^o(r;k_{bd})$ as well as to the residue of $\psi^o(r;E)$ as k approaches k_{bd}, with the constant of proportionality chosen to fix the L^2 norm of the bound states to one.

Now the resolvent operator $(H^o-E)^{-1}$ can be written as the Green's function

$$G_+^o(r,r';k) = R^o(r_<;k)\varepsilon_+^o(r_>;k)/W_+^o(k) \quad 10)$$

$$= \sqrt{\pi/2k} \; \psi^o(r_<;E)\varepsilon_+^o(r_>;k)$$

for outgoing wave boundary conditions. Finally, performing a Cauchy integral[4] in the complex k plane on the resolvent integral and removing the poles in G_+^o at the bound states yields the completeness of the set of scattering and bound states:

$$\int_0^\infty dE\psi^o(r;E)\psi^{o*}(r',E) + \sum_{bd} \psi bd(r)\psi_{bd}^*(r') = \delta(r-r') \text{ or} \quad 11a)$$

$$2/\pi \int_0^\infty dE \; k \; R^o(r;k)R^o(r';k)/W^o(k)^2 \quad 11b)$$

$$- 2\pi i \sum_{\text{Res at } k=k_{bd}} 2k/\pi R^o(r;k)R^o(r';k)/W_+^o(k)W_-^o(k) = \delta(r-r')$$

The completeness property will be used in the basis set analysis in Part III to establish the orthogonality of the Sturm sequence polynominals on an weight function closely related to the spectral density $2k/\pi W^o(k)^{-2}$ of the $R^o(r;k)$ of Eqn. (11b).

B. Addition of a Short Range Potential.

Now consider adding a potential $\Delta V(r)$ which is dominated by $V^{o}(r)$ both at r=0 and as $r\to\infty$ to give a potential of $V=V^{o}+\Delta V$ and a total Hamiltonian

$$H = H^{o} + \Delta V. \qquad 12)$$

Clearly, one can apply the same analysis to H as to H^{o} above to obtain solutions $R(r;k)$, $\varepsilon_{\pm}(r;k)$ and $\psi(r;E)$ with the Jost functions $W_{\pm}(k)$ directly. Of more interest, however, are the relations between the solutions with the Hamiltonians H and H^{o}, which depict how the scattering continuum of H^{o} is perturbed by the additional interaction $\Delta V(r)$. These relations can be obtained by writing $\Delta V\psi$ on the right hand side of the Schrödinger equation as an inhomogeneity,

$$(H^{o}-E)\psi(r;E) = -V(r)\psi(r;E) \qquad 13)$$

and integrating using the Green's function (Eqn. (10)) to give the Lippmann-Schwinger equation

$$\psi(r;E) = \psi^{o}(r;E) - \int dr' G^{o}_{+}(r,r';k)\Delta V(r')\psi(r';E), \qquad 14)$$

which ensures that $\psi\sim\psi^{o}$ if ΔV is turned off, thereby removing any arbitrariness in going from Eqn. (13) to Eqn. (14).

Using the dominance of $V^{o}(r)$ over $\Delta V(r)$ at infinity to require $\varepsilon_{\pm}(r;k)\sim\varepsilon^{o}_{\pm}(r;k)$ as $r\to\infty$ and some algebra[4] yields two important relations. First, the Fredholm determinant of the integral kernel $-G^{o}_{+}\Delta V$ in Eqn. (14) can be expressed in terms of the ratio of the H and H^{o} Jost functions as

$$D_{+}(k) \equiv \mathrm{Det}(1+G^{o}_{+}V) = \mathrm{Det}\left(\frac{H-E}{H^{o}-E}\right) = W_{+}(k)/W^{o}_{+}(k). \qquad 15)$$

Secondly, a comparison of the asymptotic form of the physical wave function $\psi(r;E)$ on the left and right hand sides of Eqn. (14) reveals

$$\psi(r;E) \sim \sqrt{2/\pi k}\; i/2\; [\varepsilon^{o}_{-} - W_{-}/W_{+}\; \varepsilon^{o}_{+}] \qquad 16)$$

$$\sqrt{2/\pi k}\; i/2\; [\varepsilon^{o}_{-} - W^{o}_{-}/W^{o}_{+}\; S(k)\varepsilon^{o}_{+}] \quad \text{where}$$

$$S(k) = e^{2i\delta(k)} = W_{-}/W_{+}\; W^{o}_{+}/W^{o}_{-} = 1-2\pi i T(k) \qquad 17)$$

and $\delta(k)$ is the phase shift, or the difference in phase of the two Jost functions

$$\delta(k) = \phi(k) - \phi^{o}(k), \text{ and} \qquad 18)$$

$$T(k) = \int_0^\infty dr\psi^o(r;E)\Delta V(r)\psi(r;E) \qquad 19)$$

is the transition matrix element giving the probability amplitude that the additional potential $\Delta V(r)$ perturbs a wave coming in as $\psi^o(r;E)$ into one going out as $\psi(r;k)$. The square of the sum of these amplitudes for each angular momentum (different $V^o(r)$) then gives the measurable scattering cross section.

III. A. Solutions of the Matrix Schrödinger Equation for H^o.

Now, after this lengthy introduction on the known form of the solutions of the differential equation, consider how the same problem looks in a complete orthonormal basis of functions ($X_n(r)$; $n=0,1,2,...$)

$$\int_0^\infty drX_n(r)X_{n'}(r) = \delta_{nn'} \qquad 20)$$

in which H^o is tridiagonal

$$\int X_n(r)H^oX_{n'}(r) \equiv H_{nn'} = 0 \text{ unless } n=n' \text{ or } n'\pm 1. \qquad 21)$$

Since bases can be found for physically relevant model Hamiltonians H^o (see Ref. 2 for the radial Coulomb Hamiltonian and the Appendix below for the one dimensional Morse oscillator), the striking analogies with the results of Part II are of more than mere formal interest.

We proceed by expanding the physical wave function in the basis as

$$\psi^o(r;E) = \Sigma_n X_n(r)\psi_n^o(k),$$

which must be understood in the sense of a distribution to be integrated over, since the $X_n(r)$ must all die at large r to be L^2, while $\psi^o(r;E)$ oscillates at large r, and presumes that $X_n(r)$ is regular at the origin in the same way as ψ^o. Projecting onto the Schrödinger Eqn. (1) with a basis function X_n then gives a three term recursion relation

$$H^o_{nn+1}\psi^o_{n+1}(k)+(H^o_{nn}-E)\psi^o_n(k)+H^o_{nn-1}\psi^o_{n-1}(k) = 0 \quad n=1,2,..., \qquad 23)$$

$$H^o_{o1}\psi^o_1(k) + (H^o_{oo}-E)\psi^o_o(k) = 0. \qquad 23a)$$

Concentrating on the first N equations, which can be depicted as

$$\begin{pmatrix} H^o_{oo}-E & H^o_{o1} & 0 & 0 & 0 \\ H^o_{10} & H^o_{11}-E & H^o_{12} & 0 & 0 \\ 0 & 0 & H^o_{N-1,N-2} & H^o_{N-1,N-1}-E \end{pmatrix} \begin{pmatrix} \psi^o_o \\ \vdots \\ \psi^o_{N-1} \end{pmatrix} = - \begin{pmatrix} 0 \\ \vdots \\ 0 \\ H^o_{N-1,N}\psi^o_N \end{pmatrix} \qquad 24)$$

makes it clear that $\psi^o_N(k)$ has N zeros at the N eigenvalues of H^o, truncated to N rows and columns. But we know another solution of this homogeneous three term recursion relation with exactly the same zeros, namely the Sturm sequence polynomials $p^o_n(E)$, which start with $p^o_o(E)=1$. Since the solution is unique up to an n-independent factor, this implies that

$$\psi^o_n(k) = \psi^o_o(k)p^o_n(E) \text{ for all n.} \qquad 25)$$

Now exploit the completeness of the $\psi^o(r;E)$ of Eqn. (11a):

$$\delta_{nn'} = \int_o^\infty\int drdr'X_n(r)\delta(r-r')X_{n'}(r) \qquad 26)$$

$$= \int_o^\infty dE\psi^o_n(k)\psi^{o*}_{n'}(k) - 2\pi i \sum_{\text{Res at } k=k_{bd}} \psi^o_n(k)\psi^{o*}_{n'}(k)$$

or using Eqn. (15),

$$\delta_{nn'} = \int_o^\infty dE\rho(E)p^o_n(E)p^o_{n'}(E) + \sum_{bd}\rho_{bd}p^o_n(E_{bd})p^o_{n'}(E_{bd}) \qquad 27)$$

$$= dE\rho(E)p^o_n(E)p^o_{n'}(E) \quad \text{where}$$

$$\rho(E) = |\psi^o_o(k)|^2 \quad \text{and} \qquad 28)$$

$$\rho_{bd} = -2\pi i \text{ Res } \psi^o_o(k)\psi^o_o(k^*) \quad \text{as} \quad E \to E_{bd}. \qquad 29)$$

Thus, the Sturm sequence polynomials are orthogonal with respect to the weight function $\rho(E)$ on the positive real axis augmented with ρ_{bd} at the bound states. They play the role in n space which the regular solutions $R^o(r;k)$ play in r space, while $\psi^o_o(k)$ corresponds to $\sqrt{2k/\pi}\, W^o_+(k)$.

Before exploiting further such analogies, let us look at the most important consequences of the orthogonality of the Sturm sequence polynomials, namely that they can be used to define a Gauss quadrature[6] tailor made to approximate matrix elements of operator functions of H^o. Suppose $O(H^o)$ is an operator function of H^o. Then a matrix element of $O(H^o)$ between some initial state i and some final state f can be written using the completeness relation Eqn. 11a) as

$$\int_0^\infty dr\; i^*(r)O(H^o)f(r) \equiv (i,O(H^o)f) \qquad 30)$$

$$= dE \int_0^\infty dr\; i^*(r)\psi^o(r;E)\; O(E)\int_0^\infty dr'\psi^{o*}(r';E)f(r')$$

where the sum over the bound states and integral over the continuum are abbreviated as in Eqn. (27). Introducing the weight function $\rho(E)$ in numerator and denominator then allows a quadrature of, say, degree N:

$$(i,O(H^o)f) = \sum_{j=1}^{N} w_j^{o(N)}/\rho^o(E_j^{o(N)})(i,\psi^o(E_j^{o(N)})O(E_j^{o(N)}) \times (\psi^o(E_j^{o(N)}),f) \qquad 31)$$

where the $w_j^{o(N)}$ are the quadrature weights, $E_j^{o(N)}$ the abscissas, i.e. $p_N^o(E_j^{o(N)}) = 0$ for j=1,2,...,N, and the radial integrals are denoted by scalar products in the obvious way. If, in addition, an N term separable expansion in the basis set of the scalar products:

$$(i,\psi^o(E)) \approx \sum_{m=0}^{N-1} (i,X_m)\psi_m^o(k) \qquad 32)$$

is approximately correct, the quadrature can be expressed as

$$(i,O(H^o)f) \approx \sum_{j=1}^{N} w_j^{o(N)} \sum_{m=0}^{N-1}(i,X_m)\psi_m^o(k_j^{o(N)})\; O(E_j^{o(N)}) \times \sum_{n=0}^{N-1} \psi_n^{o*}(k_j^{o(N)})(X_n,f) \qquad 33)$$

This can in turn be greatly simplified by recalling the eigenvalue equation satisfied by the $\psi_n^o(k)$ and $p_n^o(E)$ at the zeros of $p_N^o(E)$, for the normalized eigenvectors $\psi_{nj}^{o(N)}$ can be written as

$$\psi_{nj}^{o(N)} = p_n^o(E_j^{o(N)})/\left(\sum_{m=0}^{N-1} p_m^o(E_j^{o(N)})\right)^{1/2} = p_n^o(E_j^{o(N)})/(w_j^{o(N)})^{1/2} \qquad 34)$$

using the Christoffel[6] sum definition of the weights. Then using Eqn. (25) to relate ψ_n^o to p_n^o and Eqn. (28) for the weight function, the matrix element becomes

$$(i,O(H^o)f) \approx \sum_{j=1}^{N} (i,\psi_j^{o(N)})O(E_j^{o(N)})(\psi_j^{o(N)},f) \text{ where} \qquad 35)$$

$$\psi_j^{o(N)}(r) \equiv \sum_{n=0}^{N-1} X_n(r)\psi_{nj}^{o(N)} \qquad 36)$$

are the radial pseudostate eigenfunctions in the truncated basis.

Eqns. (34) through (36) display clearly how the eigenfunctions of H^{o} in the truncated basis (X_n;n=0,1,...,N-1), or pseudostates, represent the full bound and continuum spectral density: up to radii where N basis functions suffice, the pseudostates differ from the scattering wave function evaluated at the pseudostate eigenvalues by a normalization constant of just the form to make a sum over pseudostates (Eqn. (35)) equivalent to a Gauss quadrature (Eqn. (33)). The approximation scheme will be successful when (a) the separable expansion in Eqn. (32) is good, and when (b) O(E) is a smooth function.

Unfortunately, the physically most interesting operator function of H^{o}, its resolvent $(H^{o}-E)^{-1}$, is not a smooth function. As may be expected in analogy with the form of the Green's function in $\underline{r}$ space, Eqn. (10), an expression for the resolvent requires a second solution of the Schrödinger equation. Since the exponential like solutions, $\varepsilon^{o}_{\pm}(r;k)$, are irregular at the origin, while the basis functions behave like the regular solution there, an expansion of the form of Eqn. (22) for $\varepsilon^{o}_{\pm}(r;k)$ would not be appropriate. Instead, we can look directly for a second solution of the three term recursion relation Eqn. (23), but with a non-zero initial condition:

$$H^{o}_{nn+1}q^{o}_{n+1}(k)+(H^{o}_{nn}-E)q^{o}_{n}(k)+H^{o}_{nn-1}\,q^{o}_{n-1}(k) = 0 \quad n=1,2,... \qquad 37)$$

$$H^{o}_{01}q^{o}_{1}(k)+(H^{o}_{00}-E)q^{o}_{0}(k) = 1 \qquad 37a)$$

Based on some experience with the classical orthogonal polynomials, one can immediately write down an integral representation of the second solution:

$$q^{o}_{n}(k) = dE'\rho^{o}(E')p^{o}_{n}(E')\;/(E'-E), \qquad 38)$$

which, on the positive real energy axis in the $E+i0^{+}$ limit becomes

$$q^{o}_{n}(k) = P\;\; dE'\rho^{o}(E')p^{o}_{n}(E')/(E'-E)+i\pi\rho^{o}(E)p^{o}_{n}(E). \qquad 38a)$$

Furthermore, $q^{o}_{n}(k)$ can be shown to be an analytic function in the upper half k plane except for poles at bound states on the positive k axis. Now we examine the relation between the two solutions of the recursion relation to develop analogies to the Jost functions and Wronskians of Part II. From Eqn. (38a), the regular solution of the recursion, $p^{o}_{n}(E)$, can be expressed as

$$p_n^o(E) = [q_n^o(k) - q_n^o(k^*)]/2\pi i\rho(E) \tag{39}$$

This can be put in the form of Eqn. (5), if we let

$$w_+^o(k) = \sqrt{2k/\pi}\ \psi_o^o(k) \tag{40}$$

to put Eqn. (27) in line with Eqn. (11b) and define

$$e_n^o(k) = q_n^o(k)w_+^o(k) \tag{41}$$

and use Eqn. (28) for $\rho(E)$ to give

$$p_n^o(E) = [e_n^o(k)w_+^o(k^*) - e_n^o(k^*)w_n^o(k)]/2ik \tag{42}$$

Furthermore, if a Wronskian is defined in n-space as

$$Wn(f,g) \equiv H_{nn+1}^o(f_{n+1}g_n - f_n g_{n+1}), \tag{43}$$

then it follows from the recursion relations, Eqns. (23) and (37), that $Wn(q^o,p^o)=1$, and hence from the definition of e_n^o, that

$$Wn(e^o,p^o) = H_{nn+1}^o(e_{n+1}^o p_n^o - e_n^o p_{n+1}^o) = w_+^o, \tag{44}$$

so that the role of the Jost function as a Wronskian of the regular and the exponential solution is maintained. Finally, the e_n^o so defined can be used to construct a second function in r-space:

$$\tilde{\varepsilon}_+^o(r;k) \equiv \sum_n X_n(r)e_n^o(k). \tag{45}$$

Since this second function is regular at the origin and is linearly independent of the regular solution, it cannot satisfy the Schrödinger equation exactly. Indeed, from Eqns. (37), it follows that

$$(H^o-E)\tilde{\varepsilon}_+^o(r;k) = X_o(r)w_+^o(k) \tag{46}$$

which can be solved using the Green's function to give

$$\tilde{\varepsilon}_+^o(r;k) = w_+^o(k)\int dr' G_+^o(r,r';k)X_o(r) \tag{47}$$

or, using Eqn. (10) and taking r very large,

$$\tilde{\varepsilon}_+^o(r;k) \sim \varepsilon_+^o(r;k) \quad \text{as } r \to \infty\ ^* \tag{48}$$

---

*$\tilde{\varepsilon}_+^o(r;k)$ is then a function with the asymptotic behavior of the exponential solution regularized at the origin, and is hence perhaps the ideal choice for use in the Kohn variational method in this basis.

Now it is straightforward to construct the resolvent, or Green's matrix. Taking a hint from Eqn. (10) for the Green's function, we try, using Eqns. (40), (41) and (25),

$$G^{o}_{nn'}(k) = p^{o}_{n<}(E)q^{o}_{n>}(k) = p^{o}_{n<}(E)e^{o}_{n>}(k)/w^{o}_{+} = \sqrt{\pi/2k}\,\psi^{o}_{n<}(k)\,e^{o}_{n>}(k), \qquad 49)$$

which can be shown to be the inverse of the matrix $H^{o}-E$ using Eqns. (23) and (47). As a alternative, $G^{o}_{nn'}(k)$ can be expressed using the integral representation of q^{o}_{n}, Eqn. (38), and the orthogonality of the p^{o}_{n}, Eqn. (27), as

$$G^{o}_{nn'}(k) = dE'\rho^{o}(E')p^{o}_{n}(E')p^{o}_{n'}(E')/(E'-E), \qquad 49a)$$

which corresponds to the spectral representation of the resolvent matrix.

In addition, it is illuminating to examine the quadrature approximations to the singular integrals $q^{o}_{n}(k)$ and $G^{o}_{nn'}(k)$, after splitting off the singularity. Adding and subtracting $p^{o}_{n}(E)$ to the integrand (in Eqn. 38) gives

$$q^{o}_{n}(k)= dE'\rho^{o}(E')(p^{o}_{n}(E')-p^{o}_{n}(E))/(E'-E)+p^{o}_{n}(E)\ dE'\rho^{o}(E')/(E'-E) \qquad 50)$$

Since the integrand in the first term is a polynomial of degree n-1, a quadrature of degree n will be exact, giving

$$q^{o}_{n}(k) = p^{o}_{n}(E)[q^{o}_{n}(k) - \sum_{j=1}^{n} w^{o(n)}_{j}/(E^{o(n)}_{j}-E)], \qquad 51)$$

where the definition of q^{o}_{o}, Eqn. (38), has been used. This could also be interpreted to say that the difference between $q^{o}_{o}(k)$ and its quadrature approximation is $q^{o}_{n}(k)/p^{o}_{n}(E)$. In the complex k plane, away from the zeros of $p^{o}_{n}(E)$, this difference should be small[6], while on the real axis the smooth but multivalued structure of $q^{o}_{n}(k)$ is approximated by a set of poles and zeros.

Performing the same subtraction on the integrand of Eqn. (49a) for the Green's matrix and performing a quadrature of degree $N \geq n$ and n' yields

$$G^{o}_{nn'}(k)=\frac{\int dE'\rho^{o}(E')p^{o}_{n}(E')p^{o}_{n'}(E')-p^{o}_{n}(E)p^{o}_{n'}(E)+p^{o}_{n}(E)p^{o}_{n'}(E)}{E'-E} \qquad 52)$$

$$= \sum_{j=1}^{N} w^{o(N)}_{j}\, p^{o}_{n}(E^{o(N)}_{j})\,p^{o}_{n'}(E^{o(N)}_{j})/(E^{o(N)}_{j}-E)$$

$$+ p_n^o(E)p_{n'}^o(E)[q_n^o(E) - \sum_{j=1}^{N} w_j^{o(N)}/(E_j^{o(N)}-E)]$$

$$= \sum_{j=1}^{N} \psi_{nj}^{o(N)}\psi_{n'j}^{o(N)}/(E_j^{o(N)}-E) + p_n^o(E)p_{n'}^o(E)q_N^o(E)/p_N^o(E) \quad ,$$

where Eqns. (34) and (51) were used to get the last line. The first term is exactly the approximate resolvent which would be obtained by inverting the matrix H^o-E truncated to N rows and columns, while the second term corrects the error and recovers the smooth but multivalued analytic structure of the resolvent from the rational approximation of the first term. As noted above for q_n^o, this correction should be small for complex k. The complex coordinate method[7] exploits this by introducing complex r to remove the pseudostate eigenvalues from the real axis and thereby vastly improve the accuracy of approximating $G_{nn'}^o(k)$ by the pseudostate sum evaluated at real energies. With the explicit form of the error term obtained here, the convergence properties of that approximation, at least in the special basis X_n in which H^o is tridiagonal, can be examined.

Finally, a new computational approach to the generalized Gauss quadrature discussed above can be extracted from Eqns. (38a) and (51) for $q_n^o(k)$. At a zero of $p_N^o(E)$,

$$q_N^o(E_j^{o(N)}) = w_j^{o(N)} p_{N'}^o(E_j^{o(N)}) \quad , \tag{53)}$$

which is clearly real. Hence, if we define the phase of $q_N^o(k)$ as

$$\phi_N^o(E) = \arg q_N^o(k) \quad , \tag{54)}$$

this phase goes through at each zero of $p_N^o(E)$:

$$\phi_N^o(E_j^{o(N)}) = \pi j. \tag{55)}$$

This is reminiscent of Levinson's rule for the number of bound states, but appearing here for pseudostates. Moreover, the derivative of this function which maps the quadrature abscissas onto their cardinal number, evaluated at the abscissas gives

$$d\phi_N^o/dE \quad E_j^{o(N)} = \rho^o(E_j^{o(N)})/w_j^{o(N)} \quad , \tag{56)}$$

which is just the factor needed in Eqn. (34) to relate the pseudostate eigenvectors to the expansion coefficients of the actual scattering wave functions. Numerical experiments[8] on known quadratures have shown that interpolating the abscissas by some smooth function and evaluating its derivative at the abscissas to be a reasonably accurate and very cheap method of generating $w_j^{o(N)}/\rho(E_j^{o(N)})$.

B. Addition of a Potential of Finite Rank in the Basis.

Nothing of the form of the results obtained above is lost, if we add a potential of finite range, not in r, but in n, i.e.

$$\Delta V_{nn'} = \int_0^\infty X_n(r)\Delta V(r)V_{n'}(r)dr \neq 0 \text{ for n and } n' \geq N-1 \tag{57}$$

Performing a Householder transformation on the first N rows and columns then gives a tridiagonal Hamiltonian matrix $H_{nn'} = H^o_{nn'} + \Delta V_{nn'}$ albeit with the first N basis functions transformed among one another, of the same form as for $H^o_{nn'}$. We again obtain a set of coefficients $\psi_n(k)$, $p_n(E)$, $q_n(k)$, $e_n(k)$ and a Jost function $w_+(k)$, denoted now without a superscript zero, and as with the solutions of the differential equation, examining the perturbations introduced by the potential ΔV reveals the physics of interest.

Since $H_{nn'}$ and $H^o_{nn'}$ are identical for n or $n' \geq N$, enforcing the condition $\tilde{\varepsilon}_+(r;k) \sim \tilde{\varepsilon}^o_+(r;k) \sim \exp(ikr)$ as $r \to \infty$ implies

$$e_n(k) = e^o_n(k) \quad \text{for } n \geq N \tag{58}$$

Then, as in Part II, Eqn. (18), the phase shift δ can be identified with the difference in phase between w^o_+ and w_+ and finally expressed in terms of the Sturm sequence polynomials of H and the two solutions of the H^o problem at the boundary in n-space where the potential ceases to act:

$$\delta(k) = -\arg\left(H^o_{N-1,N}(p_{N-1}q^o_N - p_N q^o_{N-1})\right) \tag{59}$$

Similarly, the ratio of the two Jost functions is given explicitly as

$$D_+(k) = w^o_+(k)/w_+(k) = H^o_{N-1,N}(p_{N-1}q^o_N - p_N q^o_{N-1}) \tag{60}$$

That $D_+(k)$ is also the Fredholm determinant of the Lippmann Schwinger kernel now requires no generalization of the concept of determinant to operators, for, since $H^o_{nn'} = H_{nn'}$ for n or $n' \geq N$,

$$\mathrm{Det}(\mathbb{1} + G^o_+V) = \mathrm{Det}((H-E)/(H^o-E)) = \mathrm{Det}_N((H-E)/(H^o-E)), \tag{61}$$

where the subscript N denotes restriction to the first N rows and columns.

In a few algebraic steps, this can be shown to be identical to Eqn. (60).

On the other hand, the phase of the Fredholm determinant can be expressed through Eqns. (41), (53), (59) and (60) as

$$\delta(k) = -\arg(w_+^o(k)/w_+(k)) = \arg(q_N/q_N^o) = \phi_N(k)-\phi_N^o(k) \qquad 62)$$

which relates, through Eqn. (64), the phase shift directly to the relative spacing of the pseudostate eigenvalues of H and H^o.

IV Generalization to Other Bases and Many Channel Scattering.

In Part II, we were able to show explicitly how a complete L^2 basis can be used to represent the scattering solutions, and how a truncation of the basis to N terms generates an N term Gauss quadrature approximation to integrals over the spectral density. Unfortunately, all the proofs depend on knowing a complete basis in which some appropriate H^o is tri-diagonal, which is tantamount to being able to solve the H^o Schrödinger equation exactly in the first place. The possibility developed in Part IIIB of adding a finite rank approximation to an arbitrary additional potential is only a good start, since the results require variational correction to get the correct Born approximation high energy behavior.

As a first approximation for potential scattering calculations in arbitrary basis sets, however, one could conjecture that Gauss-like quadratures are generated by diagonalizing H and H^o truncated to N rows and columns and that the phase shift $\delta(k)$ can be obtained from interpolations of the eigenvalues in accordance with Eqn. (72). Similarly, the Stieltjes imaging work of Langhoff[9] conjectures that the imaginary part of the Fredholm determinant can be approximated by a reference function times a polynomial in the energy, just as in Eqn. (70) above. Furthermore, the Stieltjes imaging technique[9] applied to many partical systems assumes the existence of an effective one particle density, which is constructed by using total energy moments of the many particle density in a finite L^2 basis to determine the quadrature weights and abscissae belonging to the density.

In the closing coupling approach to many channel scattering, only one particle is allowed to go to large distances, leaving the target in one of several discrete states[4]. In this approximation, the ideas discussed in Part II can be generalized to allow calculation of multichannel scattering in an L^2 basis[10]. This has been done for certain cases and applied with success to electron-hydrogen scattering[11] in a basis set where the radial kinetic energy is tridiagonal. In addition, the splitting of the Green's matrix into the resolvent of the truncated Hamiltonian and a correction which is small away from the pseudostate eigenvalues still holds in the multichannel case, which helps explain the success of the dilitation transformation[7] in calculating resonances in electron-molecular scattering.

Unfortunately, at energies near or above the three particle break-up threshold, the close coupling approach is in trouble, for the only representation of the target's continuum is through a small number of pseudostates. This generates non-physical, resonance-like, effects even in the two particle channels much as what would happen if the correction to the Green's matrix were left out of Eqn. (72). Clearly, the quadrature of the target continuum generated by the pseudostates must be understood, meaning in some complicated sense a two dimensional quadrature overall. In the Appendix, we develop the pseudostate Gauss quadrature of the Morse oscillator to contribute to understanding Dennis Diestler's pseudostate approach to 3 body break-up. Yet, there is still another difficulty, since even a two dimensional quadrature would constitute a separable approximation to an integral kernel which is known from the Fadeev analysis to be non-compact.[4] Perhaps a detailed analysis of the L^2 representation of the three body problem in the spirit of Fadeev will reveal how to smooth the pseudostates of the target and allow accurate calculations in the break-up region.

Appendix: An Application to the One Dimensional Morse Oscillator

Dennis Diestler's approach[3] to calculating the probability of collision induced dissociation of a Morse oscillator turns out to include a direct application of the equivalent quadrature formalism. Diestler approximated the Morse oscillator continuum by a finite number of pseudostates and even chose an L^2 basis in which the Hamiltonian is tridiagonal. In this Appendix, I will delineate how the pseudostates represent the oscillator continuum without touching on the larger 3-body problem.

In appropriately scaled coordinates, the Morse oscillator potential takes the form

$$V(x) = b^2(e^{-2x} - 2e^{-x}) \text{ with } x \quad (-\infty,\infty), \tag{A1}$$

while the Schrödinger becomes

$$-d^2\psi/dx^2+V(x)\psi(x) = k^2\psi(x). \quad (A2)$$

Changing the coordinate to

$$z = 2be^{-x} \qquad z\varepsilon(0,\infty) \quad (A3)$$

allows the differential equation (A2) to be solved exactly[12] in terms of confluent hypergeometric functions. The two exponential solutions analogous to Eqn. (3) are

$$\varepsilon_{\pm}(x;k) = e^{\pm ikx} e^{-z/2} (1/2-b\mp ik;1\mp 2ik;z) \quad (A4)$$

which behave asymptotically as $\exp(\pm ikx)$ as $x\to\infty$ or $z\to 0$. The solution which is regular at the other end of the interval ($x\to-\infty$, or $z\to\infty$) is

$$R(x;k) = e^{-z/2} z^{\pm ik} (1/2-b\pm ik;\ 1\pm 2ik;z), \quad (A5)$$

which can be expressed as a linear combination of $\varepsilon_{\pm}$ as in Eqn. (6) through the Wronskian

$$W_{\pm}(k) = Wr(\varepsilon_{\pm},R) = \Gamma(1\mp 2ik)(2b)^{\pm ik}/\Gamma(1/2-b\mp ik) \quad (A6)$$

The physical wave function $\psi(x;k)$ is given then by Eqn. (9).

The bound states appear at the zeros of $W_+(k)$ on the positive imaginary k axis where the argument of the gamma function in the denominator of Eqn. (A6) is a negative integer $\bar{n}$ at

$$k_{\bar{n}} = i(b-\bar{n}-1/2) \quad \text{for} \quad 0 \le \bar{n} < b-1/2. \quad (A7)$$

At the bound state energies $E_{\bar{n}} = k_{\bar{n}}^2/2$, R and ε_+ become proportional and the confluent hypergeometric functions become Laguerre polynomials giving normalized bound states

$$\psi_{\bar{n}}(x) = (-1)^{\bar{n}} \sqrt{\frac{(2b-2\bar{n}-1)\Gamma(\bar{n}+1)}{\Gamma(2b-\bar{n})}}\ z^{b-1/2-\bar{n}} e^{-z/2} L_{\bar{n}}^{2b-2\bar{n}-1}(z) \quad (A7)$$

We now turn to the basis set used by Diestler, which was chosed to represent the bound states well, but still be complete:

$$X_n(x) = z^{b-\bar{N}+1/2} e^{-z/2} L_n^{2b-2\bar{N}} \sqrt{\Gamma(n+1)/\Gamma(2b-2\bar{N}+n+1)} \quad n=0,1,2,\ldots \quad (A9)$$

where $\bar{N}$ is the number of bound states, or according to Eqn. (A7), the largest integer less than b+1/2. The basis functions X_n are orthonormal, while the Hamiltonians $-d^2/dx^2+V(x)$ is tridiagonal symmetric with the non-zero elements

$$H_{nn} = (n+1-\bar{N})(2b-2\bar{N}+2n+1)-n-(b-\bar{N}+1/2)^2 \tag{A10}$$

$$H_{nn-1} = -(n(2b-2\bar{N}+n))^{1/2}(n-\bar{N})$$

Note that the Hamiltonian matrix splits into two blocks: $0\leq n\leq\bar{N}-1$ and $n\geq\bar{N}$. This occurs because the $\bar{N}$ bound states $\psi_{\bar{n}}$ are exactly representable in terms of the first $\bar{N}$ basis functions $X_n(0\leq n\leq\bar{N}-1)$. Since the scattering wave functions are orthogonal to the bound states, this also means that the formal expansion in Eqn. (22) of the physical wave function at positive energies starts at $n=\bar{N}$:

$$\psi(x;k) = \sum_{n=\bar{N}} X_n(x)\psi_n(k) \tag{A11}$$

Using the known solution from Eqns. (A5) and (A6) and the orthonormality of the basis functions, it requires only straightforward algebra to express the coefficients $\psi_n(k)$ in a form analogous to Eqn. (25) (some modification is required because $\psi_o(k)$ vanishes)

$$\psi_n(k) = D_+(k)p_n(E) \text{ for } n=\bar{N}, \bar{N}+1,... \tag{A12}$$

with

$$D_+(k) = \sqrt{k/\pi}\,|\Gamma(b-\bar{N}+1/2+ik)|^2/W_+(k) \tag{A13}$$

and

$$p_n(E) = \pi_{\bar{N}}(E)(-1)^{\bar{N}} \sqrt{\Gamma(n+1)\Gamma(2b-2\bar{N}+n+1)}/(\Gamma(2b-\bar{N}+1)\Gamma(\bar{N}+1)) \tag{A14}$$

$$\times \; {}_3F_2(-n+\bar{N},b+1/2+ik,b+1/2-ik;2b-\bar{N}+1,\bar{N}+1;1) \quad ,$$

where

$$\pi_{\bar{N}}(E) = \prod_{\bar{n}=0}^{\bar{N}-1} (k^2-k_{\bar{n}}^2) \tag{A15}$$

has zeros at all the bound state energies and the ${}_3F_2$ can be shown to be a polynomial of degree $n-\bar{N}$ in E. The polynomials $p_n(E)$ are orthogonal on the weight function

$$\rho(E) = |D_+(k)|^2 = k/\pi|\Gamma(1/2-b+ik)\Gamma^2(1/2+b-\bar{N}+ik)/\Gamma(1+2ik)|^2 \tag{A16}$$

The second function $q_n(k)$ is best expressed by recurring down to $q_{\bar{N}}$ (recursion to q_o is not possible because $H_{\bar{N}-1,\bar{H}}$ vanishes) and then evaluating numerically the integral representation analogous to Eqn. 38)

$$q_{\bar{N}}(k) = \int dE'\; \rho(E')p_{\bar{N}}^2(E')/(E'-E) \tag{A17}$$

For the case of one bound state with $b=\bar{N}=1$ at $k=i/2$, the integral in Eqn. (A17) is expressible in terms of a known polygamma function[6] as

$$q_1(k) = 1-(k^2+1/4)G^{(1)}(1/2-ik) \tag{A18}$$

Explicit knowledge of the q_n then gives the correction term to the pseudostate approximation to the Green's matrix in Eqn. (52), with which it should be possible to determine how good Diestler's approach to 3 body break-up is. Work is in progress on this problem.

REFERENCES

1) For an overview through 1978, see W.P. Reinhardt, Comp. Phys. Comm. 17, 1 (1979) and references therein, as well as Ref. 2,7,9 and 10 below.

2) E.J. Heller, W.P. Reinhardt and H.A. Yamani, J. Comput. Phys. 13, 536 (1973); E.J. Heller, T.N. Rescigno, and W.P. Reinhardt, Phys. Rev. A 8, 2946 (1973); H.A. Yamani and W.P. Reinhardt, Phys. Rev. A11, 1156 (1975); J.T. Broad, Phys. Rev. A18, 1012 (1978).

3) E.-W. Knapp and D.J. Diestler, J.Chem.Phys. 67, 4969 (1977); D.J. Diestler, this volume.

4) R.G. Newton, Scattering Theory of Waves and Particles (McGraw-Hill, New York, 1966), especially Chapter 12.

5) H. Weyl, Math. Ann. 68, 220 (1910); H. Hehenberger, H.V. McIntosh and Erkii Brändas, Phys. Rev. A10, 1494 (1974.

6) Higher Transcendental Functions, Bateman Manuscript Project, A. Erdeýli ed. (McGraw-Hill, New York 1953), Vols I and II.

7) Int.J.Quantum Chem. 14, No. 4 (1978) is devoted entirely to complex scaling; C.W. McUrdy and T.N. Rescigno, Phys. Rev. A21, 1499 (1980) and references therein.

8) H.A. Yamani and W.P. Reinhardt, Phys. Rev. A11, 1156 (1975).

9) C.T. Corcoran and P.W. Langhoff, Chem. Phys. Lett. 41, 609 (1976).

10) J.T. Broad and W.P. Reinhardt, J. Phys. B9, 1491 (1976).

11) J.T. Broad and W.P. Reinhardt, Phys. Rev. A14, 2159 (1976).

12) L.D. Landau and E.M. Lifschitz, Quantum Mechanics, Non-Relativistic Theory (Pergammon, London 2nd.ed. 1965), §23.

APPROXIMATE SOLUTION OF SCHRÖDINGER'S EQUATION FOR ATOMS

Charlotte Froese Fischer
Department of Computer Science
Vanderbilt University
Nashville, TN 37235, USA

I INTRODUCTION

The numerical solution of Schrödinger's equation for atoms, $H\psi = E\psi$, where

$$H = \sum_{i=1}^{m} \left\{ -\frac{1}{2}\Delta_i + \frac{Z}{r_i} \right\} + \sum_{i \neq j} \frac{1}{r_{ij}}$$

presents several difficulties, the most serious one being the high dimensionality of the equation, even for moderately sized atoms. For example, for iron which has 26 electrons, the number of space variables is already 78, a number well beyond the capability of conventional methods. A second, though less severe difficulty, arises from the singularities that occur when either r_i (the distance of the ith electron from the nucleus) is zero, or r_{ij} (the distance between the ith and the jth electron) is zero. Fortunately, Schrödinger's equation also has several properties which often make it possible to obtain reasonably accurate, approximate solutions.

First, the Hamiltonian H is "almost" separable. In fact, in the limit as the nuclear charge Z tends to infinity with the number of electrons remaining fixed, the repulsion between the electrons becomes negligible and the dominant term of the Hamiltonian is separable. Second, the terms in the Hamiltonian never involve the co-ordinates of more than two electrons. Thus, by using a separable approximation (to be described later) a fairly accurate zero-order reference state may be obtained. This approximation neglects the correlation in the motion of the electrons and so the error in this solution is referred to as the "correlation error". Using the second property, the solution may be corrected to first-order, in a perturbation theory sense, by a series of problems of similar complexity as that of a two-electron system. In most applications the resulting solution is of sufficient accuracy. When specific atomic properties are desired, the correlation corrections can often be kept to a minimum[1].

Before considering the separable approximation, let us review results for two special cases.

(i) $m = 1$. In this case there is no r_{ij} term and the equation is separable in spherical co-ordinates with

$$\psi(r, \theta, \phi) = R(nl; r)\, Y_{lm}(\theta, \phi) \tag{1}$$

The radial *factor* $R(nl; r)$ is usually expressed in terms of a radial *function* $P(nl; r)$, where

$$R(nl; r) = (1/r)\, P(nl; r)$$

and

$$\left(\frac{d^2}{dr^2}+\frac{2Z}{r}-\frac{l(l+1)}{r^2}-\frac{1}{n^2}\right)P(nl;r)=0$$

$$P(nl;0)=0;\quad P(nl;r)\to 0 \text{ as } r\to\infty$$

The functions $Y_{lm}(\theta,\phi)$ present in Eq. (1) are well-known spherical harmonic functions. Each electron also has a spin co-ordinate and since H is independent of spin, the function multiplied by a spin function will still be a solution of Schrödinger's equation. One electron functions of this form are referred to as "spin-orbitals".

(*ii*) $m = 2$. In this case there are six space variables though the problem may be reduced to one with only three variables[2]. For spherically symmetric solutions one may think of the nucleus and the two electrons as being in a plane with the co-ordinates of the problem being either (r_1, r_2, r_{12}) or (r_1, r_2, θ) where θ is the angle between the two position vectors. Some results for this case are presented in Table 1.

Table 1. Comparison of energies in atomic units for He

Method	$1s^2\ ^1S$	$1s2s\ ^3S$	$1s2s\ ^1S$
Series[3]	—2.9037	—2.1752	—2.1460
Finite Difference[4]	—2.9036	—2.1741	
Hartree-Fock[1]	—2.8617	—2.1743	—2.1699

The first are very accurate results obtained by Pekeris from variational series expansions in powers of r_1, r_2, and r_{12}. (Only the first few decimal places are quoted here.) They are generally considered to be as accurate as experiment. The finite difference results were obtained by Hawk and Hardcastle using the r_1, r_2, and θ variables. The $1s^2\ ^1S$ calculation required 30,525 unknowns and 2,000 iterations for convergence whereas the $1s2s\ ^3S$ case required 62,049 unknowns and 1,700 iterations. The amount of computer time was not specified in their paper, but if the improvement of a single unknown took a microsecond, the first calculation would have required about 17 hours of computer time and the second about 35 hours. Note that their result is accurate for the first case but not as accurate for the second. The latter requires an even larger grid with more unknowns to achieve similar accuracy. The Hartree-Fock method is based on the separable approximation and the above calculations required only a few seconds of CPU time on an IBM 370/168. For the $1s2s\ ^3S$ case the results are more accurate than the those obtained from the lengthier finite difference calculations. This is an example of a case where the solution is such that the separable approximation produces remarkably good results. For the ground state $1s^2\ ^1S$ case, the two electrons occupy similar regions of space and hence the correlation error is greater.

II THE SEPARABLE APPROXIMATION

In order to focus on the essence of the Hartree-Fock method and its extension to the multi-configuration Hartree-Fock (MCHF) method, let us consider the simpler model problem,

$$\begin{gathered} -\Delta u + f(x,y)\,u = \lambda\, u; \quad -1 < x, y < 1 \\ u(\pm 1, y) = 0; \quad u(x, \pm 1) = 0 \end{gathered} \tag{2}$$

The solutions of the above elliptic boundary value problem are the stationary points of the functional

$$F(w) = \frac{\int_{-1}^{1}\int_{-1}^{1}\left[w_x{}^2 + w_y{}^2 + f(x,y)w^2\right]dx\,dy}{\int_{-1}^{1}\int_{-1}^{1} w^2\,dx\,dy}, \quad w \in \mathcal{H} \tag{3}$$

where $\mathcal{H}$ is a Hilbert space of twice differentiable functions on the domain $-1 < x, y < 1$.

The above model problem can readily be extended to m dimensions by letting Δ be the Lapacian over m dimensions and replacing $f(x,y)$ by $f(x_1, \ldots, x_m)$. In order that our model problem have similar properties as those of the Schrödinger equation we will also require that

$$f(x_1, \ldots, x_m) = \sum_{i>j} g(x_i, x_j)$$

The elliptic boundary value problem of Eq. (2) may be solved by a variety of methods[5,6]. Each requires a grid or mesh. Let

$$h = 2/(n+1); \quad x_i = -1 + i\,h; \quad y_j = -1 + j\,h$$

Then the finite difference approximation, for example, leads to a system of n^2 equations (n^m in general) of the form

$$A\,U = \Lambda\, S\, U$$

where U is a vector of unknowns approximating the solution $u(x_i, y_j)$ and $\Lambda \approx \lambda$. Though the matrices A and S are sparse and the problem possibly could be solved efficiently by Rayleigh-Quotient iteration, the dimension of the system grows too quickly with m to be practicable.

Suppose now that

$$f(x,y) = f_1(x) + f_2(y)$$

Then the problem is *separable* in that a solution exists of the form

$$\begin{gathered} u(x,y) = u_1(x)\,u_2(y) \\ \lambda = \lambda_1 + \lambda_2 \end{gathered}$$

where

$$\begin{gathered} -u_1'' + f_1(x)\,u_1 = \lambda_1\,u_1, \quad -1 < x < 1, \; u_1(\pm 1) = 0 \\ -u_2'' + f_2(y)\,u_2 = \lambda_2\,u_2, \quad -1 < y < 1, \; u_2(\pm 1) = 0 \end{gathered}$$

Numerical methods of solution now lead to a pair of eigenvalue problems

$$A_i\,U_i = \Lambda_i\,S_i\,U_i\,, \quad i = 1 \text{ or } 2$$

where all matrices A_i or S_i are band matrices. Again there is an obvious extension to higher dimensions and both the number of unknowns as well as the number of arithmetic operations per iteration (assuming each system goes through an iteration) is $O(mn)$. This represents a more tractable rate of growth in the problem size.

The separable approximation assumes that

$$f(x, y) \approx f_1(x) + f_2(y)$$

However, it is not known how best to define $f_1(x)$ and $f_2(y)$ *a priori* so the variational formulation is used to find two functions $u_1(x)$ and $u_2(y)$ such that the functional of Eq. (3) over a subspace of $\mathcal{H}$ of product functions, say $w_1(x)\, w_2(y)$, is stationary or possibly a minimum. That is, $F(w_1 w_2)$ must be stationary with respect to all allowed variations in the functions w_1 and w_2 which leads to a coupled system of non-linear integro-differential equations for the solution, namely

$$\begin{aligned} -u_1'' + \left\{ \int_{-1}^{1} f(x, y)\, u_2{}^2\, dy \right\} u_1 &= \lambda_1 u_1, \quad -1 < x < 1 \\ -u_2'' + \left\{ \int_{-1}^{1} f(x, y)\, u_1{}^2\, dx \right\} u_2 &= \lambda_2 u_2, \quad -1 < y < 1 \end{aligned} \tag{4}$$

with

$$u_i(\pm 1) = 0, \quad \int_{-1}^{1} u_1{}^2\, dx = 1 \text{ and } \int_{-1}^{1} u_2{}^2\, dy = 1$$

The above system of equations is solved iteratively. Let

$$\begin{aligned} g_1(x; u, v) &= \int_{-1}^{1} f(x, y)\, u(y)\, v(y)\, dy \\ g_2(y; u, v) &= \int_{-1}^{1} f(x, y)\, u(x)\, v(x)\, dx \end{aligned}$$

Then the iteration process may be defined as follows:

$$\begin{aligned} -u_1^{(k+1)''} + g_1(x; u_2^{(k)}, u_2^{(k)})\, u_1{}^{(k+1)} &= \lambda_1^{(k+1)} u_1^{(k+1)} \\ -u_2^{(k+1)''} + g_2(x; u_1^{(k+1)}, u_1^{(k+1)})\, u_2^{(k+1)} &= \lambda_2^{(k+1)} u_2^{(k+1)} \end{aligned}$$

Note that this is now an uncoupled system of equations of the same form as Eq. (4).

Several theorems can easily be proved.

Theorem 1. When $f(x, y) = f_1(x) + f_2(y)$, then $u_1^{(1)}(x) = u_1(x)$ and $u_2^{(1)}(y) = u_2(y)$.

Theorem 2. The "best" separable approximation such that $F(w)$ is stationary over the subspace of product functions is

$$f_1(x) = g_1(x; u_2, u_2); \quad f_2(y) = g_2(y; u_1, u_1)$$

The first theorem assures us that in the case of a separable problem the functions will have converged after one iteration. The second provides an *a posteriori* separable approximation to $f(x, y)$. Note however that the approximation depends on which of the many solutions of the boundary value problem is being sought.

In practice, the above "coupling" iteration is combined with an iterative procedure for finding a single eigensolution of a boundary value problem such as Rayleigh-Quotient iteration. At the same time, if a finite difference approximation is used to solve the linearized boundary value problem, a deferred difference correction based on the current estimates of the solution can be included[7].

In general, since the integrals g_i must be evaluated for n different values of x or y, each integral involving n data points, the amount of computation per iteration becomes $O(n^2)$. However, in Schrödinger's equation the function $f(x, y)$ is such that the integral can be expressed as a sum of a few solutions of differential equations for which the amount of computation is again $O(n)$.

Because of the special form of the function $f(x_1, \ldots, x_m)$ for $m > 2$, the dimension of the integrals never exceeds one, but the number of pairs is $O(m^2)$.

III IMPROVING THE SEPARABLE APPROXIMATION

The separable approximation may not always yield sufficiently accurate results and an improvement scheme is needed.

The subspace of product functions suggests that one consider the expansion of the exact solution in terms of a tensor product basis. Let

$$\{u_{1i}(x),\ i = 1, 2, \ldots\} \text{ and } \{u_{2j}(y),\ j = 1, 2, \ldots\}$$

each be an orthonormal, complete set of basis functions over the interval $[-1, 1]$ satisfying the boundary conditions. Then the tensor product forms a basis for the solution space and

$$u(x, y) = \sum_i \sum_j a_{ij}\, u_{1i}(x)\, u_{2j}(y)$$

or, in matrix vector form

$$u(x, y) = U_1{}^T A\, U_2$$

where

$$U_1{}^T = (u_{11}(x),\ u_{12}(x), \ldots)$$
$$U_2{}^T = (u_{21}(y),\ u_{22}(y), \ldots)$$

The above basis, of course, is not unique since any orthogonal transformation of a basis is also a basis. The "best" basis could be defined as one for which the number of non-zero coefficients a_{ij} is a minimum.

For a general matrix A there exist two orthogonal matrices O_1 and O_2 and a diagonal matrix D such that

$$A = O_1{}^T D\, O_2$$

Then, for a particular solution, there exists a transformation such that

$$\begin{aligned} u(x, y) &= \tilde{U}_1{}^T D\, \tilde{U}_2 \\ &= \sum_i d_i\, \tilde{u}_{1i}(x)\, \tilde{u}_{2i}(y) \end{aligned}$$

where $\tilde{U}_k = O_k U_k$, $k = 1$ or 2. An important special case is the one where $f(x, y) = f(y, x)$. Then the solution of our problem must be either symmetric or antisymmetric. If we use the same basis functions for the two dimensions, say u_i, the matrix A is similarly symmetric or antisymmetric. In the first case, there exists a single orthogonal transformation such that

$$u(x, y) = \sum_i d_i \tilde{u}_i(x) \tilde{u}_i(y)$$

whereas in the second

$$u(x, y) = \sum_i d_i \{\tilde{u}_{2i-1}(x) \tilde{u}_{2i}(y) - \tilde{u}_{2i}(x) \tilde{u}_{2i-1}(y)\}$$

In each case the transformation has the effect of reducing a double sum to a single sum. By using the variational formulation of the problem it is possible to proceed directly to an approximation in the reduced form. Let us consider an example.

Suppose $f(x, y)$ is symmetric and

$$u(x, y) \approx w(x, y) = d_1 u_1(x) u_1(y) + d_2 u_2(x) u_2(y)$$

We wish to find functions u_1 and u_2 and constants d_1 and d_2 such that $F(w)$ is stationary with respect to variations in these quantities. As before, let us assume the basis functions are orthonormal and that our solution is normalized as well. Then the variation is subject to constraints and Lagrange multipliers must be introduced, one for each constraint.

Define

$$F_{ij} = \int_{-1}^{1} \int_{-1}^{1} u_i(x) u_i(y) [-\Delta + f(x, y)] u_j(x) u_j(y) \, dx \, dy$$

Then

$$F(w) = d_1^2 F_{11} + 2 d_1 d_2 F_{12} + d_2^2 F_{22}$$

Introducing Lagrange multipliers and constraints we find that

$$\begin{aligned} W(d_1, d_2, u_1, u_2) = F(w) &+ \lambda_{11} \int_{-1}^{1} u_1^2 \, dx + \lambda_{12} \int_{-1}^{1} u_1 u_2 \, dx \\ &+ \lambda_{22} \int_{-1}^{1} u_2^2 \, dx - \lambda(d_1^2 + d_2^2) \end{aligned}$$

must be stationary. The condition, $\partial W / \partial d_i = 0$, leads to the eigenvalue problem

$$\begin{bmatrix} F_{11} - \lambda & F_{12} \\ F_{12} & F_{22} - \lambda \end{bmatrix} \begin{bmatrix} d_1 \\ d_2 \end{bmatrix} = 0$$

whereas the condition that $\delta W = 0$ for variations in u_1 and u_2 leads to

$$\begin{bmatrix} -u_1'' \\ -u_2'' \end{bmatrix} + \begin{bmatrix} g(x; u_1, u_1) - \lambda_{11} & (d_2/d_1)\{g(x; u_1, u_2) - \lambda_{12}\} \\ (d_1/d_2)\{g(x; u_1, u_2) - \lambda_{12}\} & g(x; u_2, u_2) - \lambda_{22} \end{bmatrix} \begin{bmatrix} u_1(x) \\ u_2(x) \end{bmatrix} = 0$$

Because of symmetry the subscript on g has now been omitted.

Thus it is necessary to solve an eigenvalue problem coupled to a system of coupled integro-differential equations. Note that the entries F_{ij} in the matrix depend of the functions u_1 and u_2, and that certain coefficients in the integro-differential equations depend on the eigenvector of the matrix F.

At this stage a brief comparison with the finite element method is of interest. In the latter, a basis with finite support is selected so that in this basis, many integrals F_{ij} are zero. The basis is not varied and only the eigenvalue problem

$$(F - \lambda I)\, d = 0$$

need be solved. The dimension of the problem is considerably larger but with the matrix $F = (F_{ij})$ sparse and the eigenvector d generally dense. In the above extension to the separable approximation, we determine a basis such that the eigenvector is sparse. When $m > 2$, certain matrix elements may also become zero because of orthogonality conditions, but generally the matrix is not exceedingly sparse.

IV APPLICATION TO SCHRÖDINGER'S EQUATION

Shrödinger's equation has the property that the Hamiltonian H commutes with th total angular momentum operator of the system. Thus the wavefunction ψ must also be a eigenfunction of this operator. A very general and powerful algebra has been developed by Racah[8] for generating these eigenfunctions from sums of products of the one-electron spherical harmonics or spin-functions and for evaluating the angular integrals that arise in the quantum mechanics of atoms. Thus it is both convenient and efficient to use these functions to define the spin-angular dependence of the solution.

Let the eigenfunctions of angular momentum and spin be designated by $|\gamma\ {}^{2S+1}L>$ where γ is a label describing the coupling of the one-electron functions and where ${}^{2S+1}L$ designates a spectroscopic term. Basis functions or "configuration state functions" are the defined as

$$\Phi(\gamma\ {}^{2S+1}L) = \mathcal{A}\,[\prod_{i=1}^{m} R(n_i l_i; r)]\ |\gamma\ {}^{2S+1}L>$$

where $\mathcal{A}$ is a antisymmetrizing operator. Thus the unsymmetrized function is a product of radial factors similar to those of the one-electron problem multiplied by a spin-angular factor.

In the above basis, the wavefunction for the $1s^2\ \ {}^1S$ ground state of helium has an expansion of the form

$$\psi = \sum_{l=0}^{\infty} \left\{ \sum_{n} \sum_{n'} a_{n,n'}^{(l)}\, R(nl; r_1)\, R(n'l; r_2) \right\} |ll\ {}^1S>$$

which, in the reduced form, becomes

$$\psi = \sum_{l=0}^{\infty} \left\{ \sum_{n=l+1}^{\infty} d_n^{(l)} \tilde{R}(nl; r_1) \tilde{R}(nl; r_2) \right\} |ll\, ^1S>$$

$$= \sum_{l=0}^{\infty} \sum_{n=l+1}^{\infty} d_n^{(l)} \Phi(nl^2\, ^1S)$$

In general, the multi-configuration Hartree-Fock approximation assumes

$$\psi \approx \sum_{i=1}^{M} c_i\, \Phi(\gamma_i\, ^{2S+1}L)$$

where the coefficients c_i and the radial functions that enter into the definition of the configuration state functions are such that the energy is stationary subject to orthormality constraints. The special case where $M = 1$ is known as the Hartree-Fock approximation.

Table 2. Some MCHF results for the ground and first excited state of He

$1s^2\ ^1S$				$1s2s\ ^3S$			
M	l	Configuration	E	M	l	Configuration	E
1	0	$1s^2$	-2.86168	1	0	$1s2s$	-2.17425
2	0	$+2s^2$	-2.87800	2	0	$+3s4s$	-2.17426
3	0	$+3s^2$	-2.87887	3	1	$+2p3p$	-2.17517
4	0	$+4s^2$	-2.87899	4	1	$+4p5p$	-2.17517
5	1	$+2p^2$	-2.89855	5	2	$+3d4d$	-2.17522
6	1	$+3p^2$	-2.90015			E^{exact}	-2.17523
7	1	$+4p^2$	-2.90040				
8	2	$+3d^2$	-2.90218				
9	2	$+4d^2$	-2.90252				
10	3	$+4f^2$	-2.90291				
11	4	$+5g^2$	-2.90303				
		E^{exact}	-2.90372				

Some numerical results for helium are presented in Table 2. These show how the energy decreases as more and more terms are included in the sum, first those with $l = 0$, then $l = 1$, and so on. Unfortunately the sum on l is rather slowly convergent and adding more terms becomes less and less rewarding. Bunge[9] has used systematic extrapolation procedures to estimate the remaining errors in the energy, but no research has been performed on the feasibility of extrapolation when other atomic properties are of interest.

More details on the calculations for helium can be found elsewhere[10].

When more than two electrons are present, the reduced form cannot always be employed as simply since now the basis functions involve products of more than two radial factors. However, because the Hamiltonian only includes terms with the co-ordinates of at most two electrons, the Hartree-Fock approximation can be improved appreciably by including only single and double orbital replacements that lead to expansions like those for helium. This approach was used successfully in a study of a four-electron system, *Be* $1s^2 2s^2\ {}^1S$ [11].

V NUMERICAL SOLUTION OF THE MCHF EQUATIONS

Numerical methods for this problem have been published[12] and a general program MCHF77[13] is available. Only a few special problems will be mentioned here.

Like the eigenvalue problem, the MCHF and the HF equations have many solutions. Spectroscopists label the observed states according to the characteristics of the dominant configuration state in the expansion of the exact wavefunction. In the Hartree-Fock approximation, the radial functions to a large extent are like those for hydrogen; that is, the lowest eigenfunction of a given symmetry type has no node, the next one node, and so on. Thus node counting is a simple method for obtaining the desired solution but must be used with care since solutions to the Hartre-Fock equations may have additional small oscillations for large values of r. Cases also have been found where a small oscillation occurred near the origin, at least during the intial phases of the iterative process.

In some cases the energy functional is invariant under a transformation in the form of a rotation of a pair of radial functions constrained through orthogonality. These cases must be detected and the corresponding Lagrange multiplier set to zero as shown by Koopmans[14]. In other cases, the radial basis can be rotated during each cycle of the iterative process so as to attain a stationary energy. Pairwise rotation has been found to be sufficient[12].

Finally, it should be noted that the coefficients in the expansion are the components of an eigenvector of a matrix, the energy the corresponding eigenvalue. Only one eigensolution is required in an MCHF iteration and as the iterations proceed, good initial estimates of the eigenvector will be available. Again, the desired eigenvector is not specified in terms of the energy but rather in terms of the dominant component. By setting this component to unity, computing an energy estimate from a Rayleigh-Quotient, and then correcting the remaining components by solving the system of equations obtained from the eigenvalue problem by omitting the equation for the dominant component, a rapidly convergent procedure is obtained. Occasionally degeneracy effects arise and several large components are present. In such cases a procedure for finding the eigenvector with a specified relative phase in these components would be ideal. Instead, when difficulties arise, MCHF77 uses a damping procedure which tends to prevent the eigenvector components from changing sign as a sequence of eigenvalue problems are solved, each with slightly different matrix elements.

VI CONCLUSION

Schrödinger's equation differs from many other equations occuring in science or engineering in that it is a partial differential equation with many variables. By taking advantage of its properties and using the separable approximation reasonably accurate solutions can be obtained even for fairly large atoms.

The above discussion has assumed that the MCHF equations were solved numerically. A common alternative approach is to expand each radial function in terms of an analytic basis, usually of the type $r^q e^{-\varsigma r}, q = l, l+1, \ldots$, an approach that leads to the matrix MC-SCF method[15]. This avoids the necessity of solving differential equations but introduces an uncertainty about the adequacy of a given basis. Frequently the exponents ς are optimized. This is a nonlinear optimization problem that adds considerably to the computation time of the method. In order to avoid the necessity of optimization Gilbert[16] considered the use of a spline basis but finally concluded that this approach was not suitable for atomic structure calculations[17].

ACKNOWLEDGMENTS

This research was supported in part by a US Department of Energy grant.

REFERENCES

1. Froese Fischer, C., **The Hartree-Fock Method for Atoms**, Wiley Interscience, New York, (1977).
2. Bhatia, A. K., and Temkin, A., Symmetric Euler-angle decomposition of the two-electron fixed-nucleus problem, **Rev. Mod. Phys. 36** (1964) 1050-1064.
3. Pekeris, C. L., $1\ ^1S, 2\ ^1S$, and $2\ ^3S$ states of H^- and He, **Phys. Rev. 126** (1962) 1470-1476.
4. Hawk, I.L., and Hardcastle, D.L., Finite-difference solution to the Schrödinger equation for the ground state and first-excited state of Helium, **J. Comput. Phys. 21** (1976) 197-207.
5. Fox, L., Finite difference methods for elliptic boundary value problems, **The State of the Art in Numerical Analysis** (Edited by D. Jacobs) Academic Press, New York (1977).
6. Wait, R., Finite element methods for elliptic problems, **The State of the Art in Numerical Analysis** (Edited by D. Jacobs) Academic Press, New York (1977).
7. Froese Fischer, C., The deferred difference correction for the Numerov method, **Comput. Phys. Commun. 2** (1971) 124-126.
8. Racah, G., The theory of complex spectra II, **Phys. Rev. 62** (1942) 438-462; also III, **Phys. Rev. 63** (1943) 367-382.
9. Bunge, C., Accurate determination of the total electronic energy of the Be ground state, **Phys. Rev. A14** (1976) 1965-1978.
10. Froese Fischer, C., The solution of Schrödinger's equation for two-electron systems by the MCHF procedure, **J. Comput. Phys. 13** (1973) 502-521.
11. Froese Fischer, C., and Saxena, K.M.S., Correlation study of Be $1s^2 2s^2$ by a separated pair numerical multiconfiguration Hartree-Fock procedure, **Phys. Rev. A9** (1974) 1498-1506.
12. Froese Fischer, C., Numerical solution of general Hartree- Fock equations for atoms, **J. Comput. Phys. 27** (1978) 221-241.

13. Froese Fischer, C., A general multiconfiguration Hartree-Fock program, **Comput. Phys. Commun.** 14 (1978) 145-153.

14. Koopmans, T. A., Über die zuordnung von Wellenfunktionen und Eigenwerten zu den einzelnen Electronen eines Atoms, **Physica** 1 (1933) 104-113.

15. Hinze, J. and Roothaan, C.C.J., Multiconfiguration Self-consistent field theory, **Prog. of Theor. Phys. Suppl.** 40 (1967) 37-51.

16. Gilbert, T. L., The spline representation, **J. Chem. Phys.** 62 (1975) 1289-1298.

17. Altenberger–Siczek, A. and Gilbert, T. L., Spline bases for atomic calculations, **J. Chem. Phys.** 64 (1976) 432-433.

NUMERICAL INTEGRATION OF LINEAR INHOMOGENEOUS ORDINARY DIFFERENTIAL EQUATIONS APPEARING IN THE NONADIABATIC THEORY OF SMALL MOLECULES

L. Wolniewicz
Institute of Physics
Nicholas Copernicus University
87-100 Torun, Poland

1. Introduction

Sets of coupled ordinary differential equations appear in various branches of physics and chemistry. In consequence many special numerical techniques have been developed to deal efficiently with particular physical problems. However, only very few papers have been published on the problem of molecular bound states described by sets of ordinary differential equations. For the computation of eigenvalues two efficient methods exist: Gordon's method [1,2] and the method due to Johnson [3]. Yet these methods are not very efficient if the eigenfunctions rather than eigenvalues are needed. Also neither of these methods is directly applicable to inhomogeneous equations which arise, e.g. when a coupled channel problem is perturbed and the perturbation theory is used to deal with the perturbation. This is a situation that is encountered in one of the approaches to the nonadiabatic theory of diatomic molecules [4]. In this theory the solutions of the inhomogeneous equations are needed for a subsequent evaluation of integrals representing the nonadiabatic energy corrections.

The origin of difficulties in the numerical integration of the inhomogeneous set is the same as in the case of homogeneous equations: some of the solutions of the corresponding homogeneous equations grow very fast and in consequence they swamp completely the desired solution of the homogeneous set. Below we present a method [5] that can be used to overcome this difficulty.

2. Specification of the problem

Since it is practically impossible to design an efficient numerical method without making some assumptions about the elements of the Jacobi matrix we give below the relevant equations for a molecule to specify the orders of magnitude that we are interested in. Let μ be the reduced mass of the two nuclei and $r \in [0,\infty)$ the scalar internuclear distance. If the remaining coordinates in the center of mass system are denoted by x, the nonrelativistic hamiltonian of the molecule can be written as (see, e.g. [6])

$$H = -\frac{1}{2\mu r}\frac{\partial^2}{\partial r^2} r + H_m(x;r) \tag{1}$$

where the operator $H_m(x;\ r)$ commutes with r. The explicit form of H_m can be found in [6] but we will not need it here.

To avoid unnecessary complications in the presentation we omit here the physical aspects of the problem and refer the interested reader to [4, 5].

Our numerical problem consists in evaluating integrals of the form

$$E'' = \int \psi\theta dx dr \tag{2}$$

where $\theta(x,r)$ is a given function and ψ a bounded solution of the equation

$$(H-E)\psi = \rho(x,r) \tag{3}$$

with E being a constant and ρ a given function vanishing at $r = 0$ and at $r \to \infty$.

Approximate solutions of (3) can be sought in the form of a finite expansion

$$\psi = r^{-1}\sum_{i=1}^{N} \phi_i(x;r)\chi_i(r) \tag{4}$$

where $\{\phi_i(x;r)\}$ is a given basis set that we will assume to be real and orthonormal for all r:

$$\int\phi_i(x,r)\ \phi_k(x,r)dx = \delta_{ik} \tag{5}$$

Below we will use the notation A^T for a transposition of a matrix or vector A.

A substitution of (4) into (3) leads to the coupled equations for the column vector $\chi = (\chi_1,\ \chi_2,\ldots,\chi_N)^T$:

$$\{\frac{d^2}{dr^2} + B\frac{d}{dr} + \lambda - V\} \chi = \phi \qquad (6)$$

with the boundary conditions

$$\chi(0) = 0 \quad , \quad \chi(r_{max}) = 0. \qquad (7)$$

$\lambda = 2\mu E$, B and V are r-dependent NxN matrices with elements

$$B_{ik} = 2\int \phi_i \frac{\partial}{\partial r} \phi_k \, dx$$

$$V_{ik} = \int \phi_i [2\mu H_m \phi_k - \frac{\partial^2}{\partial r^2} \phi_k] \, dx \qquad (8)$$

and Φ is a vector

$$\Phi_i = 2\mu r \int \phi_i \, \theta dx$$

The integral (2) reads now

$$E'' = (2\mu)^{-1} \lambda''$$

with

$$\lambda'' = \int \chi^T \Phi dr \qquad (9)$$

Clearly, since the hamiltonian (1) is hermitian,

$$V^T = V + dB/dr . \qquad (10)$$

In the molecular bound state problems one has usually: $r_{max} \approx 10$, $\mu > 1000$, $\|\Phi\|$, $\|B\| \approx 1$, $\|\lambda - V\| \approx 200 - 1000$ with $\|\cdot\|$ being a norm: $\|A\| = \max |A_{iu}|$.

Moreover, for small and large values of r all eigenvalues of $\lambda - V$ are negative while they are both positive and negative for intermediate r. Thus the homogeneous equations corresponding to (6) have some solutions that grow very fast. If no precautions were taken these solutions would swamp completely the desired solution of the inhomogeneous problem. This difficulty can be in principle overcome either by making use of the imbedding technique [7] or by employing some kind of stabilizing transformation related to Gordon's stabilization [1]. Here we will make use of the latter approach.

Since the elements of V are quite large it is advantageous to transform the equations (6) by a linear transformation

$$\chi = S \cdot f \tag{11}$$

where the NxN matrix S satisfies the differential equation

$$2dS/dr + BS = 0 \tag{12}$$

with the initial condition

$$S(r_o) = I \ . \tag{13}$$

Here I is the indentiy transformation and r_o is arbitrary but fixed. Clearly, since B is skewsymmetric, S is orthogonal. Hence, the numerical integration of (12) creates no problems and can be performed by any standard method.

Now we get instead of (6), (7), (9) the problem

$$\frac{d^2f}{dr^2} + Qf = g \tag{14}$$

$$f(0) = f(r_{max}) = 0 \tag{15}$$

$$\lambda'' = \int f^T g dr \tag{16}$$

with

$$g = S^T \Phi$$

and

$$Q = -\frac{1}{2} S^T \{V + V^T - \frac{1}{2} B^T B - 2\lambda\} S \tag{17}$$

Naturally Eqs. (14) have all the inherent instabilities that are typical for Eqs. (6) and so (14) cannot be integrated in a straightforward manner. This point is discussed in the next section.

3. The algorithm

Let us consider the boundary value problem (14), (15) with f and g being N-dimensional vectors and Q a real and symmetric r-dependent matrix. By F^{out} we will denote a NxN matrix whose columns represent N linearly independent solutions of the corresponding homogeneous equations

$$d^2F/dr^2 + Q\cdot F = 0 \tag{18}$$

satisfying

$$F(0) = 0 \; . \tag{19}$$

The solutions of Eq. (18) satisfying

$$F(r_{max}) = 0 \tag{20}$$

will be denoted by F^{in}.

Similarly, f^{out} and f^{in} are solutions of Eq. (14) satisfying $f^{out}(0) = 0$, and $f^{in}(r_{max}) = 0$. The solution of (14), (15) is now

$$f = f^{out} + F^{out}\, c = f^{in} + F^{in}\, d \tag{21}$$

where c and d are constant vectors that can be in principle easily determined e.g. by using Eq. (21) twice, for two different values of r, and solving the resulting linear equations. However, in physical applications Q has both large and small eigenvalues and therefore the solutions forming F become practically linearly dependent and f becomes proportional to a fast growing solution of the homogeneous equations, when one proceeds with the integration of Eqs. (18) and (14), respectively. In such a case Eq. (21) does not hold and it is difficult to determine f. Fortunately, the difficulties arising from the linear dependencies in F can be overcome by using some sort of an orthogonalization process whenever F threatens to become linearly dependent [1, 8] and thus assuring the linear independence of the solutions. This, however, does not suffice if one lost the small inhomogeneous solution in f^{out} and f^{in} because it was swamped by large homogeneous solutions. Therefore, one should try to subtract from f, in the course of integration, linear combinations of the homogeneous solutions in order to keep f possibly small. Below we apply this idea to get an algorithm for the solution of Eqs. (14) - (15) and for the evaluation of the integral (16).

To begin with let us convert Eqs. (14) and (18) into a discrete problem by using the Numerov method (see [9,3]). For a given number of $n + 1$ grid points we write $h = r_{max}/n$ and $r_k = k \cdot h$ $(k = 0,1,\dots, n)$.

Defining

$$T_k = -\frac{h^2}{12} Q(r_k) , \qquad G_k = \frac{h^2}{12} g(r_k) , \tag{22}$$

we get the following recurrence relations for $f_j = f(r_j)$, $F_j = F(r_j)$:

$$Y_{j+1} - U_j Y_j + Y_{j-1} = G_{j+1} + 10\, G_j + G_{j-1} , \tag{23}$$

$$Y_o = Y_n = 0 , \tag{24}$$

$$Z_{j+1} - U_j Z_j + Z_{j-1} = 0 \tag{25}$$

where

$$Y_j = (I - T_j) f_j , \tag{26}$$

$$Z_j = (I - T_j) F_j , \tag{27}$$

$$U_j = (2 \cdot I + 10 \cdot T_j) \cdot (I - T_j)^{-1} . \tag{28}$$

Let Z^1_j be a solution of Eq. (25) satisfying

$$Z^1{}_0 = 0 , \quad Z^1{}_1 = I . \tag{29}$$

We define now

$$Z^s{}_j = Z^1{}_j \, (Z^1{}_s)^{-1} , \; s > 0, \qquad j = 0,1,\dots \tag{30}$$

i.e. $Z^s{}_j$ is a solution of Eq. (25) with the boundary conditions

$$Z^s{}_0 = 0 \;,\; Z^s{}_s = I \;. \tag{31}$$

If Y_j is any solution of Eq. (23) satisfying $Y_0 = 0$, and c a constant vector,

$$\bar{Y}_j(c) = Y_j - Z^1{}_j \; c \tag{32}$$

is also a solution and satifies $\bar{Y}_O = O$. We will use the notation

$$Y^k{}_j = \bar{Y}_j(c_o) \tag{33}$$

where c_o is a vector that minimizes

$$\delta = (\bar{Y}_k)^T \; \bar{Y}_k + (\bar{Y}_{k-1})^T \; \bar{Y}_{k-1} \;. \tag{34}$$

Note that in view of $Z_O = O$, and $\bar{Y}_O = O$ Eq. (34) leads to $Y^1{}_1 = O$.

It follows from the above definitions that the vectors that form $Z^s{}_j$ are orthogonal for j = s, and so they are certainly linearly independent for indices j close to s. At the same time $Y^s{}_j$ is small.

To avoid the problem of testing $Z^s{}_j$ for the linear independence of columns, it is advantageous to change the independent basis by going over from $Z^s{}_j$ to $Z^{s+1}{}_j$ after each integration step. Thus we can proceed as follows: Suppose Eq. (23) and (25) have been solved for $j \leq k$ and we have $Y^k{}_j$ and $Z^k{}_j$. Now, computing $Z^k{}_{j+1}$ from Eq. (25), we get for $j \leq k+1$:

$$Z^{k+1}{}_j = Z^k{}_j \; (Z^k{}_{k+1})^{-1} \;. \tag{35}$$

It follows from Eq. (31):

$$Z^k{}_{k+1} = Z^1{}_{k+1} \; (Z^1{}_k)^{-1} = R_k \tag{36}$$

where R_k is the ratio matrix introduced recently by Johnson [3]. Thus Eq. (35) reads now

$$Z^{k+1}{}_j = Z^k{}_j \cdot R_k{}^{-1} \tag{37}$$

Equation (23) yields $Y^k{}_j$ for j = k+1 and the new vector is:

$$Y^{k+1}{}_j = Y^k{}_j - Z^{k+1}{}_j \, c^k \tag{38}$$

with c^k satisfying according to Eq. (34):

$$(R_k + R_k{}^{-1}) \, c^k = Y^k{}_k + R_k Y^k{}_{k+1} \, . \tag{39}$$

In (39) use has been made of (31) and of the symmetry of R_k [3].

So far we have constructed the outward solutions related through Eqs. (26), (27) to f^{out} and F^{out} appearing in Eq. (21). The inward solutions, $\hat{Y}^k{}_j$ and $\hat{Z}^k{}_j$ can be constructed in a similar manner if we start from

$$\hat{Y}^{n-1}_n = \hat{Y}^{n-1}_{n-1} = 0 \, , \; \hat{Z}^{n-1}_n = 0 \text{ and } \hat{Z}^{n-1}_{n-1} = I \, ,$$

and use (23) and (25) for decreasing indices. Instead of Eqs. (36)-(39) we get for the inward solutions:

$$\hat{Z}^k{}_{k-1} = \hat{Z}^{n-1}{}_{k-1} \, (\hat{Z}^{n-1}{}_k)^{-1} = \hat{R}_k \tag{40}$$

$$\hat{Z}^{k-1}{}_j = \hat{Z}^k{}_j \, \hat{R}_k^{-1} \tag{41}$$

$$\hat{Y}^{k-1}{}_j = \hat{Y}^k{}_j - \hat{Z}^{k-1}{}_j \cdot d^k \tag{42}$$

with d^k satisfying:

$$(\hat{R}_k + \hat{R}_k^{-1}) \, d^k = \hat{Y}^k{}_k + \hat{R}_k \hat{Y}^k{}_{k-1} \, . \tag{43}$$

Now, having both the outward solution, $Y^m{}_j$, for $0 \leq j \leq m+1$ and the inward solution $\hat{Y}^{m+1}{}_j$ for $m \leq j \leq n$, we can use Eq. (21) to match them.

In analogy with Eq. (21) we write:

$$Y_j = Y^m{}_j - Z^{m+1}{}_j \, c = \hat{Y}^{m+1}{}_j - \hat{Z}^m{}_j \, d \tag{44}$$

and for j = m and j = m+1 we get, respectively:

$$Y^m{}_m - R_m{}^{-1} \, c = \hat{Y}^{m+1}{}_m - d \, , \tag{45}$$

$$Y^m{}_{m+1} - c = \hat{Y}^{m+1}{}_{m+1} - \hat{R}_{m+1}{}^{-1} \, d \, , \tag{46}$$

i.e.

$$c = Y^{m}_{m+1} - \hat{Y}^{m+1}_{m+1} + \hat{R}_{m+1}^{-1} \cdot d \, , \tag{47}$$

and d is given by

$$(R_m - \hat{R}_{m+1}^{-1}) \, d = Y^{m}_{m+1} - \hat{Y}^{m+1}_{m+1} - R_m (Y^{m}_{m} - \hat{Y}^{m+1}_{m}) \tag{48}$$

If the matrix $R_m - \hat{R}_{m+1}^{-1}$ is nonsingular, we get from Eqs. (48), (47), and (44) a unique solution Y_j. If the matrix is singular, the homogeneous problem corresponding to (23), (24) has [3] a nontrivial solution Yo_j and in consequence Y_j is not unique, similarly as is the case with the differential equations (14).

It is clear from the definitions (30) and (36) that

$$Z^{k}_{j} = \begin{cases} R_{j-1} R_{j-2} \cdots R_k & \text{for } j > k \\ R_j^{-1} R_{j+1}^{-1} \cdots R_{k-1}^{-1} & \text{for } j < k \end{cases} \tag{49}$$

and similar relations hold for $\hat{Z}^{k}_{j}$. Hence, it suffices to compute Y^{k}_{j}, $\hat{Y}^{k}_{j}$ and the ratio matrices R_j , $\hat{R}_j$ to solve our problem. As was shown by Johnson [3], these matrices can be obtained conveniently from the equations:

$$R_k = U_k - R_{k-1}^{-1} \, , \quad R_0^{-1} = 0 \tag{50}$$

and

$$\hat{R}_k = U_k - \hat{R}_{k+1}^{-1} \, , \quad \hat{R}_n^{-1} = 0 \, . \tag{51}$$

Although the method that was outlined above can be used for a step by step integration without the danger of instabilities connected with the initial rapid growth of the solutions, the computation of the wavefunction, for large systems, is still a practical problem, similarly as in the case of homogeneous equations. For relatively weakly bound states, as e.g. those considered by Dunker and Gordon [2], one can - by using repeatedly the formulas(23),(37-43) and (50), (51) - compute the proper initial values Y_1 and Y_{n-1} and then obtain Y_j from Eq. (23)

in analogy with [2]. However, for stiff problems this method must fail because of inherent instabilities, and therefore we will not go into any detail of this approach.

An other possibility is to compute and store the matrices R_j and $\hat{R}_k$ and construct the solution with the aid of Eqs. (40)-(49). This method is stable and was successfully used by Johnson [3] for homogeneous equations. However, it has the disadvantage that it requires a very large amount of storage to store the nxN^2 numbers forming the ratio matrices. Therefore, it cannot very well be used even for moderately large systems. Fortunately, in most cases in practice we need the solutions of Eqs. (14)-(15) only to compute integrals of the form

$$\int f^T \bar{P}\, dr = h \cdot J_1 \tag{52}$$

or

$$\int f^T \bar{A}\, f\, dr = h \cdot J_2 \tag{53}$$

where the vector $\bar{P}$ and the symmetric matrix $\bar{A}$ are given functions of r. An example of Eq. (52) is the second order energy in the perturbation theory.

Below we present a stable computational scheme for the evaluation of (52) that does not require storing of the ratio matrices.The evaluation of (53) can be performed in a similar way [5] but we will not give the detailed formulas here.

If we use the trapezoidal rule we get in view of the boundary conditions, Eq. (24), the expressions:

$$J_1 = \sum_{j=1}^{n-1} (Y_j)^T P_j \quad , \tag{54}$$

where

$$P_j = (I - T_j)^{-1} \bar{P}_j \ . \tag{55}$$

Now we define the auxiliary quantities:

$$M_s = \sum_{j=1}^{s} (Z^{s+1}{}_j)^T P_j \ , \tag{56}$$

$$E_s = \sum_{j=1}^{s} (Y^s{}_j)^T P_j \tag{57}$$

Obviously, E_s is scalar and M_s a vector.

By making use of Eq. (50) we get the following recurrence relations

$$M_{s+1} = R_{s+1}^{-1} (M_s + P_{s+1}) , \tag{58}$$

$$E_{s+1} = E_s - (c^s)^T M_s + (Y^{s+1}_{s+1})^T P_{s+1} \tag{59}$$

Thus, starting with $E_0 = 0$, $M_0 = 0$ and using (58) - (59) simultaneously with the step-by-step integration described above we get at the matching point M_m and E_m.

If we define

$$\hat{M}_s = \sum_{j=s}^{n-1} (Z^{s-1}_j)^T P_j \tag{60}$$

and similarly $\hat{E}_s$ we get [5] for the inward integration formulas quite analogous to Eqs. (58) - (59) and $\hat{M}_{m+1}$, $\hat{E}_{m+1}$ can be easily computed.

Now using (44) we write with c and d given by (47) - (48):

$$J_1 = \sum_{j=1}^{m} (Y^m_j - Z^{m+1}_j c)^T P_j + \sum_{j=m+1}^{n-1} (Y^{m+1}_j - Z^m_j d)^T P_j$$

$$= E_m - c^T M_m + \hat{E}_{m+1} - d^T \hat{M}_{m+1} \tag{61}$$

It is worth noting that our method of simultaneous integration of the differential equations and evaluation of the necessary integrals can be easily applied to the case of homogeneous equations. If one uses Johnson's method [3] for the homogeneous equations, one can get integrals of type (52) (53) involving the wavefunction [5].

4. Numerical example

As an illustration of the method a two-dimensional problem

$$L(r)y = g(r)$$
$$y(a) = y(b) = 0 \tag{62}$$

with

$$L(r) = -\frac{1}{2000}(d/dr)^2 + V - E \tag{63}$$

was solved [5] and the integral

$$J = \int_a^b y \cdot g \, dr \qquad (64)$$

was computed with the aid of the formulas given in the preceding section.

In (82) V is a symmetrix 2x2 matrix with elements:

$$V_{11} = -0.66 - 0.035 \times \{1 - \exp(-r + 2)\}^2 \qquad (65)$$

$$V_{22} = -0.72 - 0.095 \times \{1 - \exp[-0.7(r - 3)]\}^2 \qquad (66)$$

$$V_{12} = 0.0005 \exp[-5.8(r - 3.125)^2] \qquad (67)$$

and the right-hand side in (62) was given as

$$g(r) = L(r)y_o(r) \qquad (68)$$

with

$$y_o = \begin{pmatrix} 1 \\ 2 \end{pmatrix} \exp[-5(r - 3)^2] \quad . \qquad (69)$$

The remaining constants were: a = 0 and b = 7. Equation (62) was solved and J computed by the method described above for several different values of E chosen in such a way that the lowest value used was below the lowest eigenvalue of L y = 0 and the highest E used was well above the asymptotic values of V for large r. Thus the lowest E corresponds to a situation were the solution of the homogeneous problem grows very fast and one could expect numerical instabilities. On the other hand for high E the solutions of Ly = 0 oscillate relatively fast which again could lead to numerical problems.

The results obtained with different integration steps, h, are given in the table together with the exact results. It is seen that even for a relatively large integration step h = 0.07 the results are accurate to within 10^{-7} and they rapidly converge to the exact results when h decreases.

The method was also tested for stability on a real, HD^+ molecule, problem [10]. Several sets of up to 21 coupled second order equations

Table 1. Results obtained for the integral J

h \ E	-0.6	-0.65	-0.725
0.07	-0.28330654	-0.14318176	0.067005408
0.04	-0.28330678	-0.14318200	0.067005172
0.02	-0.28330681	-0.14318203	0.067005145
0.01	-0.28330681	-0.14318203	0.067005144
0.005	-0.28330681	-0.14318203	0.067005144
exact	-0.283306807	-0.143182027	0.0670051438

of the form (6) were solved and second order energies of the form (9) computed. The results were stable, i.e. they were practically insensitive to the position of the matching point and to the step size. In view of these tests it is hoped that the method of simultaneous integration of the equations and of the integrals will be helpful in the solution of various difficult inhomogeneous and homogeneous problems appearing in molecular physics.

References

1. R.G. Gordon, J.Chem.Phys. 51, 14 (1969).
2. A.M. Dunker and R.G. Gordon, J.Chem.Phys. 64, 4984 (1976).
3. B.R. Johnson, J.Chem.Phys. 69, 4678 (1978).
4. L. Wolniewicz, Can.J.Phys. 53, 1207 (1975).
5. L. Wolniewicz, J.Comput.Phys. 40, 440 (1981)
6. R.T. Pack and J.O. Hirschfelder, J.Chem.Phys. 52, 521 (1970).
7. J. Casti and R. Kalaba, Imbeding Methods in Applied Mathematics, Adison-Wesley Publishing Company (1973).
8. R.E. Bellman and R.E. Kalaba, Modern analytic and computational methods in science and mathematics, §21, - American Elsevier Publishing Company, New York (1965).
9. D.R. Hartree, The calculation of atomic structures - John Wiley and Sons (1957).
10. L. Wolniewicz and J.D. Poll, J.Chem.Phys. 73. 6225 (1980).

COMPUTATION OF SOLENOIDAL (DIVERGENCE-FREE) VECTOR FIELDS*

Karl E. Gustafson
David P. Young

Department of Mathematics, University of Colorado, Boulder, Colorado 80309
Boeing Computer Services, Tukwila, Washington 98188

ABSTRACT

In many important scientific applications (e.g., incompressible fluids) the diivergence-free property is not preserved by the partial differential equations describing the flow. Accordingly projection of a vector field v onto its solenoidal (divergence-free) part plays a fundamental role and in some respects is one of the most difficult aspects in the numerical analysis of such problems.

We first survey and describe the schemes that have been devised to deal computationally with this difficulty. Relatively few have been implemented in three dimensions and even fewer for three-dimensional stationary flows.

We then present a new scheme for the direct computation of the projection of an arbitrary three-dimensional vector field v(x) onto its solenoidal (divergence--free) part. The algorithm combines finite differences before and after the calculation of a singular integral. We prove convergence for this algorithm and present illustrative numerical results for the cases tested. A number of applications are discussed.

*Partially supported by a Computing resources Grant from the National Center of Atmosperic Research.

1. INTRODUCTION

In the Navier-Stokes equations for incompressible fluids, the Helmholtz projection of a vector field onto its divergence-free, i.e., solenoidal part plays a fundamental role and in some respects is one of the most difficult aspects of the problem for numerical analysis (see section 2). The general question of the computation of the divergence-free part of a vector field arises in other problems of physical interest, for example for the electric intensity in electromagnetic theory.

Performing three-dimensional computation in such problems is an essentially open subject. A number of codes have been implemented (most of them only in two dimensions) and we survey and describe them in section 2. In section 3 we present a new algorithm for the computation of the Helmholtz projector of $(L^2(\Omega))^3$ onto the solenoidal subspace, based upon the analytic formula

$$H\vec{v}(P) = \text{curl} \int_\Omega \frac{\text{curl } v(Q)}{4\pi|P - Q|} dV_Q \qquad (1)$$

for sufficiently smooth vector fields $\vec{v}$ of compact support in the domain Ω. Stability and convergence are shown. Section 4 contains numerical results, and a number of applications are discussed in section 5.

The potential complexity and magnitude of such higher dimensional computations is nicely summarized by Morse and Feshbach [1, p. 1759]: "Naturally the calculation of vector fields, which cannot be expressed in terms of the gradient of a scalar, is a more arduous task than it is for scalar fields, since three numbers must be calculated for each point in space, rather than one."

2. SCHEMES FOR HANDLING div $\vec{v} = 0$

A number of schemes have been proposed recently for handling the divergence-free condition div $\vec{v} = 0$. Some of these are described below. Most have been implemented only in two dimensions.

The most obvious approach from the numerical point of view is to finite difference the divergence-free condition along with the basic equations of motion being considered. Let us consider for example the Navier-Stokes equations for the hydrodynamical stability of a viscous incompressible flow:

$$\vec{v}_t - \nu\Delta\vec{v} + (\vec{v}\cdot\vec{\nabla})\vec{v} = \vec{f} - \vec{\nabla}p \text{ in } \Omega \qquad (2)$$

$$\vec{v} = 0 \text{ on } \partial\Omega \qquad (3)$$

$$\vec{\nabla}\cdot\vec{v} = 0 \text{ in } \Omega \qquad (4)$$

where Ω is the vessel containing the flow, $\delta\Omega$ denotes its boundary, ν is the viscosity, $\vec{f}$ is the body force, $\vec{v}$ is the velocity excess over that of the basic flow $\vec{U}$, and p is the pressure. See Ladyzhenskaya [2] and Sattinger [3]. One may take $\vec{v} = 0$ on $\delta\Omega$ here because it is the disturbance velocity. In other flow problems the boundary condition $\vec{v} = 0$ often appears as the "no slip" or "viscous" boundary condition, and we shall in this paper for simplicity usually tacitly assume it. In section 5 we describe how our algorithm can be extended to treat non-zero boundary data.

While at first sight it may seem somewhat inocuous, the divergence-free condition (4) causes serious difficulties in solving such flow problems, both theoretically and numerically. Equation (4) is often called the continuity equation, that is $\rho_t + \rho v_{i,i} = 0$, and to which (4) is equivalent for constant densities. Roughly speaking and from the physical point of view, the difficulties theoretically and numerically come about because the incompressibility condition (4) acts as a constraint which causes the pressure to vary continuously with the flow.

If one just finite-differences the whole system (2) (3) (4) one arrives at a very large matrix in the discretized equations which is difficult to solve efficiently. Modified finite-difference methods have been devised to try to overcome this difficulty, but with only limited success. See the discussion in Temam [4, e.g., p. 64] and in Gresho, Lee, Sani, and Stullich [5].

There are two principal elements in the theoretical treatment of the Navier--Stokes equations which have some bearing here.

The first theoretical element, and about which we shall comment further only briefly in section 5, is that on a physical and modeling basis an important modification for the validity of the Navier-Stokes equations at high Reynold's number is that the viscosity ν may have a (nonlinear) dependence on $\vec{v}$. This is mentioned here only because in those considerations, whereas equation (2) is susceptible to modification with considerable justification, there are in most instances no compelling reasons to change the divergence-free condition (4).

Chorin [6,7] approaches the divergence-free condition (4) by writing (2) as

$$\vec{v}_t + \vec{\nabla}p = \vec{f} + \nu\Delta\vec{v} - (\vec{v} \cdot \vec{\nabla})\vec{v}$$

and noting that $\vec{v}_t$ is divergence-free and that curl $\vec{\nabla}p = 0$. An iterative technique is then used to calculate the divergence-free projection to obtain $\vec{v}_t$. See also Peskin [8] for an interesting application of this method to the study of the fluid flow near heart valves.

The second theoretical element which bears here is the use of the weak formulations of the Navier-Stokes equations to obtain existence proofs. See [2,3] and the references therein. These amount to variations on the Lax-Milgram Theorem, and the existence and uniqueness in the stationary, linear cases then follow from the coercivity of the induced form operator. The nonlinear nonstationary cases are usually treated by comparison through bifurcation theory with the linear nonstationary cases and the existence of flows for the latter are obtained by the Hille-Yosida semigroup theory. This seems to have started with Prodi [9]. See Sattinger [3,10]. The (weak,conditional) stability is then obtained from the linearized stationary operator L deduced from (2) (3) (4) by showing a discrete half-plane parabolic-like spectrum λ_k and a corresponding exponential falloff of the eigenfunctions. Roughly speaking one may say that a key ingredient is establishing that L_0^{-1} is a Hilbert-Schmidt (or Carleman or other compact) operator, that is, establishing the estimate

$$\sum_{k=1}^{\infty} (\lambda_k^0)^{-2} < \infty \tag{5}$$

for the eigenvalues λ_k^0 of $L_0 = -H\Delta$, where H is the Helmholtz projector onto the divergence-free subspace and where Δ is the Laplacian and principal part of the linearization of equation (2). We will propose an improved estimate similar to (5) in section 5 of this paper.

For the moment, we have described this second theoretical element and the weak formulation existence proofs for the Navier-Stokes equations because they have generated a number of corresponding numerical schemes. The latter are mostly of finite element type and have the advantages of adaptability to different boundaries $\partial\Omega$ and systematic mathematical discretization. Serious problems, which may be described as combinatorial, are encountered in actually putting the algorithms into practice, even in two dimensions. These stem in significant part from the divergence-free condition (4) in discretized form.

A thorough treatment of these finite element schemes for dealing with div $\vec{v} = 0$ may be found in Temam [4]. See also Fortin [11], Crouseix and Raviart [12], and Thomasset[13]. In these schemes the functions are divergence-free only in an average sense. For example, APX2 of [4] uses second order polynomial approximations on triangular elements and the divergence-free condition (4) is satisfied in the boundary sense

$$\int_{\partial S} \vec{v} \cdot \vec{n} = 0 \tag{6}$$

on elements. APX2´ is APX2 strengthened by a cubic perturbation. APX3 is an attempt to generalize APX2 and APX2´from two to three dimensions. APX4 and APX5 are variations of the above involving nonconforming elements.

Some of these schemes use functions satisfying

$$\int_S \operatorname{div} \vec{v} = 0 \, . \tag{7}$$

Again this is only an average divergence-free condition. On the other hand the power of nonconforming element methods is that the continuities of the flow and of the fluid may be neglected in order to overcome the difficulties of exactly fitting the condition (4). Even so, it seems difficult to construct divergence-free element bases in three dimensions.

In simulating turbulent flows, Schumann [14] also uses an average divergence-free condition. This averaging over grid volumes filters the small scale motions and gives a reasonable simulation of large scale turbulent motions. This method is one of the few that have actually been applied to the three dimensional (cylindrical) domains.

The penalty method has been employed in order to treat equations with a divergence-free condition (4), especially in case of three dimensions where many of the finite element schemes run into trouble. Penalty methods go under the names penalty-duality, augmented Lagrangian, Uzawa-Arrow-Hurwitz, and conjugate gradient, among others, and incorporate the divergence-free condition (4) as a linear constraint in the manner familiar to the calculus of variations. For application to the Navier-Stokes equations they amount physically to iterative methods in which a slight compressibility of the fluid is allowed. See Temam [4], Chorin [7], and for similar perturbative methods that have been employed Ladyshenskaya [15], Lions [16], and Temam [17]. For a recent scheme of similar type but more in the flavor of optimal control theory see Glowinski-Pironneau [18]. For an original paper on the application of the analytic penalty method to fluids problems see Fujita and Kato [19].

These and other iterative methods have severe limitations as concerns the transition from two to three dimensions. Gresho, Lee, Sani, and Stullich [5] speculate indeed that no iterative scheme can be made to really work on three dimensional flow problems. This pessimism may be traced to the discretized divergence-free condition (4) in the large matrix discretization of the whole system and its indefiniteness.

Finally, the classic "pressure trick" should be mentioned. One takes the divergence of both sides of equation (2) and thus has the Poisson problem

$$\Delta p = \text{fn. of } (f, \vec{v}) \text{ in } \Omega. \tag{8}$$

There are a number of numerical dangers inherent in this indirect approach (see Ames [20]), and in particular a significant numerical difficulty in handling p on $\partial\Omega$. These pressure trick boundary condition difficulties led to the SMAC development in Amsden and Harlow [21]. Recent algorithms (Widlund/Peskin, NYU) are essentially based on the pressure trick and use fast Poisson solvers, as do the earlier vector potential methods. However, as pointed out by Sweet [22] the associated matrix decomposition techniques based on fast Fourier transforms are not necessarily efficient for general grid size.

One of the earlier three dimensional analyses with code and actual application is a lesser-known paper by G. Williams [23]. The code there was developed for a problem of thermal convection in a rotating annulus. To address the divergence-free condition the pressure trick was employed.

Ames [20,24] gives good accounts of some of the earlier approaches such as the marker and cell method, the stream function vorticity methods in two dimensions, the vector potential methods in three dimensions, and the work of Chorin [6,7] and its later extensions. Large bibliographies are also provided in [20,24].

Most of the above methods were designed with time-dependend problems in mind, and many are iterative in nature. We were led, from the three dimensional stationary case as encountered in specific problems in bifurcation theory for general flows in general domains, to develop an algorithm for the direct computation of the projection given by equation (1). Although our preliminary code is far from optimal, convergence was demonstrated and rather good accuracy obtained on rather sparce grids. Moreover there is a certain intrinsic beauty involved in (1) that should eventually lead to an efficient general algorithm for the Helmholtz projection. For in (1) one really has a form $H = L^{1/2}L^{-1}L^{1/2}$ which may be computed all at once (we know the square root of the vector Laplacian is not really the curl, nor is the square root of the scalar Laplacian even a gradient or divergence, but those technicalities are not the point here), which makes sense for a projection of a vector onto a subspace.

3. A DIRECT PROJECTION ALGORITHM

Given a bounded domain Ω in R^3 and any vector field on Ω that is continuously differentiable on $\overline{\Omega}$ and vanishing along with its derivative on $\partial\Omega$, the Helmholtz Projection H is given by (1). Because of our initial interest in its application to problems such as the Taylor Problem (see section 5) we have coded it and tested it for a three dimensional cylinder Ω and for paired nested cylinders with Ω the contained ring domain. General domains can in principle be handled in much the same way and in fact cylindrical domains and coordinates introduce an additional computational difficulty into the calculation, as will be discussed below. On the other hand for general domains Ω a technical problem is the efficient numerical generation of the limits of integration in the integral.

Non-zero boundary conditions can be handled by using the more general form of (1) involving a surface integral (see section 5).

In the algorithm development we used the notation $\vec{F}$ for vector fields $\vec{v}$ and P and Q for spatial points $\vec{x}$ and we shall do so in the following. In the computation of the divergence-free part of $\vec{F}$ according to (1), that is, in the computation of the quantity

$$H\vec{F}(P) = \text{curl} \iiint_{\Omega} \frac{\text{curl } \vec{F}\ (Q)}{4\pi |P-Q|} dV_Q \tag{9}$$

the code has P as an input parameter, Q is the variable of integration, $|P-Q|$ is the Euclidean distance. Both curls are computed by second order centered finite differences (we assume P is not on $\partial\Omega$) with the mesh spacings as input parameters. These are called Δx_I, Δy_I, Δz_I for the inside curl and Δx_0, Δy_0, Δz_0 for the outside curl. As long as $\vec{F}$ is a three times continuously differentiable vector field, the inside curl, curl $\vec{F}$, can be calculated with as much accuracy as desired by taking small mesh spacings. Because of this and the fact that the discretized integral is a bounded operator we were led to include an option for using an analytically calculated exact curl $\vec{F}$ for our test cases so that we could test the integral, where the real difficulties arise. This option also cut processing time when employed in those cases.

The integral has an infinite discontinuity at Q = P, but a change of coordinates can make this singularity more tractable. If we change to spherical coordinates, the integrand becomes as smooth as curl $\vec{F}$. For a general code, spherical coordinates thus offer obvious advantages. However, all of our test cases involved a cylindrical ring domain Ω with inner radius r_1 and outer radius r_2, top at z_2 and bottom at z_1 (see Figure 1). In this domain the calculating of the limits of integration is significantly easier in cylindrical coordinates. The price that is paid is that the integrand has a jump discontinuity at the origin. Any limit between $\vec{0}$ and curl $\vec{F}(0)$ can be obtained by approaching the origin from some direction. However a convergent quadrature rule for such a function was not hard to find.

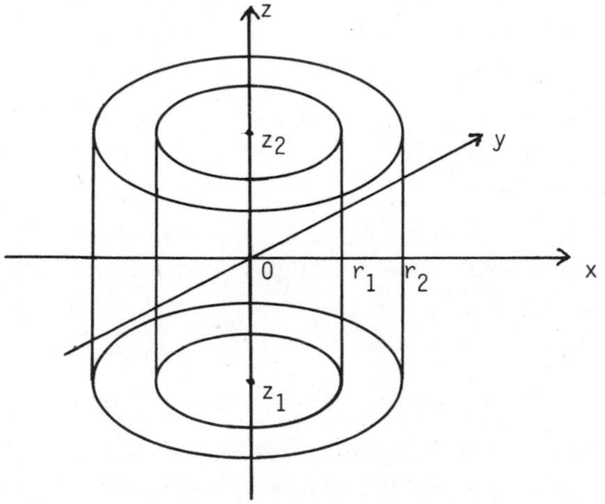

Figure 1. The domain Ω.

If P has coordinates (x_0, y_0, z_0), the integrand is of the form

$$\frac{r \cdot \text{curl}\,\vec{F}(r\cos\theta + x_0,\ r\sin\theta + y_0, z + z_0)}{\sqrt{r^2 + z^2}} \tag{10}$$

The integral (9) is evaluated as the iterated integral

$$H\vec{F}(P) = \int_{z_1-z_0}^{z_2-z_0} \int_0^{2\pi} \int_0^{r_0\cos(\theta-\theta_1) + \sqrt{r_0^2\cos^2(\theta-\theta_1) - r_0^2 + r_2^2}} \left[\frac{r \cdot \text{curl}\,\vec{F}\,(r\cos\theta + x_0,\ r\sin\theta + y_0,\ z + z_0)}{\sqrt{r^2 + z^2}}\right] dr\,d\theta\,dz$$

$$- \int_{z_1-z_0}^{z_2-z_0} \int_{\theta_1-\varphi}^{\theta_1+\varphi} \int_{r_0\cos(\theta-\theta_1) - \sqrt{r_0^2\cos^2(\theta-\theta_1) - r_0^2 + r_1^2}}^{r_0\cos(\theta-\theta_1) + \sqrt{r_0^2\cos^2(\theta-\theta_1) - r_0^2 + r_1^2}} \left[\frac{r \cdot \text{curl}\,\vec{F}\,(r\cos\theta + x_0,\ r\sin\theta + y_0,\ z + z_0)}{\sqrt{r^2 + z^2}}\right] dr\,d\theta\,dz \tag{11}$$

where (r_0, θ_0, z_0) are the cylindrical coordinates of P, $\theta_0 = \theta_1 - \pi$ and $\varphi = \cos^{-1}\frac{\sqrt{r_0^2 - r_1^2}}{r_0}$. If $r_1 = 0$, the second integral is omitted. A mesh in θ and z is set up for the outer double integral which is done by an IMSL routine that interpolates the data with a two dimensional spline and then integrates the spline. At each point of this mesh, an integral with respect to r must be computed. This r integral can be done by Simpson's rule or by an IMSL routine using cautious adaptive Romberg extrapolation. n_z, n_r, and n_θ are input parameters controlling the number of mesh intervals in z, r, and θ respectively and thus specifying the maximum distances between mesh points. Several options for computing the meshes, including equally spaced meshes and meshes that are finer near the jump discontinuity, are available.

$D(r,z) = \frac{r}{\sqrt{r^2 + z^2}}$ contains the finite jump discontinuity. At $r = 0$, $z = 0$, the value of $D(r,z)$ is taken to be an average value of D near $r = 0$, $z = 0$.

The outside curl is computed with second order centered finite differences. It should be noted that if the mesh spacing for these differences are too small for the accuracy of the integral, very large errors will result. Thus, Δx_0, Δy_0, Δz_0 should be chosen conservatively. The actual calculations summarized in section 4 give some idea of what are reasonable values.

The stability of the integral with respect to errors in the values of the curl $\vec{F}$ is clear since the integrand is bounded. In particular if $\vec{A}$ is an approximation to curl $\vec{F}$ such that $\|\vec{A} - \text{curl } \vec{F}\|_2 < \varepsilon$ where $\|\vec{f}\|_2 = (\iiint_\Omega \|\vec{f}(Q)\|_2^2 dV_Q)^{\frac{1}{2}}$ is the standard L_2 norm , then

$$\left\| \iiint_\Omega \frac{\vec{A}(Q)}{4\pi|P - Q|} dV_Q - \iiint_\Omega^\Omega \frac{\text{curl } \vec{F}(Q)}{4\pi|P - Q|} dV_Q \right\|_2$$

$$= \left\| \iiint_\Omega (\vec{A} - \text{curl } \vec{F})(r\cos\theta + x_0, r\sin\theta + y_0, z + z_0) \cdot \frac{r}{\sqrt{r^2 + z^2}} \, dr d\theta dz \right\|_2 \tag{12}$$

$$\leq \iiint_\Omega \| (\vec{A} - \text{curl } \vec{F})(r\cos\theta + x_0, r\sin\theta + y_0, z + z_0)\|_2 \cdot \left|\frac{r}{\sqrt{r^2 + z^2}}\right| dr d\theta dz$$

$$\leq \varepsilon \cdot (2\pi)(R)(z_2 - z_1) ,$$

where Ω is contained in the cylinder of radius R with top at z_2 and bottom at z_1.

The stability of our discretization of the integral follows similarly. That is, for any given $\vec{g}(r, \theta, z)$, the integral $\iiint_\Omega \vec{g}(r, \theta, z)D(r, z)dr d\theta dz$ is first discretized to $\int_{Z_1-z_0}^{Z_2-z_0} \int_0^{2\pi} [\sum_K W(k)\vec{g}(Rk, \theta, z)D(Rk, z) \frac{\Delta r(k)}{2}] d\theta dz$, where $\Delta r(k)$ is the length of the $k^{\underline{th}}$ mesh interval in r, W(k) is the Simpson's rule weight, and Rk is $k^{\underline{th}}$ mesh point, and then to

$$\sum_{i,j} SW(i,j)[\sum_k W(k)\vec{g}(Rk,\theta j,Zi)D(Rk,Zi) \frac{\Delta r(k)}{2} \tag{13}$$

where SW(i,j) are the weights for the IMSL quadrature code, θj is the $j^{\underline{th}}$ mesh point in θ and Zi is the $i^{\underline{th}}$ mesh point in Z. Since all the weights are bounded and $|D(r,z)| \leq 1$, stability of the integral discretization for variations in $\vec{g}$ is clear.

Convergence of the whole algorithm is not hard to see. As n_z, n_r, and $n_\theta \to \infty$, and as Δx_0, Δy_0, Δz_0, Δx_I, Δy_I and $\Delta z_I \to 0$, one has FHF(P) $\to$ HF(P) for any $P \in \Omega$, where FHF(P) is our discrete approximation to the Helmhotz projection. This follows, provided that Δx_0, Δy_0 and Δz_0 do not outstrip the accuracy of the integral approximation. To see this, consider $I_1 = \iiint_{\Omega'} \frac{\text{curl } \vec{F}(Q)}{4\pi|P - Q|} dV_Q$ where Ω' is Ω with a vertical cylinder of radius δ around P removed. The integrand is smooth on Ω' and so our discretization process for I_1 converges to the value I_1 as the meshes are refined. If the first mesh interval in r is $[0,\Delta r(1)]$ and $\Delta r(1) = \delta$, this process of discretizing I_1 corresponds to leaving out the first term in the r sum in our discretization of $I = \iiint_\Omega \frac{\text{curl } \vec{F}(Q)}{4\pi|P - Q|} dV_Q$. Let $ID_{\Omega'}$ be the discretized integral on Ω' . Since the first mesh point in r is on the boundary $\delta(\Omega \sim \Omega')$, we have $ID_\Omega = ID_{\Omega'} + ID_{\Omega\sim\Omega'}$. Then

$$\|ID_\Omega - ID_{\Omega'}\|_2 \;=\; \|ID_{\Omega\sim\Omega'}\|_2 \;\rightarrow\; 0 \quad \text{as} \quad \Delta r(1) \;\rightarrow\; 0 \tag{14}$$

This follows from the description of ID in (13) above. Using the L_2 norm $\| \;\|_2$, smoothness of F on $\Omega\sim\Omega'$ implies

$$\|I_1 - ID_{\Omega'}\|_2 \;\rightarrow\; 0 \tag{15}$$

as n_z, n_r and $n_\theta \rightarrow \infty$. By boundedness of the integrand after the change to cylindrical coordinates

$$\|I_1 - I\|_2 \;\rightarrow\; 0 \tag{16}$$

as $\delta \rightarrow 0$. For the cases of not necessarily equal mesh spacing, note that one still has $\Delta r(1) \rightarrow 0$ as $n_r \rightarrow \infty$. In those cases in which $\Delta r(1)$ is proportional to $(\frac{1}{n_r})^2$, the convergence in $\Delta r(1)$ is quadratic rather than linear. Thus from (14), (15) and (16) we have that

$$\Big\|\iiint_\Omega \frac{\text{curl } \vec{F}(Q)}{4\pi|P - Q|} dV_Q - ID_\Omega\Big\|_2 \;\leq\; \|I - I_1\|_2 + \|I_1 - ID_{\Omega'}\|_2$$

$$+ \|ID_{\Omega'} - ID_\Omega\|_2 \;\rightarrow\; 0 \tag{17}$$

as n_z, n_r, $n_\theta \rightarrow \infty$.

The error ε in the inner curl, as already observed, introduces an error in the discretized operator ID that goes to 0 as $\varepsilon \rightarrow 0$. Thus, if Δx_0, Δy_0, and Δz_0 do not become too small for the accuracy of the integral, FHF(P) $\rightarrow$ HF(P) in the vector L_2 norm for any fixed $P \in \Omega$.

Even though convergence rates for iterated integral approximations cannot be obtained for general domains, the convergence rate of the approximation to the r integral is determined by what the mesh looks like near r = 0. Empirically, the results of the next section indicate convergence that is almost quadratic in $\max\{\frac{K_1}{n_z}, \frac{K_2}{n_r}, \frac{K_3}{n_\theta}\}$, where the K_i are constants depending on the vector field $\vec{F}$.

The same considerations will apply for a general domain . By use of spherical coordinates the discussion of the singularity may be obviated. However a price may have to be paid in the computation of the limits of integration. The employing of better multiple quadrature rules should increase the accuracy.

4. NUMERICAL RESULTS

We used our code on several test cases, all of which showed good convergence of the approximation. Optimality of the approximation seems to depend on the interplay between the mesh spacings chosen for the differencing and those for the integral as discussed in the previous section.

All test cases reported on below were defined in the cylinder Ω: $0 \leq r \leq 2$, $-\frac{1}{2} \leq z \leq \frac{1}{2}$. The parameters Δx_0, Δy_0, Δz_0, Δx_I, Δy_I, Δz_I, n_z, n_r, and n_θ are described in the previous section. In all the tables of this section, $\Delta x_0 = \Delta y_0 = \Delta z_0 = 0.1$ and $\Delta x_I = \Delta y_I = \Delta z_I = 0.05$ unless otherwise indicated. An even quadrature mesh spacing in z and spacing in r that is quadratically closer near $r = 0$ was used. The inner curl was computed with finite differences in some cases but using an analytic formula in others. The X error is the error in computing the first component of $H\vec{F}$, etc. and L_2 error is the L_2 error in the vector approximation of $H\vec{F}$.

The first test case was a divergence free vector field defined as follows:

$$\vec{F} = F_1\vec{i} + F_2\vec{j} + F_3\vec{k},$$

$$F_1 = F_2 = g(x,y)f(z),$$

$$F_3 = -\left(\int_{-\frac{1}{2}}^{z} f(\xi)d\xi\right)\left[\frac{\partial g}{\partial x} + \frac{\partial g}{\partial y}\right],$$

$$g(x,y) = \begin{cases} (x^2 + y^2 - 4)^4 & \text{if } \sqrt{x^2 + y^2} \leq 2 \\ 0 & \text{otherwise} \end{cases}, \tag{18}$$

$$f(z) = \begin{cases} (z - \frac{1}{2})^4 z & \text{if } 0 \leq z \leq \frac{1}{2} \\ (z + \frac{1}{2})^4 z & \text{if } -\frac{1}{2} \leq z < 0 \\ 0 & \text{otherwise} \end{cases}.$$

Note that $\vec{F} = \vec{0}$ on the boundary of Ω, but that $\vec{F}$ is nonzero in the interior except at the origin. Thus, $H\vec{F}$ is $\vec{F}$ and we have an analytic test of the accuracy of the code. The following table shows convergence of the approximation at one point in Ω.

TABLE I

Values at the point $r = 0.3$, $\theta = 3.0$, $z = -0.15$ for the first test case. The inner curl was computed from an analytic formula. The analytic values at this point are $H\vec{F} = (-.5261E0, -.5261E0, -.1548E - 1)$ and $\|H\vec{F}\|_2 = .7442E0$. These results were obtained using the Cray I.

n_z	n_r	n_θ		X error	Y error	Z error	L_2 error
5	3	5		.33E0	.33E0	.17E-1	.46E0
10	6	10		-.22E0	-.22E0	-.15E-1	.32E0
20	12	20	$\Delta x_0 = \Delta y_0 = \Delta z_0 = 0.05$	-.41E-1	-.41E-1	-.43E-2	.58E-1
40	24	40	$\Delta x_0 = \Delta y_0 = \Delta z_0 = 0.02$	-.87E-2	-.87E-2	-.11E-2	.12E-1
80	48	80	$\Delta x_0 = \Delta y_0 = \Delta z_0 = 0.01$	-.20E-2	-.20E-2	-.28E-3	.29E-2

This table shows that the L_2 error in the approximation goes down by a factor of roughly 160 when n_z, n_r, and n_θ are all increased by a factor of 16. The last line shows a relative L_2 error of about 0.4%, and the approximation has two good digits in each of the three components. Convergence to the analytic answer is clear. Convergence tests for this case were run at three other points with similar results.

The second test case is the same as the first except that

$$g(x,y) = \begin{cases} (x^2 + y^2 - 4)^{10} & \text{if} \quad \sqrt{x^2 + y^2} \le 2 \\ 0 & \text{otherwise} \end{cases} \tag{19}$$

This gives more smoothness near the boundary at $r = 2$.

TABLE II

Values at the point $r = 0.3$, $\theta = 2.0$, $z = -0.15$ for the second test case. The inner curl was computed from an analytic formula. The analytic values at this point are $H\vec{F} = (-.1880E4, -.1880E4, -.1383E3)$ and $\|HF\|_2 = .2662E4$. These results were obtained using the Cray I.

n_z	n_r	n_θ		X error	Y error	Z error	L_2 error
5	3	5		1.0E3	1.0E3	.92E2	.14E4
10	6	10		-.77E3	-.77E3	-.49E2	.11E4
20	12	20	$\Delta x_0 = \Delta y_0 = \Delta z_0 = 0.05$	-.14E3	-.14E3	-.16E2	.19E3
40	24	40	$\Delta x_0 = \Delta y_0 = \Delta z_0 = 0.02$	-.28E2	-.28E2	-.42E1	.40E2
80	48	80	$\Delta x_0 = \Delta y_0 = \Delta z_0 = 0.01$	-.65E1	-.65E1	.11E1	.93E1

This table shows the L_2 error in the approximation goes down by a factor of about 150. The last line shows a relative L_2 error of about 0.3% and the result has two good digits in each component.

The third test case is a divergence-free vector defined as follows:

$$\vec{F} = F_1\vec{i} + F_2\vec{j} + F_3\vec{k} ,$$

$$F_1 = F_3 = \alpha(x)\beta(z)f(y) ,$$

$$F_2 = -\left(\int_{-1}^{y} f(\xi)d\xi\right)[\alpha(x)\beta'(z) + \alpha'(x)\beta(z)] ,$$

$$\alpha(x) = \begin{cases} (x-1)^3(x+1)^3 & \text{if } -1 \le x \le 1 \\ 0 & \text{otherwise} \end{cases} ,$$

$$\beta(z) = \begin{cases} (z-\tfrac{1}{4})^3(z+\tfrac{1}{4})^3 & \text{if } -\tfrac{1}{4} \le z \le \tfrac{1}{4} \\ 0 & \text{otherwise} \end{cases} , \qquad (20)$$

$$f(y) = \begin{cases} y(y-1)^2 & \text{if } 0 \le y \le 1 \\ y(y+1)^2 & \text{if } -1 \le y \le 0 \\ 0 & \text{otherwise} \end{cases} .$$

$\vec{F}$ has compact support in Ω.

TABLE III

Values at the point $r = 0.5$, $\theta = 1.0$, $z = 0.15$ for the third test case. The inner curl is computed from an analytic formula. The analytic values at this point are $H\vec{F} = (.7198E-5, -.4530E-4, .7198E-5)$ and $\|HF\|_2 = .4643E-4$. These results were obtained on a CDC 640C.

n_z	n_r	n_θ		X error	Y error	Z error	L_2 error
4	4	12		.98E-5	-.60E-4	.71E-4	.94E-4
10	6	25		-.32E-5	.15E-5	-.10E-4	.11E-4
20	12	50		.16E-5	.86E-5	.20E-5	.90E-5
30	18	96	$\Delta x_0 = \Delta y_0 = \Delta z_0 = 0.01$	-.20E-6	-.10E-5	-.59E-6	.12E-5

This table shows that the L_2 error in the approximation went down by a factor of roughly 80 when n_z, n_r, and n_θ are all increased by roughly a factor of 6. The last line shows a relative L_2 error of about 4%. Convergence tests were run at other points with similar results.

The fourth test case is a gradient. $\vec{F} = \text{grad}\ \Phi$ where $\Phi = \alpha(x)\alpha(y)\beta(z)$, α and β as in the third test case. For this test case, $H\vec{F} = \vec{0}$. Because of the stability of the quadrature scheme, the accuracy of our scheme for such test codes will depend on how close the values of curl $\vec{F}$ are to $\vec{0}$. Thus, the parameters Δx_I, Δy_I, and Δz_I are the controlling ones for this test case. Thus by itself it is of limited interest for purposes of testing the integral approximation.

The fifth test case has both divergence-free and gradient parts. If $\vec{G}_1$ is the vector field of the first test case and $\vec{G}_2$ the vector field of the fourth test case, then $\vec{F} = C_1 \cdot \vec{G}_1 + C_2 \cdot \vec{G}_2$ where C_1 and C_2 are code input parameters. For a meaningful test, analytic curl $\vec{F}$ should not be used in this case. For all the tables below, $C_1 = C_2 = 1.0$. For this case, we give results at two points.

TABLE IV

Values at the point $r = 0.5$, $\theta = 1.0$, $z = 0.15$ for the fourth test case. The inner curl was computed by finite differences. The analytic values at this point are $H\vec{F} = (.4451E0, .4451E0, -.6379E-1)$ and $\|H\vec{F}\|_2 = .6327E0$. These results were obtained using a Cray I.

n_z	n_r	n_θ		X error	Y error	Z error	L_2 error
5	3	5		-.22E0	-.22E0	.10E0	.33E0
10	6	10		.15E0	.15E0	-.48E-1	.21E0
20	12	20	$\Delta x_0 = \Delta y_0 = \Delta z_0 = 0.05$ $\Delta x_I = \Delta y_I = \Delta z_I = 0.001$	.18E-1	.18E-1	-.80E-2	.26E-1
40	24	40	$\Delta x_0 = \Delta y_0 = \Delta z_0 = 0.02$ $\Delta x_I = \Delta y_I = \Delta z_I = 0.001$	.65E-2	.65E-2	-.41E-2	.10E-1
80	48	80	$\Delta x_0 = \Delta y_0 = \Delta z_0 = 0.01$ $\Delta x_I = \Delta y_I = \Delta z_I = 0.001$	.13E-2	.13E-2	-.91E-3	.20E-2

This table shows that the L_2 error in the approximation goes down by a factor of 165. The last line shows a relative L_2 error of about 0.3% in the approximation and the approximation has two good digits in each component.

TABLE V

Values at the point $r = 0.1$, $\theta = 3.0$, $z = 0.05$ for the fifth test case. The inner curl was computed by finite differences. The analytic values at this point are $H\vec{F} = (.5197E0, .5197E0, .1990E-1)$ and $\|H\vec{F}\|_2 = .7352E0$. These results were obtained on the Cray I.

n_z	n_r	n_θ		X error	Y error	Z error	L_2 error
5	3	5		-.41E0	-.41E0	-.28E-1	.58E0
10	6	10		-.39E-1	-.39E-1	.41E-1	.56E-1
20	12	20	$\Delta x_0 = \Delta y_0 = \Delta z_0 = 0.05$ $\Delta x_I = \Delta y_I = \Delta z_I = 0.001$	.21E-1	.21E-1	.21E-2	.30E-1
40	24	40	$\Delta x_0 = \Delta y_0 = \Delta z_0 = 0.02$ $\Delta x_I = \Delta y_I = \Delta z_I = 0.001$	.54E-2	.54E-2	.88E-3	.77E-2
80	48	80	$\Delta x_0 = \Delta y_0 = \Delta z_0 = 0.01$ $\Delta x_I = \Delta y_I = \Delta z_I = 0.001$	.41E-3	.41E-3	.19E-3	.61E-3

This table shows that the L_2 error in the approximation goes down by a factor of about 950. The last line shows a relative error in the L_2 approximation of about 0.1% and the approximation has two good digits in each component. Convergence tests were run at other points with similar results.

We tried to choose the parameters Δx_0, Δy_0, and Δz_0 to match the accuracy of the integral. Taking Δx_0, Δy_0, and Δz_0 too small for the accuracy of the integral will result in very large errors. Thus we were conservative in reducing these parameters, and some experimentation was necessary to find reasonable values. We also tried to take Δx_I, Δy_I, and Δz_I small enough so that they were not the controlling factors. For the fifth test case, the values of these parameters were accordingly taken quite small. When the analytic curl $\vec{F}$ option is employed, these parameters do not enter in the calculation.

We believe that with further experience all of the grid parameters can be set automatically. The mesh ratios will depend not only on the just-discussed integral vs. differencing tradeoff but also on the domain geometry.

Overall, the results indicated that the integral approximation is second order in max $\left\{\frac{K_1}{n_z}, \frac{2K_2}{n_r}, \frac{2\pi K_3}{n_\theta}\right\}$ for some constants K_1, K_2, K_3 . This is reasonable; our mesh spacing in r is quadratic near $r = 0$. With even mesh spacing and a jump discontinuity, one would expect Simpson's rule to be first order. But our mesh technique achieves a considerable improvement.

5. APPLICATIONS

A number of applications of the algorithm are envisioned. A limitation to the extent of applicability is the dearth of three-dimensional codes for flow problems. Some of these may have to be written as the following are investigated. In so doing, code efficiency may become a factor. Our code thus far for the test cases described in section 4 was written only to test the accuracy and convergence of the algorithm described in section 2 and without regard for CPU times. We anticipate greatly increased efficiency after coding and algorithm improvements. Further cost effectiveness may be gained in certain applications by localizing and averaging the integral and by employing better quadrature methods.

An initial stimulus for this work was the Taylor Problem of flow between rotating cylinders. A great deal of work, both theoretical and numerical, has been done on this problem. See for example Eagles [25] for recent work. As proposed in Gustafson [26], one can obtain stability conclusions from the estimation of the eigenvalue spacing of linearizations such as $A = -HD$ where

$$D = \begin{bmatrix} \Delta - 1/r^2 & 0 & 0 \\ 0 & \Delta - 1/r^2 & 0 \\ 0 & 0 & \Delta \end{bmatrix} \tag{21}$$

Specifically one is interested in whether the estimate

$$\sum_{k=1}^{\infty} (\lambda_{k+1} - \lambda_k)^{-2} < \infty \tag{22}$$

holds. Such eigenvalue estimates for λ_k for general domains and for other linearizations about other (a basic Couette flow U was used in the particular case (21) given above) basic flows are of interest in their own right. Heretofore because of the divergence-free constraint one has usually been limited to an estimate such as (5), and the finiteness there follows only theoretically from the bound

$$\sum_{k=1}^{\infty} (\lambda_k^0)^{-2} \leq \sum (\mu_k^0)^{-2} < \infty \tag{23}$$

where μ_k^0 are the eigenvalues for $-\Delta$ in the whole space L^2. The finiteness of the latter bound follows from the compact domain Ω and the resulting (generalized) Hilbert-Schmidtness of Δ^{-1}. The effect of the physical restriction to the divergence-free subspace has thus not yet been measured.

In particular the approach of [26] and the estimate discussed above may answer open questions about spectra such as that mentioned in Marsden and McCracken [27, p. 326].

As also mentioned in section 2, we hope to apply the algorithm to test the recent nonlinear viscosity hypotheses of Ladyzhenskaya [28] and others. In addition to the projection algorithm, the numerical techniques of Young [29] developed for nonlinear concentration-diffusion equations should be useful, in as much as the nonlinear principal terms proposed in [28] are similar in type to those studied in [29].

Among other applications envisioned are the use of the projector H in connection with, and as a test on, the algorithms and solutions of Richtmyer [30], Beam and Warming [31], Gresho et al. [5], and the methods of Temam et al. [4]. The applicability of the projector will depend on the numerical accessibility of discretized vector fields as found in those and other works.

The experience gained with such specific applications should enable us to design a general software package for incompressible flow problems. Because of its efficient handling of the pressure terms, the method of Chorin [6] is a natural context for the use of our projection algorithm for such computations. In this regard it should be noted that the efficiency of our preliminary code can be significantly increased by using spline techniques to approximate the vector functions, by the use of singular integral techniques developed for panel methods used in solving subsonic potential flow problems around aircraft bodies and wings [32], and careful attention to coding detail. Let us note that by utilizing a more general Green's formula for the projection, nonzero boundary conditions can be dealt with in our approach simply by approximating the additional (nonsingular) boundary integral. Convergence of our projection algorithm follows for these boundary terms also. More complicated and general domains can be handled by flat panel approximations to the surface.

REFERENCES

1. P. Morse and H. Feshbach, Methods of Theoretical Physics, Parts I and II, McGraw-Hill, New York, 1953.
2. O. Ladyzhenskaya, The Mathematical Theory of Viscous Incompressible Flow, Gordon and Breach, New York, 1963.
3. D. Sattinger, Topics in Stability and Bifurcation Theory, Lec. Notes in Math. 309, Springer, Berlin, 1973.
4. R. Temam, Navier-Stokes Equations: Theory and Numerical Analysis, Elsevich-North Holland, New York, 1977.
5. P. Gresho, R. Lee, R. Sani, T. Stullich, On the time-dependend FEM solution of the incompressible Navier-Stokes Equations in two and three dimensions, Lawrence Livermore Lab. Rept. UCRL-81323 (1978).
6. A. Chorin, The numerical solution of the Navier-Stokes equation for an incompressible fluid, Bull. Amer. Math. Soc. 73 (1967), 928-931.
7. A. Chorin, A numerical method for solving incompressible viscous flow problems, J. Comp. Physics 2 (1967), 12-26.
8. C. Peskin, Flow patterns around heart valves: a numerical method, J. Comp. Physics 10 (1972), 252-271.
9. G. Prodi, Theoremi di Tipo Locale per il Sistema di Navier-Stokes e Stabilita delle Soluzione Stazionarei, Rend. Sem. Mat. Univ. Padova 32 (1962), 374-397.
10. D. Sattinger, The mathematical problem of hydrodynamical stability, J. Math. Mech. 19 (1970), 797-817.
11. M. Fortin, Approximation des Fonctions à Divergence Nulle par la Méthode des Eléments Finis, Lec. Notes in Physics 18, Springer, Berlin (1973), 99-103.
12. M. Crouzeix and P. Raviart, Conforming and Nonconforming Finite Element Methods for Solving the Stationary Stokes Equations (to appear).
13. F. Thomasset, Application d'une Méthode d'éléments finis d'ordre un à la résolution numérique des équations de Navier-Stokes, IRIA Rept. NO. 150, Le Chesnay, France, 1975.
14. U. Schumann, Subgrid scale model for finite difference simulations of turbulent flows in plane channels and annuli, J. Comp. Physics 18 (1975), 376-404.

15. O. Ladyzhenskaya and V. Rivland, On the alternating direction method for the computation of a viscous incompressible fluid flow in cylindrical coordinates, Izv. Akad. Nank. 35 (1971), 259-268.

16. J. Lions, On the numerical approximation of some equations arising in hydrodynamics, A.M.S. Symposium, Durham, April, 1968.

17. R. Temam, Une mêthode d'approximation de la solution des êquations de Navier-Stokes, Bull. Soc. Math. France 98 (1968), 115-152.

18. R. Glowinski and O. Pironneau, "Numerical methods for the 2-dimensional Stokes Problem through the stream function-voticity formulation", 1st France-Japan Colloq. on Funct. Analysis and Num. Analysis, Tokyo, 1976.

19. H. Fujita and T. Kato, On the Navier-Stokes Initial Value Problem I, Tech. Rept. 121, Stanford University, 1963.

20. W. Ames, Some computation-steeples in fluid mechanics, SIAM Review 15 (1973), 524-552.

21. A. Amsden and F. Harlow, A simplified MAC technique for incompressible fluid flow calculations, J. Comp. Physics 6 (1970), 322-325.

22. R. Sweet, A cyclic reduction algorithm for solving block tridiagonal systems of arbitrary dimension, SIAM J. Num. Anal. 14 (1977), 706-720.

23. G. Williams, Numerical integration of the three-dimensional Navier-Stokes equations for incompressible flow, J. Fluid. Mech. 37 (1969), 727-750.

24. W. Ames, Numerical Methods for Partial Differential Equations, 2nd Ed., Academic Press, New York, 1977.

25. P. Eagles, On stability of Taylor vortices by fifth-order amplitude expansions, J. Fluid Mech. 49 (1971), 529-550.

26. K. Gustafson, Estimation of eigenvalue aggregates determining hydrodynamic stability, Notices Amer. Math. Soc. 23 (1976), A-682.

27. J. Marsden and M. McCracken, The Hopf Bifurcation and its applications, Springer, Berlin, 1976.

28. O. Ladyzhenskaya, Mathematical analysis of Navier-Stokes equations for incompressible liquids, in "Annual Review of Fluid Mechanics", Vol. 7, Annual Reviews Inc., Palo Alto, California, 1975.

29. D. Young, Nonlinear Diffusion with Traveling Waves and Numerical Solutions, Thesis, University of Colorado, 1979, to appear.

30. R. Richtmyer, Invariant manifolds and attractors in the Taylor Problem, preprint, 1978.

31. R. Beam and R. Warming, An implicit finite difference algorithm for hyperbolic systems in conservation-law form, J. Comp. Physics 22 (1976), 87-110.

32. F. Johnson and L. Erickson, A general panel method for the analysis and design of arbitrary configurations in incompressible flows, NASA report, NASA CR-3079 (1979).

EFFICIENT SOLUTION OF A NONLINEAR HEAT CONDUCTION PROBLEM BY USE OF FAST ELLIPTIC REDUCTION AND MULTIGRID METHODS

Karl Solchenbach (*)
Klaus Stüben (*)
Ulrich Trottenberg (**)
Kristian Witsch (*)

ABSTRACT :

We report on numerical investigations which were made in the development of the code SIHEM (SImulation of HEating processes in Metals). In SIHEM the associated nonlinear parabolic problem (2D-space variables) is treated by a combination of simple (low order) space and implicit time discretizations (Crank-Nicolson) with a time step size control, Newton's or a Newton-like linearization and the use of Fast Elliptic Solvers for the large linear systems. In this report, the emphasis is laid upon systematic investigations and comparisons involving the use of typical Fast Solvers and well-known classical methods. As the solvers are applied at each time and each linearization step the total computing time considerably depends on their efficiency. As a result, we show that the use of a special Multigrid Solver in connection with the time step size control yields a very efficient composite algorithm.

(*) Institut für Mathematik der Gesellschaft für Mathematik und Datenverarbeitung (GMD)
Schloß Birlinghoven
D-5205 St. Augustin / West Germany

(**) Institut für Angewandte Mathematik der Universität Bonn
Wegelerstr. 6
D-5300 Bonn / West Germany

1. INTRODUCTION

In this report, we summarize numerical investigations which we have made in developing the code SIHEM. Based on the Crank-Nicolson scheme this code solves a nonlinear parabolic initial boundary value problem in rectangular domains simulating a special heating process of metal slabs (see Section 2). The demands concerning accuracy and efficiency for the algorithm were given by users from the steel industry.

SIHEM, which satisfies these demands, combines the following elements (see Section 3):

- a space discretization of the second order with suitable fixed step size h resulting in a nonlinear ODE system with respect to time (method of lines),

- an implicit time discretization of the second order with variable step size Δt (trapezoidal rule),

- step size control in time direction,

- Newton's (or Newton-like) linearization,

- use of Fast Elliptic Solvers for the occurring large linear systems, in particular use of appropriate Multigrid techniques.

The fixed space discretization is used here for simplicity (see, however, Section 5.2). Assuming this, the implicitness of time discretization and the time step size control are necessary to get an efficient algorithm for the cases considered here. Well-known explicit methods do not take sufficient account of the stiffness of the ODE system mentioned: Because of the stability limitations of explicit methods [19], a satisfactory adjustment of the time step size to the - harmlessly behaving - solution is not possible [3].

On the other hand, the use of the trapezoidal rule turned out to be sufficient for our purposes. Its lack of strong absolute stability does not essentially affect the possibility of performing a satisfactory time step size control, at least not for the moderate accuracies as required here. (For cases in which much higher accuracy is wanted, strongly absolute stable methods are preferable. In such cases we used smoothing and extrapolation techniques [25] and a "global" step size control for the initial time phase.)

In applying implicit methods, the essential numerical difficulty lies in the nonlinear elliptic difference equations which are to be solved at each time step. In particular, we here report on systematical studies with so-called Fast Elliptic Solvers (FES) which can be applied after appropriate linearization of the nonlinear systems.

We distinguish between methods for special problems (Special FES, formerly often called Fast Poisson Solvers) and more general approaches (General FES).

Among the Special FES, Buneman's algorithm [12], [26] is best known and can be implemented most easily, Hockney's FACR methods [22], [40] might be the fastest, and the methods of "Total and Alternating Reduction" by Schröder/ Trottenberg [30], [31] are distinguished by their extraordinary numerical stability. All these methods are directly applicable only in cases of simple geometries, as in the given application. On the other hand, the Multigrid techniques as proposed e.g. in [4], [5], [18], [20] belong to the General FES.

The use of FES in the parabolic case leads to rather involved efficiency discussions. The advantages of FES as known from the elliptic case (see Section 4.1) are still present; they are, however, not at all as impressive as in the elliptic case. This is due to several influences some of which we discuss in detail. In all situations considered here, the General FES of Multigrid type proved to be most suitable. Appropriate versions of the Special FES turned out to be similarly efficient for certain (linear and) nearly linear cases, but are less flexible and therefore less advantageous for more general situations. (FES, viz. Buneman's algorithm, were first applied in connection with parabolic problems by Buzbee [10], [11]. He, however, treated only linear problems and did not use a time step size control. Therefore his results cannot be compared with ours very well.)

For comparison, we additionally used several classical approaches (see Section 4). For instance SOR turned out to be satisfactory if only "small" time steps are performed. The classical ADI method [15] is known to be a suitable approximate solver for the Crank-Nicolson scheme. This was confirmed also by our investigations. On the other hand, the use of FES in connection with the implicit time discretization of parabolic problems is a general approach, independent of the special features of the problem and the other algorithmical components.

The composite algorithm proposed here consists of components which are known in principle. But their composition to an efficient method requires several specific arrangements. For example, the efficieny critically depends on the suitable application of the step size control: To perform this, one has to tune

- the required local time discretization accuracy,
- the stopping criterion for the linearization method,
- the (algebraic) accuracy of the used FES.

We want to point out that the quantitative numerical insight won here refers to the special class of problems considered and to the associated accuracy requirements. The results do not necessarily carry over to more general cases (highly nonlinear problems, strong sources etc.). In the underlying situation, however, the proposed composite algorithm turned out to be superior to several approaches which are recommended by well-known software libraries. (We used, for example, various finite element packages and ODE solvers for stiff systems. As the special structure of the given problem was not fully exploited by these approaches, they proved to be uncompetitive.)

We are currently working on several alternate approaches to parabolic problems. Section 5 gives an outlook on two important topics of this work: The application of nonlinear elliptic Multigrid techniques (to the nonlinear systems) and the development of "direct" parabolic Multigrid techniques (with respect to space and time). Furthermore, we are working on various higher order methods for the space discretization ("Mehrstellenverfahren"; defect correction [34]; τ-extrapolation [6]; discrete Newton methods [1]).

The development and the implementation of the code SIHEM was carried out under project MATAN at the Institut für Mathematik of the Gesellschaft für Mathematik und Datenverarbeitung. Special emphasis was laid on a graphic (coloured) display of the numerical results [38]. SIHEM was presented in action at the Hannover-Messe 1979.

2. THE TECHNICAL BACKGROUND AND THE MATHEMATICAL MODEL

2.1 The technical background

The technical background for the development of SIHEM is the following problem from the steel industry. Before steel is rolled, it has to be heated up to a certain fixed temperature (e.g. 1500 K). The final temperature distribution must be as homogeneous as possible. This is necessary in order to

- perform the rolling process faultlessly and to guarantee a sufficient quality of the final product.

Furthermore, one wants to

- minimize the costs of heating energy (oil or gas),
- maximize the throughput rate of the furnace by a short passage time.

The concrete situation is as follows :

Steel slabs (big blocks) are heated in hearth-type furnaces to temperatures of about 1500 K and are afterwards carried to the rolling line, being exposed to and cooled by the surrounding air for a short time. The furnace is partitioned into several zones the size and energy supply of which can be adjusted within certain limits. During their migration through the furnace, which is about 40 m long, the slabs are supported on "skids" which are kept at a temperature of about 320 K by water cooling.

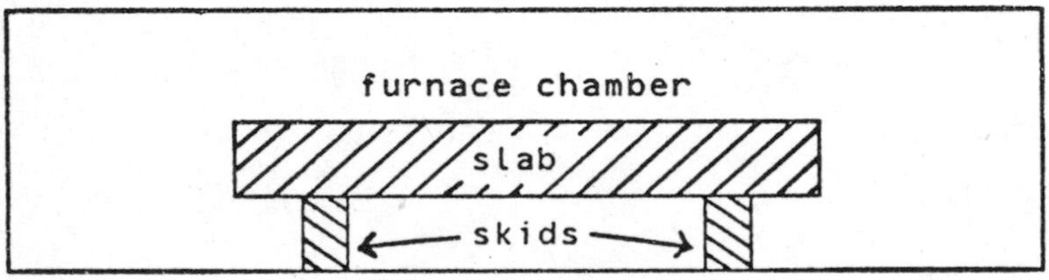

Figure 1 : Cross-section of the furnace

Now the heating should be performed in such a manner that the conditions mentioned above are met as well as possible (e.g. by appropriately fuelling the furnace, varying the zone temperatures and latency times etc.), taking into account also the cooling by the skids and by the air during the transport from the furnace to the rolling line.

In principle it is technically possible to measure the temperature in the interior of some trial slabs by thermocouple elements. This method, however, is very expensive and not economic. Therefore a mathematical model is used and the distribution of temperature is evaluated by numerical methods.

For a numerical simulation the following two possible applications are of interest :

(D) From time to time the slab surface temperature is measured pyrometrically and fed to a computer controlling the process in online and real-time mode. The temperature in the slab interior is computed stepwise via the mathematical model [44] (Dirichlet boundary conditions).

(R) Knowing everywhere the surrounding heating temperature and the kind of heat transfer at the slab boundary (here mainly by radiation), the temperature in the slab interior is computed for the complete heating process (nonlinear mixed boundary conditions).

A rough qualitative overview of the temperature distribution in the slab interior may already be supplied by a 1D-computation. More precise knowledge about the temperature at the edges and in the neighbourhood of the skids, however, can only be obtained by using a 2D- or a 3D-model [23]. The 2D-computation usually is sufficient. (The front and back ends of the slab, where eventually a 3D-computation would be necessary, are cut off for technical reasons after the milling process anyway.)

2.2 The mathematical model

The mathematical model of the heating process described above is the following well-known nonlinear parabolic initial boundary value problem for the unknown inner slab-temperature $T = T(x_1, x_2, t)$:

(2.1a) $$c_p(T)\rho(T)\cdot(\partial T/\partial t) = \mathrm{div}[k(T)\cdot \mathrm{grad} T] \qquad (x \in \Omega,\ t > 0)$$

(2.1b) $$T(x,0) = T_0(x) \qquad (x \in \overline{\Omega},\ t = 0)$$

(2.1c) in case (D): $$T(x,t) = \gamma(x,t) \qquad (x \in \Gamma,\ t > 0)$$

in case (R): $$k(T)(\partial T/\partial N) = h(T)\cdot(T_A - T) \qquad (x \in \Gamma,\ t > 0)$$

Here

T	is the temperature, $[T] = K$,
x	$= (x_1, x_2)$ are the space variables, $[x_i] = cm$,
t	is the time variable, $[t] = sec$,
Ω	is a rectangle $\Omega = (0,A)\times(0,B)$ (the cross-section of the slab), $\bar{\Omega}$ its closure,
Γ	is the boundary of Ω,
$\partial T/\partial N$	is the (outward) normal x-derivative of T,
$T_A(x,t)$	is the surrounding temperature,
$T_0(x)$	is the initial temperature of the slab,
$k(T)$	is the thermal conductivity, $[k] = W/(cm\cdot K)$,
$c_p(T)$	is the specific heat, $[c_p] = J/(kg\cdot K)$,
$\rho(T)$	is the density of the metal, $[\rho] = kg/cm^3$,
$h(T)$	is the heat transfer coefficient, $[h] = W/(cm^2\cdot K)$,
$\gamma(x,t)$	is the (known) temperature of the surface.

For reasons of easy usage we did not normalize the variables to dimensionless form. So typical temperature values in the above application, e.g., lie in the range between 280 K and 1500 K. The typical heating time is about three hours.

The fact that in the given technical situation Ω is a rectangle was the motivation for the systematical comparison of Fast Elliptic Solvers: Most of them are restricted to such simple geometries. We point out, however, that the General FES - especially our Multigrid solver - can be applied also to general domains Ω (see Section 4.1).

In SIHEM both cases (D) and (R) can be treated. But systematic numerical comparisons of the application of Fast Elliptic Solvers have been performed only for the case (D) so far. In this paper we therefore discuss only case (D) in detail.

Finally we note that in SIHEM additionally sources in the equation (2.1a) are also allowed. In practice such sources occur, e.g., if metal is heated by electro-magnetic induction. In order to make the efficiency discussion not too involved, we confine ourselves to situations without sources.

As metals are homogeneous materials, the physical data k, c_p and ρ depend only on the temperature and not on the space variable x. The dependence on T (and by that the degree of nonlinearity of the differential equation) varies with the type of steel which is to be heated [29]. For high-alloy steels the data are nearly independent of T. For low-alloy or iron-carbon-alloy steels the specific heat raises sharply at the "Curie-temperature" (see Figure 2).

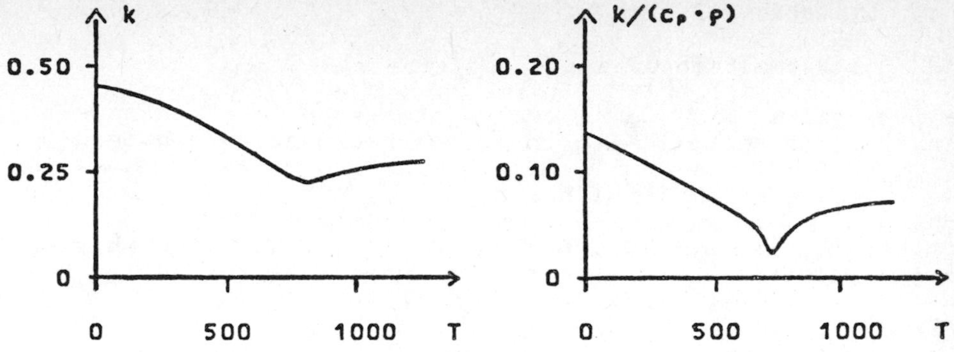

Figure 2 : Typical dependence of the data on T (in °C)

Because k does not depend on x, (2.1) is transformed by means of the well-known substitution T <—> u (cf. [26]) with some arbitrary $T_1 > 0$:

$$(2.2) \qquad u = \int_{T_1}^{T} k(\tau)d\tau .$$

By this we get the following initial boundary value problem for the unknown function $u = u(x,t)$

(2.3a) $\quad (\partial u/\partial t)(x,t) = Fu(x,t) \qquad (x \in \Omega,\ t > 0)$

(2.3b) $\quad u(x,0) = u_0(x) \qquad (x \in \overline{\Omega},\ t = 0)$

(2.3c) $\quad u(x,t) = g(x,t) \qquad (x \in \Gamma,\ t > 0)$

where F is the elliptic operator $Fu = f(u)\cdot\Delta u$ and

$$f(u) = k(T)/[c_p(T)\cdot\rho(T)] ,$$

$$u_0(x) = \int_{T_1}^{T_0(x)} k(\tau)d\tau , \qquad g(x,t) = \int_{T_1}^{\gamma(x,t)} k(\tau)d\tau .$$

In the following we usually consider this transformed problem.

3. THE COMPOSITE ALGORITHM

3.1 Discretization of space and time variables

The variables $x=(x_1,x_2)$ are discretized in a rectangular mesh G_h matching well with the given rectangular region Ω:

$$G_h = \{ x=(ih_1,jh_2) : i,j \in \mathbb{Z} \}$$

with $h=(h_1,h_2)$; $h_1=A/N_1$, $h_2=B/N_2$; N_1, $N_2 \in \mathbb{N}$. (If $h_1=h_2$, we use the notation h also for the components of the vector h.)

We denote by

$$\Omega_h := G_h \cap \Omega \; ; \qquad \Gamma_h := G_h \cap \Gamma ; \qquad \overline{\Omega}_h := \Omega_h \cup \Gamma_h$$

and by $\|.\|$ the discrete sup-norm on $\overline{\Omega}_h$

$$\|v\| := \sup \{ |v(x)| : x \in \overline{\Omega}_h \} .$$

The Laplacian Δ is approximated by the ordinary 5-point difference operator Δ_h (of order 2) with the mesh widths h_1 and h_2.

The discretization of space variables x (often called semi-discretization or method of lines) leads to a system of non-linear ODEs with respect to t:

$$(3.1a) \qquad (\partial u_h/\partial t)(x,t) = F_h u_h(x,t) \qquad (x \in \Omega_h, \; t > 0)$$

$$(3.1b) \qquad u_h(x,0) = u_0(x) \qquad (x \in \overline{\Omega}_h, \; t = 0)$$

$$(3.1c) \qquad u_h(x,t) = g(x,t) \qquad (x \in \Gamma_h, \; t > 0)$$

Here, $u_h(x,t)$ - defined for $x \in \overline{\Omega}_h$ and $t \geq 0$ - is the x-discrete approximation of $u(x,t)$ and

$$F_h u_h = f(u_h) \cdot \Delta_h u_h \; .$$

The system (3.1) is known to be highly stiff (depending on the x-mesh size). This is the reason why explicit time-discretization methods are not useful except for special situations (see Section 4.3.2). Therefore the time variable t has been discretized implicitly. We used the trapezoidal rule with $t_0=0$, $t_{n+1}=t_n+\Delta t_n$, $n=0,1,2,...$ and variable mesh sizes Δt_n, see Section 3.2.

Altogether, for each time t_{n+1}, one obtains a discrete non-linear system for the unknowns $u_{h,n+1}(x)$ $(x\in\overline{\Omega}_h)$ which denote the approximations to $u_h(x,t_{n+1})$ (Crank-Nicolson scheme):

$$\text{(3.2a)} \quad \frac{(u_{h,n+1} - u_{h,n})(x)}{\Delta t_n} = \frac{[F_h u_{h,n} + F_h u_{h,n+1}](x)}{2} \qquad (x \in \Omega_h)$$

$$\text{(3.2b)} \qquad u_{h,0}(x) = u_0(x) \qquad (x \in \bar{\Omega}_h)$$

$$\text{(3.2c)} \qquad u_{h,n+1}(x) = g(x,t_{n+1}) \qquad (x \in \Gamma_h)$$

3.2 Step size control in time direction

We want to adjust the time steps Δt_n in (3.2) automatically to some prescribed accuracy demands. For the choice of Δt_n the well-known (local) step size control described in [37] was used in a slightly modified manner (also see [19]). For completeness we outline the idea.

We assume that $u_{h,n}(x)$ for $t=t_n$ has already been calculated by (3.2). Now consider (3.1) as an initial value problem for $t \geq t_n$ with initial values $u_{h,n}(x)$. For simplicity, in the following we denote the solution of this problem by $y^*(t)$.

Let $y(t_n+\Delta t;\Delta t)$ and $y(t_n+\Delta t;\Delta t/2)$ be approximations for $y^*(t_n+\Delta t)$ evaluated from (3.2a), (3.2c) by using step sizes Δt and $\Delta t/2$, respectively. Our aim is to determine Δt_n in such a way that

$$\| y(t_n+\Delta t_n;\Delta t_n/2) - y^*(t_n+\Delta t_n) \| \approx \varepsilon_t$$

with some given ε_t prescribing the local t-discretization accuracy. The calculation of Δt_n is based on the following assumption [35]

$$\text{(3.3)} \quad \begin{aligned} y(t_n+\Delta t;\Delta t) &= y^*(t_n+\Delta t) + e(t_n+\Delta t)\Delta t^2 + O(\Delta t^4), \\ y(t_n+\Delta t;\Delta t/2) &= y^*(t_n+\Delta t) + e(t_n+\Delta t)\Delta t^2/4 + O(\Delta t^4). \end{aligned}$$

As $e(t_n)=0$ and $e(t_n+\Delta t)=e'(t_n)\Delta t+O(\Delta t^2)$, we can express the error as

$$\| y(t_n+\Delta t;\Delta t/2) - y^*(t_n+\Delta t) \| = E + O(\Delta t^4)$$

with

$$\text{(3.4)} \qquad \begin{aligned} E &:= \| y(t_n+\Delta t;\Delta t) - y(t_n+\Delta t;\Delta t/2) \|/3 \\ &= \|e'(t_n)\|\Delta t^3/4 + O(\Delta t^4) \ . \end{aligned}$$

Let Δt now be some proposed reasonable time step for which E is calculated. Then δ, given by

$$\|e'(t_n)\|\delta^3/4 = \varepsilon_t$$

with $\|e'(t_n)\|$ replaced by the approximate value $4E/\Delta t^3$, should be a good guess for the required step size Δt_n. In detail the step size control runs as shown in Figure 3.

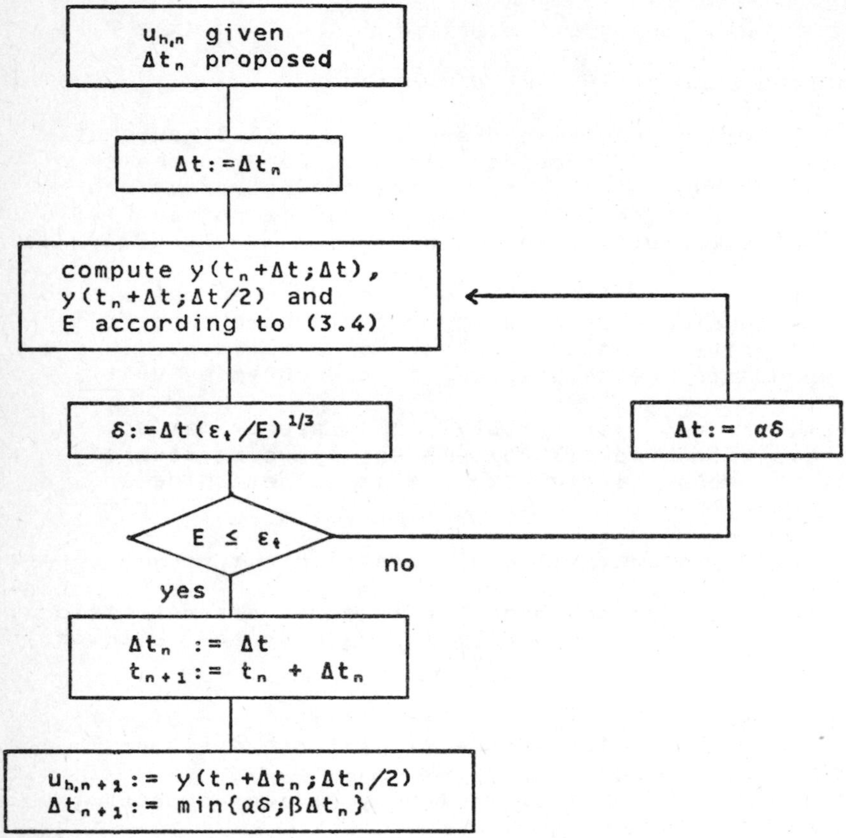

Figure 3 : Time step size control

The occurring parameters α, β ($0<\alpha<1, \beta>1$) are introduced for safety. In our program we always used $\beta=2$ and $\alpha=2^{-1/3}$. The latter value means that the local accuracy with $\Delta t_n = \alpha\delta$ should be about $\varepsilon_t/2$ rather than ε_t.

Although the algorithm requires at least 50% extra work per time step, it is very efficient for problems like the one described in Section 2.1 (compared to the choice of constant time steps): The time steps become very large for $t>>0$. This is, of course, not necessarily true in other applications. E.g. in the case of strongly varying sources the step size is limited because of physical reasons. (In practice this situation arises for example if a metal block is heated by being moved many times through an electro-magnetic coil.)

In prescribing the local accuracy parameter ε_t one should of course take into account that the global t-discretization error η_t differs from ε_t to some extent [36]. Well-known rigorous a priori estimations for η_t are usually far from being realistic. Therefore we do not state them here. We only want to mention that in the case of simple model problems the global error η_t at some fixed time behaves like $\varepsilon^{2/3}$ (more generally: $\eta_t \sim \varepsilon_t^{p/(p+1)}$, where p is the order of the time discretization method used). This has been observed in our application, too.

3.3 Linearization

In the following we consider the nonlinear system (3.2) for fixed n and Δt_n:

(3.5a) $$M_{h,n}\, u_{h,n+1}(x) = b_{h,n}(x) \qquad (x \in \Omega_h)$$

(3.5b) $$u_{h,n+1}(x) = g(x,t_{n+1}) \qquad (x \in \Gamma_h)$$

with

$$M_{h,n} = 2/\Delta t_n \cdot I_h - F_h; \quad I_h \text{ discrete identity;}$$

$$b_{h,n}(x) = (2/\Delta t_n \cdot I_h + F_h)\,u_{h,n}(x).$$

For the solution we use iterative linearization methods, replacing (3.5) for m=0,1,2,... by

(3.6a) $$H^m_{h,n}(u^{m+1}_{h,n+1} - u^m_{h,n+1})(x) + M_{h,n} u^m_{h,n+1}(x) = b_{h,n}(x) \qquad (x\in\Omega_h)$$

(3.6b) $$u^{m+1}_{h,n+1}(x) = g(x,t_{n+1}) \qquad (x\in\Gamma_h)$$

with some initial guess $u^0_{h,n+1}$ satisfying the boundary conditions (which in practice is taken by using the solution of the preceding time step or by extrapolating the solutions of two preceding time steps). $H^m_{h,n}$ is a discrete linear operator characterizing the linearization (specified below).

For well-known numerical reasons, one practically computes

$$u^{m+1}_{h,n+1} = u^m_{h,n+1} + d^m_{h,n+1}$$

with

(3.7a) $$H^m_{h,n} d^m_{h,n+1}(x) = b_{h,n}(x) - M_{h,n} u^m_{h,n+1}(x) \qquad (x \in \Omega_h)$$

(3.7b) $$d^m_{h,n+1}(x) = 0 \qquad (x \in \Gamma_h).$$

This iteration stops if $\|d^m_{h,n+1}\| \le \varepsilon_{lin}$.

In detail we consider

- Newton's method,
- a special Newton-like method.

In the case of Newtons's method we have

$$H^m_{h,n} = M'_{h,n}[u^m_{h,n+1}]$$

$$= -f(u^m_{h,n+1})\Delta_h + (2/\Delta t_n - f'(u^m_{h,n+1})\Delta_h u^m_{h,n+1})\cdot I_h.$$

Thus (3.7) is the discrete Helmholtz equation with variable c(x), which has to be solved in each time and each linearization step:

$$(3.8)\qquad \begin{aligned} -\Delta_h d + c(x)d &= r(x) \qquad (x \in \Omega_h) \\ d &= 0 \qquad (x \in \Gamma_h) \end{aligned}$$

with $d(x)=d^m_{h,n+1}(x)$ $(x\in\bar{\Omega}_h)$ and

$$(3.9)\quad c(x) = c^m_{h,n+1}(x)$$

$$= (2/\Delta t_n - f'(u^m_{h,n+1})\Delta_h u^m_{h,n+1})/f(u^m_{h,n+1}),$$

$$(3.10)\quad r(x) = r^m_{h,n+1}(x)$$

$$= \Delta_h u^m_{h,n+1} + [f(u_{h,n})\Delta_h u_{h,n} + 2(u_{h,n} - u^m_{h,n+1})/\Delta t_n]/f(u^m_{h,n+1}).$$

Some of the Fast Elliptic Solvers considered in the next section can only be applied to the discrete Helmholtz equation with constant $c(x)\equiv c$. This is the reason for introducing what we call here a Newton-like linearization ([13], also see [26], p.281f.). In this case $H^m_{h,n}$ is defined by

$$H^m_{h,n} = f(u^m_{h,n+1})\cdot(-\Delta_h + \hat{c}^m_{h,n+1}\cdot I_h)$$

with

$$\hat{c}^m_{h,n+1} = 1/2\ \{\min_{\Omega_h} c^m_{h,n+1}(x) + \max_{\Omega_h} c^m_{h,n+1}(x)\}$$

and $c^m_{h,n+1}(x)$ in (3.9). Instead of (3.8) one then has to solve in each time and each linearization step the discrete Helmholtz equation

$$(3.11)\qquad \begin{aligned} -\Delta_h d + \hat{c}d &= r(x) \qquad (x \in \Omega_h) \\ d &= 0 \qquad (x \in \Gamma_h) \end{aligned}$$

with constant $\hat{c}=\hat{c}^m_{h,n+1}$ and r(x) as in (3.10).

While - under suitable conditions - the convergence of Newton's method is of second order, the above Newton-like method converges only linearly. It nevertheless may be a reasonable approximation for Newton's method. In particular, if $c^m_{h,n+1}$ does not vary too much with respect to x and if the required accuracy is small, the Newton-like linearization is suitable.

The following Table 1 gives an impression of the convergence of Newton's and of the Newton-like method. We here treated an example with realistic physical data which are typical for the underlying technical problem, viz.

$$(P)\quad \begin{cases} k(T),\ (c_p\cdot\rho)(T) \text{ as shown in Figure 2,} \\ \Omega = (0,50)^2, \\ \gamma(x,t) = 1000\cdot(1-\exp(-t/100)) + 273, \\ T_0(x) = 273. \end{cases}$$

This example will be referred to as "problem (P)" in the following. Table 1 shows the behaviour of the linearization methods in two extreme situations (in terms of the norms of the corresponding corrections d^m computed by (3.8) and (3.11) with $N_1=N_2=32$), namely at two different times $t=t_{n+1}$ and $\Delta t_n=20$:

example (a): t = 20 ; example (b): t = 220.

In the case of example (a) (low temperatures) f(u) is nearly constant and the problem (2.3) behaves only weakly non-linear. In this case the Newton-like method is suitable compared to Newton's method: For the accuracies as typically required in our application (error in absolute values less than 0.1) the number of linearization steps is equal for both methods.

In the case (b), however, the situation is much worse for Newton-like: For t=220 the temperatures in the central area of Ω are much lower, the temperatures near the boundary Γ are much higher than the Curie-temperature (see Section 2.2). Thus (2.3) behaves stronger nonlinearly than in the case (a).

m	$\|d^m\|$ for example (a)		$\|d^m\|$ for example (b)	
	Newton	Newton-like	Newton	Newton-like
1	0.79(2)	0.79(2)	0.12(2)	0.11(2)
2	0.13(1)	0.14(1)	0.35	0.35(1)
3	0.21(-2)	0.75(-1)	0.12(-3)	0.13(1)
4	0.64(-8)	0.40(-2)	0.12(-9)	0.61
5	0.22(-13)	0.21(-3)	0.34(-12)	0.30
6	.	0.11(-4)	.	0.15
7	.	0.57(-6)	.	0.77(-1)
8	.	0.30(-7)	.	0.40(-1)
9		0.15(-8)		0.21(-1)
10		0.79(-10)		0.11(-1)

Table 1 : Convergence of the linearization methods

3.4 The composition of the components

Once a solver for the discrete equations (3.8) or (3.11) is chosen, one can compose the above components to an algorithm for solving the underlying parabolic problem. In detail several special arrangements are required to make it efficient. Mainly the following accuracy parameters have to be tuned:

ε_t : the local t-discretization accuracy for the step size control (see Section 3.2);

ε_{Lin} : the stopping parameter for the linearization method (see Section 3.3);

ε_{SOL} : a parameter prescribing the relative accuracy in solving the linear systems (3.8) or (3.11).

The time step size control with parameter ε_t works well only if a sufficient algebraic accuracy in solving the nonlinear difference equations is guaranteed, say up to an (absolute) error of $\sigma \cdot \varepsilon_t$ ($\sigma \ll 1$). In practice this accuracy is estimated by the norms $\|d^m\|$ of the Newton (-like) corrections. The stopping criterion

$$\|d^m\| \le \varepsilon_{Lin}, \quad \varepsilon_{Lin} := 0.1 \cdot \varepsilon_t$$

turned out to be sufficient for our purposes and is used in the following.

If the linear systems are solved by an iterative method, then ε_{SOL} controls the relative algebraic error for this "inner" iteration. One possibility is to adapt ε_{SOL} in each step of the "outer" iteration to the expected gain of accuracy. This is, however, complicated at least in the case of Newton's method and is not worthwile for the moderate accuracy demands given by many technical applications. We therefore preferred to choose ε_{SOL} fixed for all outer iteration steps, namely:

$$\varepsilon_{SOL} := 0.05 \ .$$

This means that Newton's method is truncated to a linearly convergent method. Practically, however, this is no essential disadvantage in our application (also see Section 5.1).

The parameter ε_{SOL} clearly is not needed if the linear equations (3.8) or (3.11) are solved by a numerically stable direct solver giving the discrete solutions "exactly" (up to neglegible roundoff). In practice, apart from iterative and direct solvers, so-called "approximative" solvers (e.g. truncated direct solvers) are of interest; see Section 4.1.

4. APPLICATION OF THE FAST SOLVERS AND NUMERICAL RESULTS FOR THE COMPOSITE ALGORITHM

The discrete elliptic problems (3.8) or (3.11), which occur in each time and linearization step, can be solved by well-known classical methods [43] or by one of the so-called Fast Elliptic Solvers developed in the last decade. Before applying these methods to the parabolic problem, we want to give a rough survey of Elliptic Solvers (see also [32]).

In all tables shown in this section the CPU-times refer to the (fairly slow) IBM/370-158 computer using the FORTRAN H-Extended Compiler with OPT(2) and single precision.

4.1 Comparison of Elliptic Solvers

In this section we consider - independently of the parabolic background - only

$$(4.1)\qquad \begin{aligned} -\Delta u + c(x)u &= f(x) \qquad (x \in \Omega) \\ u &= g(x) \qquad (x \in \Gamma) \end{aligned}$$

for rectangular domains Ω and $c(x) \geq 0$.

First we note that the matrix of the corresponding discrete problem (Δ_h and Ω_h as in Section 3) depends of course on the enumeration of the grid points in $\bar{\Omega}_h$. For the following classical methods we assume linewise enumeration.

I. Classical methods

- Gaussian block-elimination (Direct solver)
- GS (Iterative method of Gauss-Seidel; Successive relaxation method with relaxation parameter $\omega = 1$)
- SOR (Iterative solver; Successive overrelaxation method with asymptotically optimal relaxation parameter ω^*)
- ADI (Iterative solver; Alternating direction implicit Peaceman-Rachford-method with the optimal number of Wachspress-parameters [28], [42])

II. Special FES

These methods have essentially been developed for (4.1) with constant $c(x) \equiv \hat{c}$. As a matter of fact they are not directly applicable in the case of arbitrary variable c(x) or general domains.

- Buneman's algorithm (Direct solver)
  This algorithm is the well-known stabilized version of the Cyclic Reduction [9], [12].

- **TR: Method of Total Reduction (Approximative solver)**
  Due to its numerical stability, the algebraic accuracy of this method in solving the discrete equations can be prescribed by a truncation parameter, which may be used to tune efficiency and accuracy demands. Apart from the standard version TR2D01 [16], [17] which gives a relative accuracy of $\varepsilon_1=10^{-4}$, the special "strongly truncated" version TR2DA9 [17] was used in our application. In this version all difference stars occurring are approximated by 9-point-stars resulting in a relative accuracy of $\varepsilon_0=0.05$. We denote by

  TR : the version TR2D01,
  TR-A : the "strongly truncated" TR-version.

- FACR(l) (Direct solver)
  This method is a combination of a Fast Fourier Transformation with l steps of Cyclic Reduction [22], [40]. For the problem considered we used the optimal $l=l^*$ ($l^*=2$ or $l^*=3$).

III. General FES

These methods are applicable to much more general problems than (4.1). Here we used special algorithms of Multigrid type. A more detailed comparison of General FES (several Multigrid and MGR variants [18]) will be given in [39]. Within these comparisons, the Multigrid technique used here (and a similar MGR version) turned out to be most efficient.

- MG00 (Iterative solver)
  This Multigrid code - described in [39] in detail - is especially written for problems (4.1). It is a simplified version of the more general code MG01 [18]. (MG01 can easily be used instead of MG00 if the parabolic problem is given on a general domain Ω.)

  As there are many different possibilities to construct Multigrid algorithms we here want to point out the main features of our algorithm: MG00 uses grids $\Omega_h, \Omega_{2h}, \Omega_{4h}, \ldots$ up to a very coarse grid. On each of these grids the usual 5-point discretization of the Laplacian is applied. For smoothing, an iteration procedure of type Gauss-Seidel with red-black ordering of the grid-points is used, namely three relaxations at each grid per iteration step of MG00. The transfer from fine to coarse grids is done by "half weighting" [18]. Finally, linear interpolation is used as transfer from coarse to fine grids.

This "fixed" Multigrid version has a very good convergence factor. In case of Poisson's equation one can prove, for example, that the spectral radius of the iteration matrix is less than 0.038, independent of h. As the total computer work for one Multigrid iteration is roughly equivalent to five Gauss-Seidel iterations only, this method belongs to the fastest iterative methods known up to now. Experience shows that the same efficiency carries over to much more general problems, e.g. problems (4.1) on general domains. For more details, see [39].

- MGOO-F (Approximative solver)
  This code is of type Full-Multigrid as proposed by Brandt [6]. It uses MGOO in order to calculate the solution of (4.1) in a highly efficient way up to an error which is approximately equal to the h-discretization error. In particular, MGOO-F does not yield a fixed accuracy, but an accuracy which is proportional to h^2 (within a number of operations which is proportional to the number of unknowns!).

Although the magnitude of c(x) in (4.1) has essential consequences for the behaviour of some of the above methods, we present in Table 2 results only for the model problem of Poisson's equation

$$-\Delta u = f(x) \qquad (x \in \Omega = (0,1)^2)$$
$$u = g(x) \qquad (x \in \Gamma)$$

discretized on Ω_h with $h=h_1=h_2=1/N$. The given computing times refer to N=64 and N=128, i.e. to linear systems for 3969 and 16129 unknowns, respectively. For the direct and approximative solvers (second half of Table 2), the computing times are independent of the special example. For the iterative solvers, the examples have been chosen in such a way that the typical convergence behaviour of the various methods can be observed. (In the cases GS and SOR, e.g. we have chosen an example with the exact solution $u(x)=\sin\pi x_1 \cdot \sin\pi x_2$, and zero starting values.)

In order to obtain comparable results for the different methods we have chosen two fixed relative algebraic accuracies (already mentioned above in connection with TR-A and TR, respectively):

$$\varepsilon_0 = 0.5\cdot 10^{-1} \quad \text{and} \quad \varepsilon_1 = 10^{-4}.$$

For the iterative methods, computing times for both accuracies are given. The direct methods which in principle differ with respect to numerical stability [30] are arranged in such a way that the accuracy ε_1 is guaranteed in all cases (e.g. by using double precision for certain components in Buneman's algorithm). As already mentioned, MGOO-F is oriented to the discretization error rather than to a fixed algebraic accuracy: The distinction between ε_0, ε_1 is not appropriate for this method. In fact, for the example treated MGOO-F gives an algebraic accuracy much smaller than ε_1.

The last column of Table 2 shows the growth of operation count with respect to N (for fixed accuracy ε), neglecting the proportional constants (which depend logarithmically on ε for the iterative methods). For the special case of Poisson's equation the Gauss-Seidel and the block elimination method are, of course, totally inefficient.

method	computing times [sec]				operation count
	N = 64		N = 128		
	ε_0	ε_1	ε_0	ε_1	
GS	108.	332.	1730.*	5310.*	$O(N^4)$
SOR	3.2	9.8	25.8	78.4	$O(N^3)$
ADI	1.7	4.0	6.8	18.3	$O(N^2 \log N)$
MGOO	0.27	0.80	1.1	3.2	$O(N^2)$
Block-El.	-	65.	-	1040.*	$O(N^4)$
FACR	-	0.55	-	2.1	$O(N^2 \log\log N)$
Buneman	-	0.80	-	3.5	$O(N^2 \log N)$
TR	-	0.62	-	2.5	$O(N^2)$
TR-A	0.35	-	1.4	-	$O(N^2)$
MGOO-F	-	0.49	-	1.9	$O(N^2)$

Table 2 : Efficiency of the FES compared with classical methods (values marked with * are estimated)

4.2 Results for the technical problem

The relative efficiency of the different elliptic solvers given by Table 2 changes to some extent if they are used within the composite algorithm for the underlying parabolic problem. Differences may arise from the following facts:

(1) At each time step the solvers are applied to Helmholtz' equations (4.1) instead of Poisson's equation - as far as possible with c(x) variable as given in (3.9). As c(x) depends strongly on Δt, namely

$$c(x) = O(1/\Delta t) \quad (\Delta t \longrightarrow 0),$$

the values of c(x) occurring can become very small or very large depending on the size of Δt. This has consequences, e.g., on the efficiency of SOR: The smaller the step size Δt becomes the faster is the convergence of SOR (if h is kept fixed).

(2) By the replacement of c(x) by $\hat{c}$ in (3.11) the convergence of the linearization method may become considerably worse. As a consequence, the Special FES then lose much of their efficiency.

(3) The control strategy of the composite algorithm requires a certain flexibility of the solvers (adaptation of accuracy and efficiency demands). High accuracy - as given by direct solvers - usually is not needed. Thus it does not pay if it is achieved at the expense of computing time.

(4) A great part of the computer work is needed only to establish the linear systems for all time steps. In our examples this is roughly 50% of the overall work if FES are used as solvers. So the advantages of the Fast Solvers are reflected only partly by computing times.

For brevity we restrict ourselves in giving explicit results only for some typical solvers, namely

- TR, standing for those Special FES - as Buneman or FACR - which give a relative algebraic accuracy of at least 10^{-4}. Indeed, in all applications TR, Buneman, and FACR have finished similarly.

- MGOO, which could have been replaced in principle by several similarly behaved Multigrid or MGR methods (as described in [18]). MGOO is distinguished by the fact that it gives the required algebraic accuracy of about $\varepsilon_{SOL}=0.05$ in just one iteration step. (With respect to this rough accuracy required MGOO-F is less efficient and is therefore not discussed here.)

- TR-A, which should be regarded as an approximative Special FES, developed for a special situation as considered here (i.e. truncated as to give only ε_{SOL}-accuracy).

- SOR, which has been added as a widely used classical method. The required accuracy ε_{SOL} is controlled by using the spectral radius which is numerically approximated in line with the relaxation steps. The optimal relaxation parameter ω^* is approximated by simple eigenvalue estimations.

We summarize the solvers used in the following table. (Block-elimination has been added only for comparisons.) As far as possible, these methods have been used in connection with Newton's method rather than with the Newton-like method.

	relative algebr. error $\leq 10^{-4}$	relative algebr. error ≤ 0.05
Newton's linearization (3.8)	Block-elimination	MGOO; SOR
Newton-like linearization (3.11)	TR	TR-A

Furthermore we made comparisons with the classical "parabolic" ADI method [2], [15], which can be viewed as an approximation of the Crank-Nicolson scheme. As expected, for our parabolic problem this ADI version turned out to be highly preferable over the "elliptic" ADI method (cf.Section 4.1). Therefore we give numerical results only for the "parabolic" ADI. In practice we used it in connection with Newton's method, applying only one (linear) ADI-step in each Newton iteration.

As in our algorithm the accuracy is controlled by an adaptive procedure, the computing times give the main information about the efficiency of the different solvers. Besides computing times ("CPU"), Tables 3.1-3.3 show additionally the total number of outer linearization steps ("ITER") and the number of time steps needed to perform the heating process up to a fixed time t^* within which steady state is nearly reached. The given values refer to problem (P) with $t^*=9000$ sec and different accuracies, namely $h=h_1=h_2=50/N$ with $N=64$, 128 and $\varepsilon_t=1$, $1/4$, $1/16$. Values marked by $*$ are estimated. Furthermore Figure 4 shows the behaviour of Δt_n.

Let us first discuss the case of fixed ε_t (N variable). Then the number of time steps performed is independent of the solver and of N (if N is chosen sufficiently large). Therefore the total operation count is determined by the numerical effort of the specific solver used at each time step. Thus the overall work behaves like:

Block-elimination: $O(N^4)$,
SOR: $O(N^3)$,
MGOO, TR, TR-A: $O(N^2)$.

This shows that SOR is asymptotically (i.e. for $N \longrightarrow \infty$) not competitive with the FES. Indeed, for the examples given MGOO is considerably faster than SOR. The TR-methods are less efficient than MGOO, but still suitable. (Block-elimination is, of course, totally useless.)

Solver	N = 64		N = 128	
	CPU	ITER	CPU	ITER
Block-El.	161.*	149*	2548.*	147*
SOR	7.85	151	55.12	147
TR	4.70	219	18.92	222
TR-A	3.62	229	16.30	254
MGOO	2.80	149	11.77	157
par. ADI	3.85	165	16.27	174
# t-steps	40		40	

Table 3.1 : Results for ε_t = 1 (CPU-time in minutes)

Solver	N = 64		N = 128	
	CPU	ITER	CPU	ITER
Block-El.	263.*	234*	4229.*	244*
SOR	10.62	250	74.22*	244*
TR	7.98	374	33.05	389
TR-A	6.67	421	27.67	431
MGOO	4.38	234	18.28	244
par. ADI	6.35	272	24.40	261
# t-steps	60		61	

Table 3.2 : Results for ε_t = 1/4 (CPU-time in minutes)

Solver	N = 64		N = 128	
	CPU	ITER	CPU	ITER
Block-El.	496.*	458*	8337.*	481*
SOR	16.73	458	114.55*	481*
TR	15.48	730	62.72	740
TR-A	12.33	780	46.93	731
MGOO	8.58	458	36.03	481
par. ADI	11.10	476	48.88	523
# t-steps	101		101	

Table 3.3 : Results for ε_t = 1/16 (CPU-time in minutes)

However, as already pointed out, the superiority of the FES for moderate values of N is in the parabolic situation not as impressive as in Poisson's equation (cf. Table 2). This is due to the different items (1)-(4) which we mentioned at the beginning of this section. The quantitative influence of these items is very involved and considerably depends on the example. Here we want to point out that

- the overall numbers of linearization steps given for the TR-methods show the influence of the Newton-like linearization (cf. also Table 1): Because of its slower convergence these methods lose indeed much of their efficiency. TR compared to TR-A suffers from the fact that the algebraic accuracy is "too good" (i.e. part of the computer time is wasted). Therefore TR-A is faster than TR although it needs more linearization steps in most of the cases. TR has advantages only for linear problems or for problems with very weak nonlinearities (see Section 4.3.1).

- the comparison of SOR and MGOO computing times shows that the advantages of MGOO indeed are smaller in parabolic situations than in the Poisson case.

On the other hand, if ε_t decreases (N fixed), the disadvantage of the Newton-like method becomes more and more obvious. Thus the TR-methods finish worse. SOR becomes more and more efficient, because the average size of Δt_n decreases (cf. (1)). E.g. the average numbers of SOR-steps needed per Newton iteration for problem (P) and N = 64 are 15, 12, and 10 for ε_t = 1, 1/4, and 1/16, respectively.

Up to now we treated N and ε_t separately. Of course, these values should actually be coupled in a suitable manner. To get an order-of-magnitude impression of the total work in that case we assume ε_t to be chosen as

$$\varepsilon_t = O(N^{-2})$$

(the order of the space discretization). In our application the local step sizes are of order $\varepsilon_t^{1/3}$. This means that both c(x) and the number of time steps (up to a fixed heating time) are of order $\varepsilon_t^{-1/3}$. Using $\varepsilon_t = O(N^{-2})$ we get for the overall work e.g. in the case of

$$\text{MGOO:} \quad O(N^{8/3} \log N),$$
$$\text{SOR:} \quad O(N^{10/3} \log N).$$

One of the aims of our investigations was to obtain a comparison between the application of FES and the classical parabolic ADI method. At least in this special situation (5-point space discretization, rectangular domain) ADI was expected to be fairly good. This is confirmed by our results. This impression should, however, change for more general problems as far as this ADI variant is applicable at all.

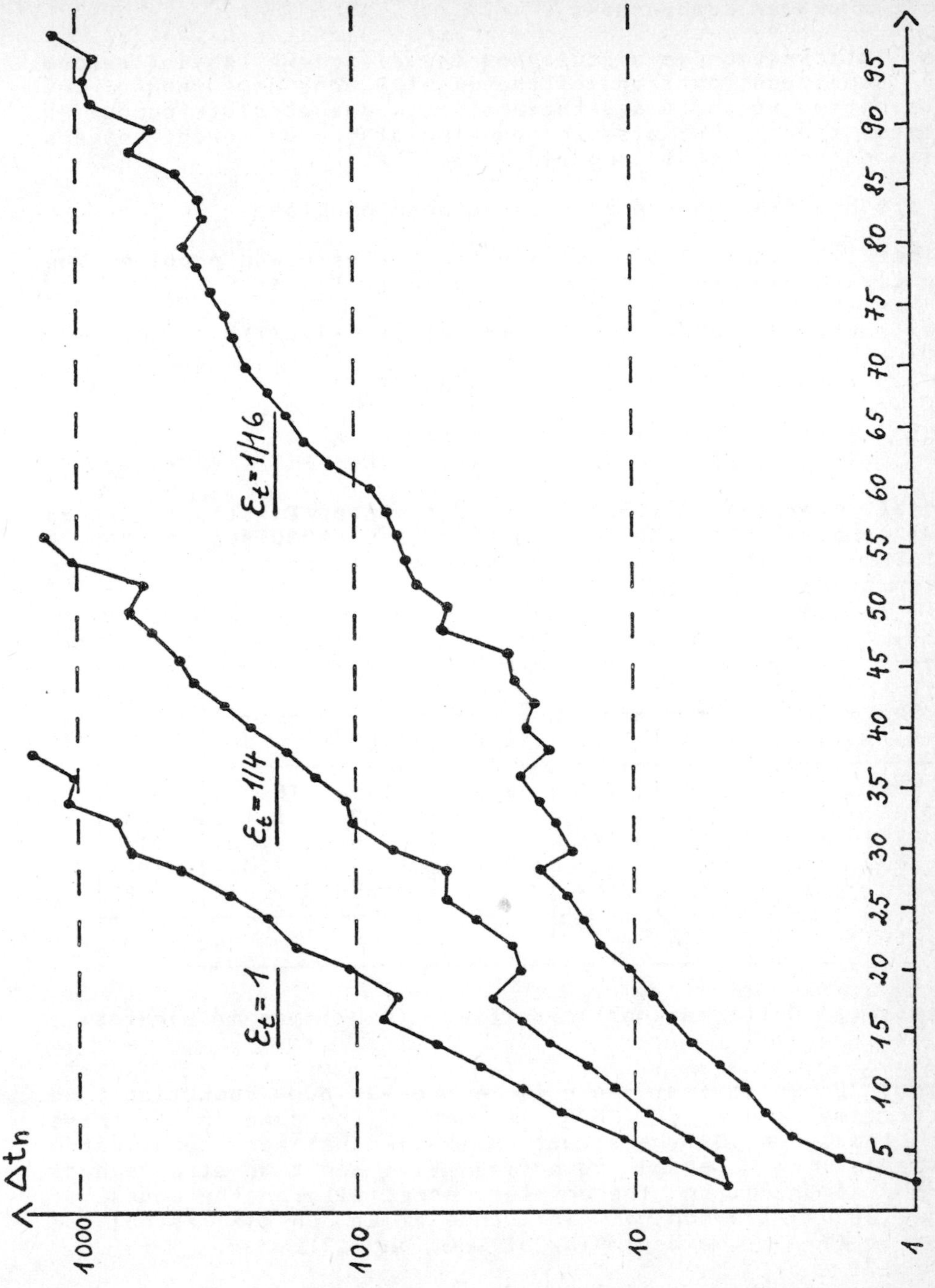

Figure 4: Δt_n as computed by the step size control. The irregular behaviour of Δt_n (in the middle of the picture) is caused by the changes of the physical data near the Curie temperature (see Figure 2).

4.3 Further comparisons

In this section we discuss how the efficiency is influenced by some modifications of the problem considered and of the algorithm. We shall see that not only the absolute computing times change, but also the ranking of the different solvers with respect to the computing times.

4.3.1 Modified nonlinearities; linear problems

First we vary the nonlinearity of the given problem. The following examples are considered in Table 4:

(a) : $k(T) = 5000/(10000+T)$, $(c_p \cdot \rho)(T) = 1/k(T)$

(b) : $k(T)$ and $(c_p \cdot \rho)(T)$ as in problem (P)

(c) : $k(T) = 1/2 - T/3892$ $(T \leq 973)$
$T/3892,$ $(T \geq 973)$, $(c_p \cdot \rho)(T) = 1/k(T)$.

In all examples $\Omega=(0,50)^2$, $N=64$, $\varepsilon_t=1$ and $\gamma(x,t)$, $T_0(x)$ as in problem (P). The heating time is $t^*=9000$ sec. As before "ITER" is the total number of outer iterations (linearization steps).

Solver	(a)		(b)		(c)	
	CPU	ITER	CPU	ITER	CPU	ITER
SOR	11.00	161	7.85	149	10.17	159
TR	3.22	144	4.70	219	6.47	305
TR-A	2.78	176	3.62	229	5.00	316
MG00	3.02	161	2.80	149	3.03	162
par. ADI	3.67	157	3.85	165	9.85	422
#t-steps	44		40		42	

Table 4: Different nonlinearities (CPU-times in minutes)

First we note that the efficient behaviour (computing time per time step) of MG00 is nearly the same in all these problems. For SOR the situation is somewhat more complicated because the average computing time per time-step depends much stronger on the problem, especially on the course of the step size control (influence of Δt_n on $c(x)$ as pointed out in (1) at the beginning of Section 4.2).

As for weakly nonlinear problems the Newton-like method converges faster than for stronger nonlinear problems, the efficiency of the TR methods decreases from (a) to (c). This is reflected by both computing times and total number of linearization steps.

The ADI method is much worse in example (c). The reason for this is that in performing the t-step size control the function c(x) in (3.8) may become slightly negative. As this is not allowed for the ADI method used the growth of the t-step size has to be limited such that $c(x) \geq 0$ is guaranteed for all t-steps. This additional limitation explains why ADI becomes worse than the other solvers.

In practice the physical data k, c_p, ρ are often supposed to be independent of the temperature (if the requirement of accuracy is small, if there is not much information about the course of these data or if the heating process is performed only within a rather small temperature interval). In this case the original nonlinear problem (2.1) becomes linear.

Our algorithm is essentially designed for nonlinear problems. Although formally applicable, it can, of course, not be recommended for linear problems. For linear problems it is more natural to use fast direct solvers - which give a sufficiently high accuracy - once in each time step. Nevertheless also fast iterative solvers as e.g. MGOO are still of a similar efficiency. As an example we treated problem (P) with k and $c_p \cdot \rho$ replaced by

$$k(T) \equiv 0.5, \quad (c_p \cdot \rho)(T) \equiv 2$$

and $N=64$, $\varepsilon_t=1$. Using TR and MGOO (with three iterations per time step which gives roughly the same accuracy as TR) we needed the computer times 2.0 and 2.5 min., respectively. SOR was more than 6 times slower.

4.3.2 Time discretization with constant time steps

If the composite algorithm is used without the time step size control, one obtains results that are very different from those in Table 3. This is demonstrated in Table 5 for problem (P) and $N = 64$.

Solver	constant time steps		step size control
	$\Delta t \equiv 5$	$\Delta t \equiv 20$	$\varepsilon_t = 1.$
SOR	48.98	22.85	7.85
TR	80.57	27.42	4.70
TR-A	56.88	20.00	3.62
MGOO	49.22	16.78	2.80
par. ADI	60.97	21.00	3.85

Table 5: Comparison of constant and variable time steps (CPU-times in minutes)

Here the computing times of our algorithm (prescribing $\varepsilon_t = 1$) are compared to those needed in case of constant time steps: $\Delta t \equiv 5$ and $\Delta t \equiv 20$. $\Delta t = 5$ is chosen to be the smallest step size computed by the step size control (cf. Figure 4). In the case of constant time steps the linearization at each time step is stopped by setting $\varepsilon_{lin} = 0.1$. (For completeness we note that for well-known stability reasons the choice of constant time steps may be problematic for the trapezoidal rule, at least in an initial phase.)

Firstly, the computing times given in Table 5 clearly show the essential advantage of the step size control for problems as treated here.

Secondly, the ranking of the different solvers with regard to their efficiency changes if (small) constant time steps are chosen. In particular SOR, which is the worst method in connection with the step size control, becomes the best (together with MGOO) in case $\Delta t \equiv 5$. The reason for this has already been mentioned in (1) at the beginning of Section 4.2.

This gives the impression that SOR is preferable over MGOO for (small) constant time steps. This is, however, no longer true if Δt and ε_{lin} are coupled to N in a natural way (for reasons of second order accuracy in space and time), namely by

$$\Delta t = O(N^{-1}) , \quad \varepsilon_{lin} = O(N^{-2}).$$

As – in each time step – $c(x)$ in (3.8) then behaves like $O(N)$, the spectral radius of the SOR method becomes $1-O(N^{-1/2})$. On the other hand the convergence factor of MGOO is independent of N. Thus we get asymptotically (i.e. for $N \longrightarrow \infty$) for the total number of operations:

$$\text{for SOR:} \quad O(N^{3.5}\log N),$$
$$\text{for MGOO:} \quad O(N^{3}\log N).$$

This shows that MGOO is (asymptotically) faster than SOR also in case of constant time steps, if only Δt, N, and ε_{lin} are adjusted properly (which is a difficult problem in practice and is usually not really done). But the gain in efficiency is less than in the case of Poisson's equation: The total operation count is smaller by a factor $O(N^{1/2})$ rather than $O(N)$ for Poisson's equation.

Finally we want to mention that many users still prefer explicit time discretization methods over implicit methods, e.g. the extremely simple Euler-method [23], [33]. Indeed, the Euler-method needs roughly 27 min. to solve the problem treated in Table 5 if $\Delta t \equiv 1.1$ is chosen (which is the maximal step size allowed by the stability condition in the example considered). So this method is fairly good in comparison to the other methods in the case of constant time steps. The preference of this explicit method is apparently no longer justified if compared to the implicit method using a time step size control (see also [3]).

Although our aim in this paper is not a comparison of implicit and explicit methods, we want to point out that there are situations, where explicit methods might be preferable to the much more complicated implicit methods. This is the case, if, for instance, only very crude accuracies are required (small N), or if the size of the time steps is extremely limited for physical reasons anyway.

5. ALTERNATE APPROACHES TO THE PARABOLIC PROBLEM

5.1 Nonlinear elliptic Multigrid methods

So far, the emphasis of this report was laid on a systematic comparison of linear elliptic solvers applied in connection with a nonlinear parabolic situation. Therefore, we discussed only "global" (Newton and Newton-like) linearization methods for the discrete nonlinear elliptic problems that arise in each time step. Of course, it is also possible to apply nonlinear iterative methods to the nonlinear problems directly (nonlinear SOR, nonlinear ADI etc.) [27], [41].

As a linear Multigrid method turned out to be the most efficient solver in our experience, we are mainly interested in the investigation of a corresponding nonlinear Multigrid solver. General nonlinear Multigrid methods have been described already by Brandt ("FAS-mode" [4]) and Hackbusch [21].

We have started systematical investigations in this area. Up to now, we have performed several experiments with various approaches. We here want to outline only the simplest technique, which, however, has already given encouraging results.

This is demonstrated by Table 6 where three methods of solving the nonlinear difference equations at each time step are compared. The results refer to an elliptic problem as it typically arises in the implicit time discretization of a parabolic situation. The numbers given are the errors to the discrete solution (which is known in this special example).

The three columns in Table 6 correspond to

(I) Newton's method combined with MG00, where the number of inner Multigrid iterations is adapted to the convergence of Newton's method. (This number is approximately doubled from one linearization step to the next.) The horizontal lines indicate that a new Newton step is performed.

(II) Newton's method combined with only one iteration of MG00 performed at each linearization step. This is what is really done in our composite algorithm (cf. Section 3.4).

(III) a nonlinear Multigrid method of FAS-type [4]. In this procedure, the Multigrid components (smoothing, fine-to-coarse, coarse-to-fine transfer, etc.) were - in principle - arranged as in MG00. In particular, for smoothing a nonlinear Gauss-Seidel relaxation method was used, resulting in a set of single nonlinear equations for each smoothing step. In our application it was sufficient to treat these single equations by just one step of a linear iteration method (using no derivatives).

The results given in the table show that the convergence properties of all three methods are very similar. Method III, however, is not only simpler but also slightly more efficient than II or I:

At first, it is simpler because no global linearization is performed at all. In particular no adaptation to any outer iteration is necessary.

Secondly it is faster: For example, within II the numerical work to establish the linear systems in all linearization steps is roughly 50% of the total work (cf. Section 4.2); whereas the numerical effort to perform one nonlinear multigrid iteration is only slightly higher than that of one linear multigrid iteration step. Altogether this means a reduction of the computing time by somewhat less than 50%, the concrete value depending on the complexity of the special problem that is to be solved.

μ	Sup-norm of the error to exact discrete solution		
	(I)	(II)	(III)
1	0.22(2)	0.22(2)	0.29(2)
2	0.60(1)	0.14(1)	0.17(1)
3	0.14(0)	0.13(0)	0.15(0)
4	0.78(-1)	0.12(-1)	0.13(-1)
5	0.27(-2)	0.93(-3)	0.92(-3)
6	0.18(-3)	0.72(-4)	0.65(-4)
7	0.20(-4)	0.56(-5)	0.47(-5)
8	0.14(-5)	0.42(-6)	0.34(-6)
9	0.49(-6)	0.20(-7)	0.22(-7)
10	0.30(-7)	0.22(-8)	0.13(-8)
11	0.17(-8)	0.14(-9)	0.77(-10)
12	0.92(-10)	0.11(-10)	0.80(-11)
13	0.62(-11)	.	.
14	0.90(-12)	.	.
15	.	.	.
16	.	.	.

Table 6: Convergence behaviour of the linear and the non-linear Multigrid method (μ number of Multigrid iterations)

We want to point out that the above remarks refer to the special class of problems discussed in this paper. In particular, it is not clear up to now, whether the very simple nonlinear smoothing procedure mentioned above is practicable for more complicated problems. But even if the smoothing procedure has to be replaced by a more sophisticated one, the whole algorithm should be at least as efficient as the corresponding one using global linearizations. It remains, however, technically simpler. A detailed report on our experience with nonlinear Multigrid methods will be given elsewhere.

5.2 Parabolic Multigrid methods

As already mentioned by Brandt ("frozen τ-technique" [6], [7]) and Dinar [14], there are possibilities of applying Multigrid methods to the parabolic problem "directly".

Here the idea is to use coarser space grids not only in the solution process at each time step but also in the time evolution process. In particular, within the FAS-mode of the Multigrid method the local truncation error τ is approximated in such a way that the accuracy of the given fine h-grid can be obtained also on coarser grids by a certain change of the right hand sides of the associated coarse grid equations. Controlling the time dependence of these changes in a suitable way, all calculations have to be carried out only on coarse grids for most of the time steps.

This idea can be applied to both implicit and explicit time discretizations. Using implicit discretizations the space and the time step sizes can in principle be varied and controlled independently of each other. In the case of explicit discretizations the stability condition can be weakened considerably, as - without loss of accuracy - the space step size may become very large.

We have started systematic investigations of these direct parabolic Multigrid-solvers. Some of these first studies - also very encouraging - are reported in [24].

5.3 Alternate computers

Let us make a last remark concerning computers. As pointed out before, our programs have been written only for "old fashioned" sequential computers and have been used only on the quite slow IBM/370-158. The applied procedures, however, namely the MG00 as well as the TR solvers, are essentially "parallel processes". This is due to the red-black ordering of the grid-points in both methods (also see [8]), which allows the grid operations to be performed simultaneously. Therefore these methods are immediately suitable for parallel and "supercomputers" [8].

References

[1] E. L. ALLGOWER, K. BÖHMER, S. F. McCORMICK, Discrete correction methods for operator equations. Proc. Conf. Numerical Solution of Nonlinear Equations: Simplicial and Classical Methods, ed. H.-O. Peitgen, Springer Lecture Notes in Mathematics, Berlin, Heidelberg, New York, 1980

[2] F. W.AMES, Nonlinear partial differential equations in engineering I. Academic Press, New York, London, 1965

[3] A. BÖRSCH-SUPAN, Über Stabilität und Schrittweitensteuerung bei der Lösung parabolischer Differentialgleichungen mit Differenzenverfahren. Diplomarbeit, Universität Bonn, 1979

[4] A. BRANDT, Multi-level adaptive solutions to boundary-value problems. Math. Comp. 31 (1977), pp. 333-390

[5] A. BRANDT, Multi-level adaptive techniques (MLAT) for partial differential equations : Ideas and software. Mathematical Software III (1977), pp. 277-317

[6] A. BRANDT, Multi-level adaptive Finite-Element methods, Special topics of applied mathematics, edts. J. Frehse, D. Pallaschke, U. Trottenberg, North-Holland Publishing Company, Amsterdam, New York, Oxford, 1980

[7] A. BRANDT, ICASE workshop on Multigrid methods. NASA Langley Research Center, Hampton VA., June 1978

[8] A. BRANDT, Multigrid solvers on parallel computers. ICASE Report No. 80-23, NASA Langley Research Center, Hampton VA., 1980
To appear in: Elliptic problem solvers, ed. Martin Schultz, Academic Press, New York, 1980

[9] O. BUNEMAN, A compact non-iterative Poisson solver. Report 294, Stanford University, Inst. for Plasma Res., Stanford, Calif., 1969

[10] B. L. BUZBEE, Application of Fast Poisson Solvers to the numerical approximation of parabolic problems. Rep. LA-4950-T, Los Alamos Scientific Laboratory, Univ. of California, Los Alamos, N.M., 1972

[11] B. L. BUZBEE, Application of Fast Poisson Solvers to A-stable marching procedures for parabolic problems. SIAM J. Numer. Anal. 14, No. 2 (1977), pp. 205-217

[12] B. L. BUZBEE, G. H. GOLUB, C. W. NIELSON, On direct methods for solving Poisson's equations. SIAM J. Numer. Anal. 7 (1973), pp. 627-656

[13] P. CONCUS, G. H. GOLUB, Use of fast direct methods for the efficient numerical solution of nonseparable elliptic equations. SIAM J. Numer. Anal. 10 (1973), pp. 1103-1120

[14] N. DINAR, Fast methods for the numerical solution of boundary value problems. Ph. D. Thesis, Weizmann Institute of Science, Rehovot, Israel, 1979

[15] J. DOUGLAS, On the numerical integration of $\partial^2 u/\partial x^2 + \partial^2 u/\partial y^2 = \partial u/\partial t$ by implicit methods. J. Soc. Indust. Appl. Math. 3, No. 1 (1955), pp. 42-65

[16] H. FOERSTER, H. FÖRSTER, U. TROTTENBERG, Modulare Programme zur schnellen Lösung elliptischer Randwertaufgaben mit Reduktionsverfahren : Programme zur Lösung der Helmholtz-Gleichung mit Dirichletschen Randbedingungen im Rechteck. Preprint No. 216, Sonderforschungsbereich 72, University of Bonn, Bonn 1978

[17] H. FOERSTER, H. FÖRSTER, U. TROTTENBERG, Modulare Programme zur schnellen Lösung elliptischer Randwertaufgaben mit Reduktionsverfahren: Algorithmische Details der Programme TR2D01 und TR2D02. Preprint No. 420, Sonderforschungsbereich 72, University of Bonn, Bonn 1980

[18] H. FOERSTER, K. STÜBEN, U. TROTTENBERG, Non-standard Multigrid techniques using checkered relaxation and intermediate grids. Preprint No. 384, Sonderforschnungsbereich 72, University of Bonn, Bonn 1980
To appear in: Elliptic problem solvers, ed. Martin Schultz, Academic Press, New York, 1980

[19] R. D. GRIGORIEFF, Numerik gewöhnlicher Differentialgleichungen, Teubner-Verlag, Stuttgart, Band 1: 1972, Band 2: 1977

[20] W. HACKBUSCH, On the multi-grid method applied to difference equations. Computing 20 (1978), pp. 291-306

[21] W. HACKBUSCH, On the convergence of multigrid iterations. Report No. 79-4, University of Cologne, Cologne 1979, to appear in: Beitr. Numer. Math. 9

[22] R. W. HOCKNEY, The potential calculation and some applications, in: Methods in Computational Physics 9 (1970), pp. 135-211

[23] H. KLAMMER, W. SCHUPE, Durcherwärmungsverhältnisse der Bramme in Stoßöfen verschiedener Bauart. Stahl u. Eisen 99, Nr. 20 (1979), pp. 1088-1093

[24] N. KROLL, Direkte Anwendung von Mehrgittermethoden auf parabolische Anfangsrandwertaufgaben. Diplomarbeit 1981, Universität Bonn, 1981

[25] B. LINDBERG, On smoothing and extrapolation for the trapezoidal rule. BIT 11 (1971), pp. 29-52

[26] TH. MEIS, U. MARCOWITZ, Numerische Behandlung partieller Differentialgleichungen. Springer, Berlin, Heidelberg, New York, 1978

[27] J. M. ORTEGA, W. C. RHEINBOLDT, Iterative solution of nonlinear equations in several variables. Academic Press, New York, London, 1971

[28] D. W. PEACEMAN, H. H. RACHFORD, The numerical solution of parabolic and elliptic differential equations. J. Soc. Indust. Appl. Math. 3, No. 1 (1955), pp. 28-41

[29] F. RICHTER, Die wichtigsten physikalischen Eigenschaften von 52 Eisenwerkstoffen. Mannesmann-Forschungsberichte, Stahleisen-Sonderbericht, Heft 8, 1973

[30] J. SCHRÖDER, U. TROTTENBERG, Reduktionsverfahren für Differenzengleichungen bei Randwertaufgaben.
I. : Numer. Math. 22 (1973), pp. 37- 68
II.: Numer. Math. 26 (1976), pp. 429-459

[31] J. SCHRÖDER, U. TROTTENBERG, K. WITSCH, On Fast Poisson Solvers and applications, Numerical treatment of differential equations, Springer Lecture Notes in Mathematics, No. 631, Berlin, Heidelberg, New York, 1978

[32] U. SCHUMANN, Computers, Fast Elliptic Solvers and applications. Proceedings of the GAMM-Workshop on Fast Solution for the discretized Poisson equation, Karlsruhe, 1977

[33] R. SEVRIN, R. PESCH, Die dreidimensionale Berechnung der Erstarrung von Stahlblöcken sowie deren Erwärmung im Tiefofen. Stahl u. Eisen 93, Nr. 18 (1973), pp. 834-837

[34] H.-J. STETTER, The defect correction principle and discretization methods. Numer. Math. 29 (1978), pp. 425-433

[35] H.-J. STETTER, Analysis of discretization methods for ordinary differential equations. Springer, Berlin, Heidelberg, New York, 1973

[36] H.-J. STETTER, Considerations concerning a theory of ODE-solvers, Numerical treatment of differential equations, Springer Lecture Notes in Mathematics, No. 631, Berlin, Heidelberg, New York, 1978

[37] J. STOER, R. BULIRSCH, Einführung in die Numerische Mathematik II. Springer, Berlin, Heidelberg, New York, 1973

[38] K. STÜBEN, U. TROTTENBERG, Numerische Software zur effizienten Lösung partieller Differentialgleichungen. GMD-Spiegel 1/79 (1979), pp. 35-39

[39] K. STÜBEN, U. TROTTENBERG, K. WITSCH, On the convergence of Multigrid-iterations; quantitative results for model problems. To appear 1981.

[40] C. TEMPERTON, On the FACR(l) algorithm for the discrete poisson equation. Internal Report 14, Research Dept., European Centre for Medium Range Weather Forecasts, Bracknell, Berks., Sept. 1977

[41] W. TÖRNIG, Numerische Mathematik für Ingenieure und Physiker, Band 1. Springer, Berlin, Heidelberg, New York, 1979

[42] E. L. WACHSPRESS, Optimum alternating-direction-implicit iteration parameters for a model problem. J. SIAM 10 (1963), pp. 339-350

[43] S. VARGA, Matrix iterative analysis. Prentice Hall, Englewood Cliffs, 1962

[44] H.-J. WICK, Erprobung eines On-line Schätzverfahrens zur betrieblichen Ermittlung des Durcherwärmungsgrades von Blöcken im Tiefofen. Stahl u. Eisen 99, Nr. 20 (1979), pp. 1083-1087

ARE THE NUMERICAL METHODS AND SOFTWARE SATISFACTORY FOR CHEMICAL KINETICS?

by

Germund Dahlquist*
Lennart Edsberg*
Gunilla Sköllermo**
Gustaf Söderlind*

1. Introduction

Let y(t) be the solution of an initial value problem for a system of ODE's,

$$dy/dt = f(t,y), \quad y(0) = c \in \mathbb{R}^s .$$

An approximate solution, $y_n \approx y(t_n)$, is obtained by *the Euler method*,

$$(y_{n+1} - y_n)/h_n = f(t_n, y_n), \quad y_0 = c, \quad h_n = t_{n+1} - t_n .$$

One might expect that if $(y(t_n+h) - y(t_n))/h$ approximates $y'(t_n)$ well enough, it would be acceptable to choose the stepsize $h_n = h$. There exist, however, differential systems where h_n has to be chosen very much smaller. The reason is that the propagation of perturbations for the difference equation does not resemble that for the differential equation, unless h_n is smaller than the *smallest time constant* of the system. Such problems are called *stiff*. A system formed by the coupling of subsystems or processes with widely differing time constants is likely to become stiff. Systems describing chemical reactions are often like that.

Note that, with our definition, a system that is stiff most of the time is not necessarily stiff all the time. Usually the initial conditions are such that the system has a fast transient, where short steps are necessary for an accurate description of the solution. After the transient the solution becomes smoother, and larger steps are desirable. Intervals of rapid changes can also take place later, either by an external action, e.g. the turning of a switch, or by an internal cause (see examples 1, 2 below). It is obvious that the control of stepsize must be an essential feature of a program for such initial value problems.

A simple stiff problem is a scalar complex differential equation,

(1.3) $$dy/dt = -ay + (a+i)e^{it}, \quad a >> 1.$$

Its solution,

$$y(t) = e^{it} + (y(0) - 1)e^{-at}, \quad a >> 1,$$

rapidly approaches the very smooth function $y(t) = e^{it}$. After the transient (for $t > 5/a$, say) a desirable step is, e.g. $h = 0.01$. For the difference $\delta y(t)$ of two solutions with different initial conditions we obtain the equations,

---

* Department of Numerical Analysis and Computing Science, Royal Institute of Technology, Stockholm

** Stockholm University Computing Centre, Stockholm

(1.3') $$d\delta y/dt = -a\delta y, \quad \delta y(t) = e^{-at}\delta y(0).$$

$\delta y(t)$ tends rapidly to zero. The difference δy_n of two solutions produced by Euler's method with different initial conditions satisfy the recurrence relation,

$$(\delta y_{n+1} - \delta y_n)/h_n = -a\delta y_n,$$

(1.4) $$\text{i.e.} \quad \delta y_{n+1} = (1 - ah_n)\delta y_n.$$

Note that the modulus of this difference grows if $|1 - ah_n| > 1$, i.e. if $ah_n > 2$ (or $ah_n < 0$). If $a > 200$ we are therefore never able to choose $h_n = 0.01$, because then perturbations would be amplified. This applies not only to a perturbation in the initial value but also, for example, to a rounding error committed during the computations. The phenomenon that errors grow when they should not is called *numerical instability* and has been the subject of an extensive literature, see e.g. Gear 1971, Lambert 1973, Dahlquist 1963, 1973, Kreiss 1978.

One can avoid the numerical instability through the use of some implicit methods. The simplest of these is *the implicit Euler method*,

(1.5) $$(y_{n+1} - y_n)/h_n = f(t_{n+1}, y_{n+1}).$$

If this method is applied to (1.3), we obtain, in analogy with (1.4),

(1.6) $$\delta y_{n+1} = (1 + ah_n)^{-1}\delta y_n.$$

Note that $|\delta y_n|$ is a decreasing sequence for any h_n when $a > 0$, as it should. Note, however, that the sequence $|\delta y_n|$ is decreasing also when $ah_n < -2$, when it should not. This may be called *numerical super-stability*, a phenomenon that has received much less attention than the numerical instability, an exception is Lindberg 1974. Suppose we have a problem, where the exact solution of the differential system enters an interval, where it becomes unstable and a rapid change takes place. If the numerical method is "super-stable" this might not be detected by the program, causing the computation to proceed with a large stepsize. The interval of rapid change can thus be ignored by the program and a smooth solution, which is completely wrong, is produced. See Example 1, where the success of the computation obtained with a widely spread library program strongly depends on the choice of a control parameter called TOL, in a manner that is hard to foresee.

It is desirable that the user is made aware of such dangers by the program documentation, for example in connection with the advice how to choose the control parameters.

For a program to be able to detect all situations like this, it seems necessary that it does much more calculation than is usually done in connection with the stepsize control. However, in these examples and probably in many other cases the user is able to provide fairly simple criteria that could help the program to avoid too large a stepsize without forcing it to use unnecessarily small stepsizes in other parts of the computation. It also motivates the following question Q, which is one of the main topics of this paper:

Q: *Is there a need for software for ODE's which can easily utilize the user's knowledge of his problem in order to improve the reliability and the efficiency of the computation?*

We shall give more reasons for asking this question, and we believe that the answer is "yes". The word "user" here means a person or a team who understands the physical background of the problem as well as qualitative and computational techniques.

In order to obtain y_{n+1} in an implicit method one has to solve a system of s simultaneous (algebraic) equations, at least approximately. One starts with a *predicted* value $y_{n+1}^{(0)}$ for example obtained by some sort of polynomial extrapolation from the past. This is then *corrected* by means of some iterative scheme, which is interrupted either after a prescribed number of iterations or when some criterion of accuracy is satisfied, see e.g. Lambert 1973 for a more complete discussion.

The scheme,

$$y_{n+1}^{(i)} = y_n + h_n f(t_{n+1}, y_{n+1}^{(i-1)}) \quad i = 1,2,\ldots$$

which we shall call the *functional substitution* scheme, is useful only if $\|h_n f'\| \ll 1$, where f' is the *Jacobian matrix*,

$$f'(t,y) = \left\{\partial f_i/\partial y_j\right\}_{i,j=1}^{s} . \tag{1.7}$$

This restricts the choice of step as severely as does the requirement of numerical stability for an explicit method. It has therefore become customary to use a *Newton-like scheme* in subroutines for stiff problems. In each time-step one solves (once or several times) a linear system, the matrix of which is an approximate Jacobian. One simplifies the work by avoiding to compute and factorize a new matrix at each step, but nevertheless the work per step is considerably larger than that with one iteration with the functional substitution scheme. This is OK as long as it is compensated for by the possibility of using much larger time-steps.

In most practical problems that we have seen, it would not have been necessary to work with the Jacobian for the whole system. It is, however, not possible to take advantage of this fact with most existing subroutines for stiff problems. This again leads to our question Q formulated above.

Stiff problems have been known for a long time under different names in many disciplines. They are really an "interdisciplinary" topic. They are often handled by a method sometimes called *pseudo-stationary approximation* or (in the theory of singular perturbations) the *reduced problem* (RP). One first introduces a parameter ε,

$0 < \varepsilon << 1$, and writes the system in the form, where both equations are vector equations,

$$(1.8) \qquad \begin{aligned} dx/dt &= f(t,x,y,\varepsilon), \quad x(0) = a, \\ \varepsilon dy/dt &= g(t,x,y,\varepsilon), \quad y(0) = b. \end{aligned}$$

We assume that the system is scaled so that none of the partial derivatives of f,g are abnormally large, nor shall all partial derivatives in a row be very small, see Example 3.

The *reduced problem* (RP) is obtained for $\varepsilon = 0$:

$$(1.9) \qquad \text{RP:} \quad \begin{aligned} dx/dt &= f(t,x,y,0), \quad x(0) = a \\ 0 &= g(t,x,y,0) \end{aligned}$$

In this "differential-algebraic system", y can be eliminated by symbolic or by numerical computation. Note that the initial condition $y(0) = b$ is in general not satisfied by the RP-solution. The celebrated Michaelis-Menten approximation in biochemical kinetics is of this type, see Lin and Segel, 1974, p 302 ff.

Comments:

1. A numerical procedure for RP usually requires the sub-Jacobian $\partial g/\partial y$, not the full Jacobian. This is an advantage of RP compared to an implicit method for the full system.
2. In some problems the accuracy of the RP-solution may be too low (under certain conditions the error is $O(\varepsilon)$ outside the transient), even if ε is sufficiently small to cause inconvenience for explicit methods.
3. In a non-linear problem the second equation of (1.9) may define several branches $y = y(x,t)$. It is important to choose that branch which corresponds to the given initial values (also $y(0)$ which is otherwise ignored in RP), and to check that this branch remains stable (see the examples).

■

The implicit Euler method can now be written in the form,

$$(1.10) \qquad \begin{aligned} x_{n+1} - x_n &= h_n f(t_{n+1}, x_{n+1}, y_{n+1}), \\ (\varepsilon/h_n) \cdot (y_{n+1} - y_n) &= g(t_{n+1}, x_{n+1}, y_{n+1}). \end{aligned}$$

Theoretically, most stiff problems can be brought into the form (1.8), (or a similar form where ε is a diagonal matrix or is variable) usually after a transformation of variables, which may sometimes be impractical for other reasons. Nevertheless, *even if this partitioning is not used in the computations, its existence simplifies the understanding of the behaviour of an implicit method.* Note in particular that when $\varepsilon << h_n$ or, more precisely, when $\|(\varepsilon/h_n)(\partial g/\partial y)^{-1}\| << 1$, the second equation of (1.10) is very similar to the corresponding equation of RP. This shows, contrary to the belief of some people, that solving a differential equation by an implicit method in an interval *where it is stiff* does not become more complicated when ε becomes smaller. The *"stiffness ratios"* used by some authors, i.e. quantities which are roughly inversely proportional to ε, may be of some use when one estimates the amount of work needed, if a stiff problem is to be solved by an *explicit* method, but they are fairly *irrelevant in connection with implicit methods*, see also Shampine 1980 (these proceedings). It is, however, also very important to note that *Comment 3 also applies to the use of the implicit Euler method and other "superstable" methods*, in the sense mentioned earlier, because we cannot trust that the computations automatically reveal, if the smooth computed solution becomes unstable (repulsive), with respect to the *differential* system.

One great advantage of implicit methods (compared to the use of RP, and other techniques based on the asymptotic expansions of singular perturbation theory) is that they are successful also if ε is not very small, or in systems of ODE's with a great number of time constants with a gap-free distribution from "small" to "large". Such systems are obtained when one makes a discretization of the space variables in a time-dependent PDE. In such cases the form (1.8) is less fruitful, and a different kind of theory better illuminates the behaviour of the methods, e.g. Dahlquist 1978, Nevanlinna and Liniger 1978, 1979, Kreiss 1978.

Quite often a fairly simple scaling transformation is sufficient to bring (1.1) into the form (1.8), see Example 3. Then it would be sufficient to use a Newton-like method to the y-system of (1.10). Functional substitution, or a *relaxation* method, where only the diagonal elements of the Jacobian are computed, would be good enough for the x-system. With most programs for stiff ODE's this desirable simplification is not possible to-day.

Concerning the fastest transients, there are two situations. If the user *knows* that there is no risk that the numerical solution goes to the wrong branch, then the implicit method can be used from the start, sometimes even with $h_0 \gg \varepsilon$. This is the case, for example, in linear problems. Otherwise, we recommend functional substitution or relaxation for the whole system, which leads to small but growing time steps until a pseudo-stationary state has been reached, after which a Newton-like method with larger times steps is to be used for the y-system. The same remark also applies to intervals of rapid changes at later stages, see Example 1 or 2. Note that if ε is very small the situation is even simpler, because then the x-variables may be kept constant during the fast transient. The program can check the validity of this assumption after, say, every tenth step.
The example

$$(1.11) \qquad \varepsilon y' = y - y^3, \quad 0 < y(0) < 1,$$

illustrates the last paragraph. RP has three "branches", two of which are stable steady states ($y = 1$, $y = -1$), and one is an unstable steady state ($y = 0$). For

$0 < y(0) < 1$ the exact solution of (1.11) rapidly and monotonically approaches 1.

If Newton-Raphson is applied to RP with $y(0)$ as initial guess, the solution converges to 1 if $y(0)^2 > 1/3$, but it converges to 0 (the unstable state) if $y(0)^2 < 1/5$. For $1/5 < y(0)^2 < 1/3$ the behaviour is more complicated. In some subintervals the solution will converge to 1 after a few oscillations. We consider this method of solution "trustworthy" only if $y(0)^2 > 1/3$. Note that $f'(y) > 0$ for $y^2 < 1/3$.

The reader may find it amusing to analyse in a similar way the solution of (1.11) by implicit Euler with y_n as the initial guess in the Newton-Raphson solution of the cubic equation to obtain y_{n+1}. We hope that the reader will accept the following conclusions. The sequence $y_0 = y(0)$, $y_1, y_2, \ldots$ will convergence monotonically to 1, after "trustworthy" Newton-Raphson iterations, if and only if,

(1.12) $$3y_n^2 > (1 - \varepsilon/h_n), \quad n = 0,1,2,3,\ldots$$

For example, if $y_0^2 > 1/3$ then $y_n \uparrow 1$, for any choice of stepsizes. Also if $h_n \le \varepsilon$ for all n, then $y_n \uparrow 1$ for all $y_0 > 0$. If (1.12) is violated, then it can happen that $y_n \to 1$. If $h_n = h > 2\varepsilon$, $y_0^2 < (1 - 2\varepsilon/h)/5$, however, then $|y_n| \downarrow 0$; global error $\to$100%.

One may object to the last case that a well-designed program is not likely to choose $h_n > 2\varepsilon$ here, if the user has set a reasonable value of the tolerance parameter for the error test, but in a *system* a poorly scaled variable with a similar behaviour may not contribute enough to the stepsize control (Ex.3). We shall also see more complicated situations, where the reduced problem has several branches. We ran them with some widely spread library programs for stiff ODE's, with well developed strategies for the choice of stepsize and order. They are based on backwards differentiation formulas which are more accurate and complicated than the implicit Euler method used here, for the sake of simplicity, in the discussion. For the questions discussed in this paper there are, however, only minor differences between the methods.

2. Examples

Example 1. The Knee Problem

The differential equation for the Knee Problem is

$$\varepsilon dy/dt = (1-t)y - y^2.$$

The solution of the reduced problem has two branches,

$$y = 1-t \quad \text{and} \quad y = 0.$$

Since $\partial g/\partial y = (1-t) - 2y$ solutions in the neighbourhood of the former are stable for $t < 1$ while solutions in the neighbourhood of the latter are stable for $t > 1$.

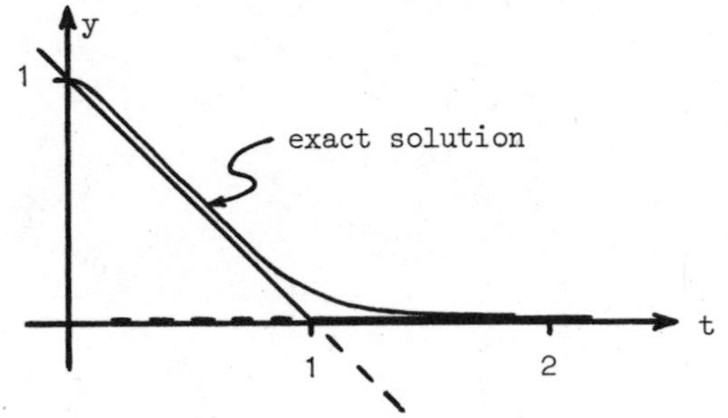

Fig. 2.1. The Knee Problem

If $y(0) > 0$ a solution rapidly approaches $y = 1-t$ if $\varepsilon << 1$ and follows that line at an $O(\varepsilon)$-distance for $t < 1-\delta$. For $t \approx 1$ a transition takes place. The solution now becomes close to the other branch, $y = 0$. The transition is smooth with respect to the time scale $O(\varepsilon^{1/2})$. Note that if we put

$$t - 1 = \varepsilon^{1/2}\tau, \quad y = \varepsilon^{1/2}\eta,$$

the parameter ε is "eliminated" from the differential equation:

$$d\eta/d\tau = -\tau\eta - \eta^2.$$

Solutions with $y(0) = 1$ were computed with the IMSL subroutine. This subroutine contains a parameter TOL. We quote from the documentation: "Single step error estimates divided by $y_{max}(i)$ will be kept less than TOL in root-mean-square norm (Euclidean norm divided by $\sqrt{s}$). The vector y_{max} of weights is computed internally and stored in work vector WK. Initially $y_{max}(i) = |y(i)|$ with a default value of 1 if $y(i) = 0$. Thereafter, $y_{max}(i)$ is the largest value of $|y(i)|$ seen so far or the initial value of $y_{max}(i)$ if that is larger." Note that this local error estimation procedure presupposes an appropriate scaling of the system.

Results: For $t \leq 1-\delta$ the solution was always accurate.

ε	TOL	
10^{-4}	10^{-4}	solution acceptable also for $t > 1$
10^{-4}	10^{-2}	solution continues along $y = 1-t$ also for $t > 1$
10^{-6}	10^{-6}	solution acceptable for $t > 1$
10^{-6}	10^{-4}	solution continues along $y = 1-t$ also for $t > 1$.

The following plots show how the stepsize depends on t in the four cases.

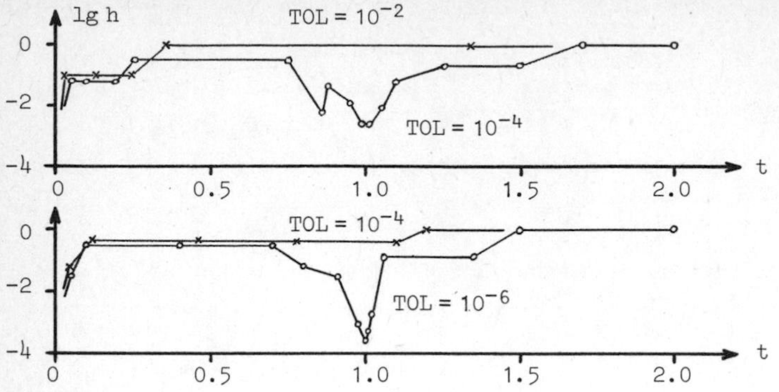

Fig. 2.2. Stepsize variation for the Knee Problem Above, $\varepsilon = 10^{-4}$; below, $\varepsilon = 10^{-6}$ (Logarithmic scale)

It seems likely that the solution would have been satisfactory also with the larger tolerances if the program had received a warning to cut down the stepsize when $\partial g/\partial y > -c\sqrt{\varepsilon}$ for some appropriately chosen c. Roughly speaking, the solution is stiff for

$$\begin{cases} \varepsilon|\log\varepsilon| < t < 1 - c\varepsilon^{1/2} & \text{and} \\ t > 1 + c\varepsilon^{1/2} . \end{cases}$$

Example 2. A stiff nonlinear oscillator.
The differential equations are

$$\begin{cases} \dot{x} = -1 - x + 8y^3 & x(0) = 0.25 \\ \varepsilon\dot{y} = -x + y - y^3 & y(0) = 0. \end{cases}$$

The solution of the reduced problem will be along the cubic curve

$$-x + y - y^3 = 0,$$

the dashed line in fig. 2.3. Since $\partial g/\partial y = 1 - 3y^2$, solutions close to the RP-solution will be stable only if $y^2 > 1/3$. The RP-solution ceases to exist when $y^2 = 1/3$.

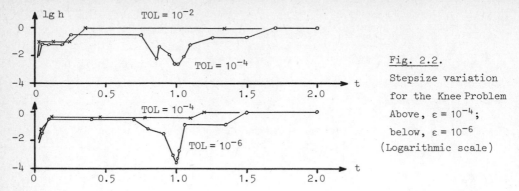

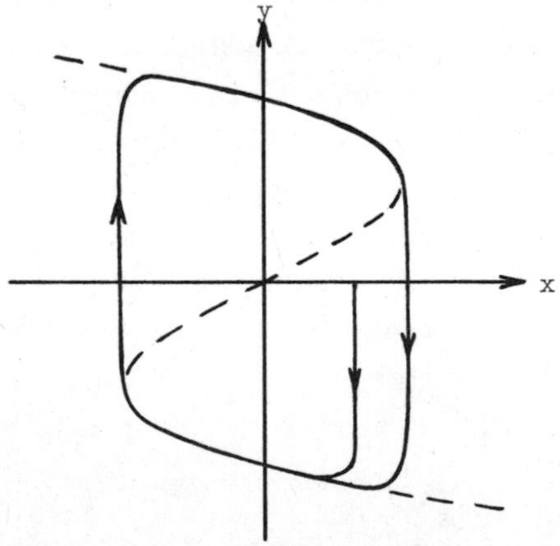

Fig. 2.3. Limit cycle for nonlinear oscillator. $\dot{y}$ is positive to the left of the cubic and negative to the right.

During a non-stiff transient interval of length $O(\varepsilon|\log\varepsilon|)$ the orbit has an almost vertical jump down to the cubic. It rapidly approaches the limit cycle indicated in the figure. The problem is stiff when the orbit is close to the cubic (at an $O(\varepsilon)$-distance). The stiffness disappears when the orbit comes close to the turning points where $y^2 = 1/3$. From the neighbourhood of the turning points the orbit has almost vertical jumps to the other stable branch during non-stiff time-intervals seemingly of length $O(\varepsilon^{2/3})$.

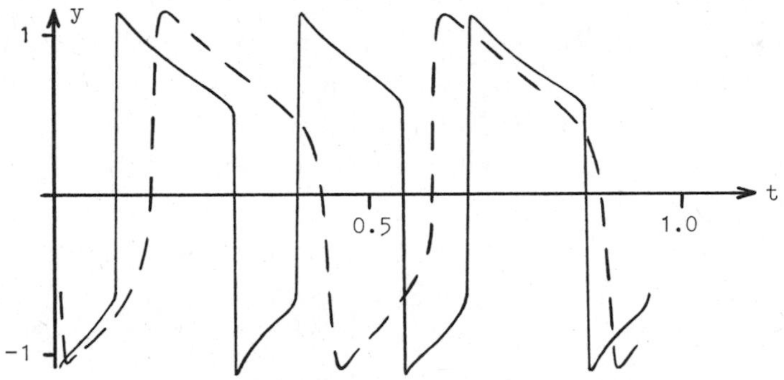

Fig. 2.4. Nonlinear oscillator solution for $\varepsilon = 10^{-4}$ and 10^{-2} (dashed)

The IMSL program performed reliably for $TOL = 10^{-4}$. It found the right branches and cut down the stepsize in the neighbourhood of the turning points. It also handled the stiff intervals efficiently, but in the neighbourhood of the turning points the efficiency is more doubtful. There were about 40 Jacobian calculations in each of the jump intervals where the problem is not really stiff. Here it would probably be more efficient to tell the program to cut down the stepsize and switch to relaxation when $\partial g/\partial y > -c\varepsilon^{1/3}$ for some appropriately chosen c. The powers of ε mentioned above were obtained by a scaling analysis of a simplified problem, reflecting the essential behaviour close to the turning points, $\varepsilon dy/dx = -c_1 x - c_2 y^2$.

It is worth mentioning that if the reduced problem is solved by determining y at $t = 0$ by Newton-Raphson's method with $y(0) = 0$ as initial approximation, one would obtain $y \approx 0.23$, which gives a point on the unstable branch of the differential system. With $TOL = 10^{-2}$, the initial stepsize was still chosen reasonably small. To begin with, the solution was fairly accurate, but after a few steps ε/h was so small that the difference equation behaved very much like the reduced problem. In fact, the numerical solution turned to the numerically stable (super-stability) but mathematically unstable branch. (Compare the equation $\varepsilon y' = y - y^3$, mentioned in the introduction.).

Example 3. Oxidation of propane

By the application of the law of mass action, Björnbom 1975 proposed the following mechanism for the oxidation of propane. See also Edsberg 1976.

$$(2.3.1)\qquad \begin{aligned} \dot{x}_1 &= -k_1x_1 - k_2x_1x_2 &&; \quad x_1(0) = 0.6\\ \dot{x}_2 &= k_1x_1 + k_3x_3 - k_4x_2x_4 - 2k_5x_2^2 &&; \quad x_2(0) = 0\\ \dot{x}_3 &= k_2x_1x_2 - k_3x_3 &&; \quad x_3(0) = 0\\ \dot{x}_4 &= -k_4x_2x_4 &&; \quad x_4(0) = 0.4\\ x_5 &= 1 - x_1 - x_2 - x_3 - x_4 . \end{aligned}$$

$k_1 = 10^{-4}$, $k_2 = 2.9\cdot 10^4$, $k_3 = 5\cdot 10^3$, $k_4 = 10^4$, $k_5 = 6.7\cdot 10^{10}$.
The variables are concentrations and are thus non-negative.

We shall now illustrate a scaling procedure, appropriate (in particular) after the fast transient. Our scaling procedure has so far been used only manually, but it is fairly systematic so it can be programmed (for interaction between user and computer), but our experience is not yet large enough to motivate that effort.

Scaling is useful for several reasons:

a) it gives a better insight into the problem,

b) it gives appropriate weights in the norm for the measuring of the local error,

c) it makes it possible to write the system in the partitioned form (1.8).

In this system x_1 may be considered as the variable which is consumed during the reaction, so it is therefore natural to introduce $\tau = k_1 t$ as a new time variable, and hence all equations are divided by k_1. Note that $k_i/k_1 \approx 5\cdot 10^7$ for $i = 2,3,4$, while $k_5/k_1 \approx 10^{15}$. Therefore, let

$$\varepsilon^{-1} = 5\cdot 10^7,\ \kappa_i = \varepsilon k_i/k_1,\ (i = 2,3,4)\quad \kappa_5 = \varepsilon^2 k_5/k_1 .$$

Now all κ_i are of normal size (see below). Then,

$$(2.3.2)\qquad \begin{aligned} \frac{dx_1}{d\tau} &= -x_1 - \kappa_2\varepsilon^{-1}x_1x_2\\ \frac{dx_2}{d\tau} &= x_1 + \kappa_3\varepsilon^{-1}x_3 - \kappa_4\varepsilon^{-1}x_2x_4 - 2\kappa_5\varepsilon^{-2}x_2^2\\ \frac{dx_3}{d\tau} &= \kappa_2\varepsilon^{-1}x_1x_2 - \kappa_3\varepsilon^{-1}x_3\\ \frac{dx_4}{d\tau} &= \kappa_4\varepsilon^{-1}x_2x_4 . \end{aligned}$$

We shall now scale the variables by appropriate powers of ε,

$$x_i = \varepsilon^{\alpha_i} x_i' .$$

The general principle is that in each equation we shall find "leading terms" of opposite sign, in the sense that if they are of order ε^{β_i}, the other terms will be, say, of order ε^{β} where $\beta \geq \beta_i$. Note that all variables are positive, and that $\frac{dx_i}{d\tau}$ and x_i are of the same order of magnitude, unless x_i is relatively constant in an interval where τ is increased by 1 or x_i would decay (or explode) very rapidly.

We make the hypothesis that x_1 needs no new scaling, i.e. $\alpha_1 = 0$. Then the first equation implies that $\alpha_2 \geq 1$. Then look at the simplest equation, i.e. the fourth.

If x_4 is not relatively constant it follows that $-1+\alpha_2=0$, i.e. $\alpha_2=1$. Moreover $\alpha_4=0$, since $x_4(0)=O(1)$. The third equation is simplest. If x_3 is not relatively constant, it follows that the two terms on the right hand side must balance each other, i.e. $\alpha_3=\alpha_2=1$. The second equation then offers no contradiction. We see no reason to test alternative hypotheses.

Since $\alpha_2=\alpha_3=1$, there will be ε-terms on the left hand side of the second and the third equations. In order to bring the equations to the form (1.8) we change the notation slightly:

$$x_1'=x_1,\ x_2'=x_4,\ y_1'=\varepsilon x_2,\ y_2'=\varepsilon x_3 .$$

Then

(2.3.3)
$$\begin{aligned}
\frac{dx_1'}{d\tau} &= -x_1'-\kappa_2 x_1'y_1' &&, \ x_1'(0)=0.6\\
\frac{dx_2'}{d\tau} &= -\kappa_4 y_1'x_2' &&, \ x_2'(0)=0.4\\
\varepsilon\frac{dy_1'}{d\tau} &= x_1'+\kappa_3 y_2'-\kappa_4 y_1'x_2'-2\kappa_5(y_1')^2 &&, \ y_1'(0)=0\\
\varepsilon\frac{dy_2'}{d\tau} &= \kappa_2 x_1'y_1'-\kappa_3 y_2' &&, \ y_2'(0)=0
\end{aligned}$$

$$\varepsilon=0.2\cdot 10^{-7},\ \kappa_2=5.8,\ \kappa_3=1,\ \kappa_4=2,\ \kappa_5=0.268 .$$

Note that it would have been possible to keep x_1',x_2' constant in the transient phase since ε is so small, but this was not done in the actual computation.

One can show (numerically or theoretically) that x_1',y_1',y_2' tend to zero almost exponentially. A quantity of interest is $x_5(\infty)$. Since $x_5 = 1-\sum_1^4 x_i$, we have to estimate

$$\lim_{t\to\infty} x_2'(t) = x_2'(0)\exp\left(-\int_0^\infty y_1'(t)dt\right).$$

Note that since all terms on the right hand side of (2.3.3) are zero at ∞, the steady state is not determined simply by setting the right hand side equal to zero. (We obtained numerically $x_2'(\infty)\approx 0.1521$).

We have, with the notations from (1.8)

$$\frac{\partial g}{\partial y} = \begin{bmatrix} -\kappa_4 x_2'-4\kappa_5 y_1' & \kappa_3\\ \kappa_2 x_1' & -\kappa_3\end{bmatrix}.$$

An eigenvalue of the full Jacobian is equal to ε^{-1}*(an eigenvalue of $\frac{\partial g}{\partial y}$) + $O(1)$. The eigenvalues of $\frac{\partial g}{\partial y}$ have negative real parts (because the variables have to be non-negative) iff

$$\kappa_4 x_2'+4\kappa_5 y_1'-\kappa_2 x_1' > 0.$$

Note that for $t=0$ the left hand side equals -2.68. The transient is therefore "locally unstable" in the beginning. If the Newton-Raphson method is used in the reduced problem, with $y_1'=y_2'=0$ as a first approximation, the iterations converge to $y_1 \approx -0.2$, $y_2 \approx -0.7$ which is on a branch which is unphysical (negative values) as well as unstable.

It was mentioned in the introduction that for $\varepsilon/h << 1$, the difference equation is very much like the equations for the reduced problem. One may thus ask whether the solution computed with a super-stable method will approach this unphysical solution.

In fact this happened when the subroutine DGEAR in the IMSL library was applied to the unscaled version of these equations, for $TOL = 10^{-3}$. The initial step was short enough, and the first steps were satisfactory, but the step soon increased rapidly and before $t = 1$ the solution had jumped to the unphysical branch.

With the same tolerance the solution of the scaled version (2.3.2) was satisfactory (but with $TOL = 10^{-2}$ overflow was obtained after a few steps).

With $TOL = 10^{-4}$ the subroutine treated both versions satisfactorily, except that some variables were negative instead of negligibly positive close to the steady state. Actually, the scaled version was more carefully treated than necessary during the transient and it therefore cost 75 function evaluations and 9 Jacobians more than the unscaled version. (In our opinion the functional substitution scheme should have been used on the whole system there).

3. Reflections and suggestions

The programs used in our examples work reliably and efficiently under the following conditions:

1) if the system is stiff all the time, except during a fast transient, that takes only a small fraction of the total computing time,

2) if the stiffness cannot be attributed to relatively few equations in the system,

3) if the user understands how to set the weight factors, tolerances etc. for the tests on the local error.

These conditions are probably satisfied in most of the situations for which these programs were designed. The simulation of chemical reactions is an important area of applications. The questioning of the reliability and efficiency in our examples is not a criticism of these particular programs, but indications of factors to be taken into account in further development. Names were mentioned, mainly to make it possible for a reader to reproduce and improve our experiments.

One of the purposes of this paper is to call attention to and make suggestions concerning some difficulties, due to the facts that many stiff problems are not stiff all the time, and that the special methods for stiff problems are "super-stable". See also Lindberg 1974 and Kreiss 1978.

We also believe that it is possible to bring a system into a *partitioned form* (perhaps a more sophisticated variant of (1.8)), so frequently that it would be worthwhile to design programs which accept a problem in this form and are able to switch between different modes: relaxation for the whole system; Newton-like solution of the whole system: relaxation for the x-system and Newton-like solution of the y-system. Many points need discussion. Should it be possible to keep the x-vector fixed in intervals of very rapid changes? Should it be possible to change the partitioning during the computation, to let ε be a diagonal matrix, to let its elements be variable, and to let some of its elements be zero (including "differential-algebraic" systems and the RP-method)? Two of us (G.Sö.,G.D.) work with a laboratory implementation of a class of methods called one-leg collocation methods, where we make experiments along these lines. We are fully aware that much flexibility can make a muddle within the program and/or for the user.

The amount of computation for the *factorization of a Jacobian* is proportional to the cube of the order of the matrix, unless the matrix has a special structure, e.g. band structure. For large stiff systems, matrix factorizations can be a substantial part of the computation. In the partitioned form the size of the matrix to be factorized is reduced.

It seems that also the frequency of those factorizations which are due to step size changes can be decreased. In the partitioned case, the matrix for which an *approximate* LU-factorization is needed is

(3.1) $$J = \varepsilon/h + \partial g/\partial y \, .$$

We note that when

(3.2) $$\varepsilon/h \, \| (\partial g/\partial y)^{-1} \| \ll 1 ,$$

the variation of h will not necessitate a new factorization. Some programs that work with the full Jacobian refactorize $I + h\partial f/\partial y$ every time h is changed. (We do not know if this is the case with the programs used in our examples.)

Automatic stepsize control is an important tool for the efficiency and also for the reliability. So far it has been mainly based on the measurement of local errors, but our examples show that also the local stability properties of the *differential system* have to be taken into account. (Remember again the super-stability notion!) This is most obviously seen in Example 1, but also in the other examples, although the failures there were partly due to inappropriate values for the parameters of the local error tests.

More research is needed concerning the *local error tests* in the hope for some standardization in the future, which the education in numerical methods can be adapted to. We are in favour of the mixed absolute-relative error test, contained for example in some NAG programs, but the user has to be trained more how to apply it to make the computation reliable and efficient. Sometimes some *scaling analysis* of a problem beforehand (or after a first trial run) may be *necessary* for a reliable and efficient performance of the error tests, see Example 3. The user may either present his original equations or a scaled version to the computer. In the former case the result of the scaling analysis is to be given to the computer in the choice of weight factors in the weighted ℓ_2-norm or ℓ_∞-norm of the error vector used in the tests.

"Componentwise relative error tests" also need scaling analysis, because they have additional rules when a variable comes close to zero (e.g. "default values of 1"), which do not make much sense unless the variable is appropriately scaled. (How many users are sufficiently informed about this?) Other programs measure the error of a component relative to the largest modulus it has had so far (again with additional rules for zero-starters). They are no good in a problem like the Belousov-Zhabatinski reaction, see e.g. Field, Noyes 1974, where a component may have the following logarithmic graph (in fact, the peaks are much thinner).

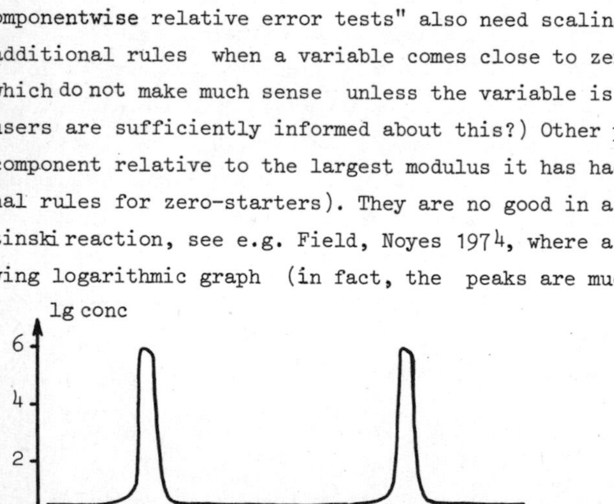

In the neighbourhood of A and B the stepsize may not be cut down enough, because the error in this component is divided by 10^6 in the test, where 10 would have been a more adequate divisor. The other components are larger, but they have a smoother behaviour there.

As was pointed out in Example 3 this is not the only purpose of scaling. It gives a better insight into the problem as a whole, and it may sometimes lead to a partitioning that ought to be useful. See Lin and Segel 1974 for further information about scaling.

Let us now turn to the most dramatic aspect of the local stability properties of the differential system, namely the existence of intervals of rapid changes, Examples 1 and 2.

In our small examples it was easy to give criteria for the *detection that an interval of rapid changes is approached* or has been entered. It ought to be possible for a user to supply a subprogram for the calculation of h_{max} as a function of x,y,t. What can otherwise be done in a larger system within a general purpose program without the user's interaction? Let J(t) be the matrix defined in (3.1) with $x = x(t)$, $y = y(t)$, $h = h(t)$, and assume that (3.2) has been satisfied in the past. Since an (approximate) factorization of J(t) is available, it is easy to detect if *a point where* $\partial g/\partial y$ *is singular* is approached or has been passed. This is one generalization of the situations in examples 1 and 2. One can, for example, look for a zero of the polynomial defined by quadratic interpolation using the values of $\det(-J(t))$ at the three most recent computations of J(t). Hopefully, the convergence of the Newton-like scheme with a several steps old Jacobian will become so poor close to a point where $\partial g/\partial y$ is singular, that the program will automatically offer sufficiently fresh Jacobians for this scheme to work.

It seems to be harder to make sure that a rapid change will not take place due to eigenvalues of $\partial g/\partial y$ that cross the imaginary axis outside the origin. This is essentially a question of showing that the spectral abscissa of a slowly varying sequence of matrices remains negative. Probably this will be solved by a combination of different tools varying from simple sufficient conditions of Gershgorin type to the computation of the eigenvalues (as the last resort). One can also from selected points compute a few short steps of length Δt with (say) the explicit Euler method. This is approximately equivalent to estimating the spectral radius of $I + \Delta t \partial g/\partial y$ by the power method, Here $\|\Delta t \partial g/\partial y\| \ll 1$. We hope to be able to return to this question in more detail. A modified form of this question is relevant for obtaining bounds for the global error.

Hopefully, our reflections and suggestions can be extended to cases where there are chemical reactions as well as diffusion and convection. Instead of question Q we then have Q': how can we merge the knowledge in ODE software about stepsize control and bells and whistles with the user's knowledge about structure and stiffness?

References

Björnbom, P., 1975, *The kinetic behaviour of the liquid phase oxidation of propane in chloro-bensene solution*, Dissertation, Dept of Chem Techn, Roy Inst of Techn, Stockholm.

Dahlquist, G., 1963, *A special stability problem for linear multistep methods*, BIT 3, 27-43.

Dahlquist, G., 1973, *Problems related to the numerical treatment of stiff differential equations*, in International computing symposium 1973, pp 307-314, North-Holland Publ Co (eds A Günther a.o.), Amsterdam.

Dahlquist, G., 1978, *G-stability is equivalent to A-stability*, BIT 18, 384-401.

Edsberg, L., 1976, *Numerical methods for mass action kinetics*, in L. Lapidus and W.E. Schiesser (eds), Numerical methods for differential systems, Acad.Press, N.Y.

Field, R. and Noyes, R., 1974, *Oscillations in chemical systems IV*, The Journal of Chemical Physics, Vol 60, No 5, pp 1877-1884.

Gear, C.W., 1971, *Numerical initial value problems in ordinary differential equations*, Prentice Hall, Englewood Cliffs, New Jersey.

Gladwell, I., 1979, *Initial Value Routines in the NAG Library*, ACM Trans on Math Software, 5, pp 386-400.

Kreiss, H.O., 1978, *Difference methods for stiff ordinary differential equations*, SIAM J. Numer Anal 15, 21-58.

Lambert, J.D., 1973, *Computational methods in ordinary differential equations*, Wiley, New York.

Lin, C.C. and Segel, L.A., 1974, *Mathematics applied to deterministic problems of natural sciences*, MacMillan, New York.

Lindberg, B., 1974, *A dangerous property of methods for stiff differential equations*, BIT 14, 430-436.

Nevanlinna, O. and Liniger, W., 1978, 1979, *Contractive methods for stiff differential equation, Part I*, BIT 18, 457-474, *Part II*, BIT 19, 53-72.

Shampine, L.F., 1980, *The stiffer the problem, the harder, is this true?* (in these Proceedings).

Optimization of Nonlinear Kinetic Equation Computation

W. F. Ames

School of Mathematics
Georgia Institute of Technology
Atlanta, Georgia 30332 U.S.A.

1. Introduction.

For certain classes of kinetic equations the complexity and amount of computation can be significantly reduced by various analytical devices. In this expository paper four of these procedures will be discussed. Each area will be illustrated with a reasonably general example followed by a discussion of generalizations.

Analytical and numerical methods which utilize the group properties of the equations will be called group methods. These will be used in illustration of exact shooting, which converts boundary value to initial value problems, and to solve classes of nonlinear eigenvalue problems.

A considerable number of problems in fluid mechanics, diffusion and reaction possess invariant (similar) solutions whose operator equations give rise to iterative monotonically convergent two sided bounds. Thus automatic error estimates are immediately available. Such a procedure will be illustrated for a classical birth and death process.

Thirdly, a reduction in complexity can be achieved by first obtaining a canonical form and then eliminating unnecessary equations. A real reaction is examined to illustrate the procedure.

Lastly, the utilization of generating functions (Z transform) to transform an infinite number of reaction equation to partial differential equations is described. The most illustrative example is a certain polymerization process.

2. Group Methods.

Perhaps the first to utilize group properties in computation, but not in the present context, was Toepfer [1]. He treated the Blasius equation of boundary layer theory $f''' + ff'' = 0$ and found that if $F(\eta)$ is a solution so then is $f = \lambda F(\mu\eta)$, $\lambda = \mu$. This permitted him to convert the boundary value problem $f(0) = f'(0) = 0$, $f'(\infty) = 1$ into two initial value problems - the predecessor of what is called "exact shooting". Among the many who exploited this idea the papers of Klamkin [2,3] and Na[4,5] stand out. They realized that the Toepfer transformation was a one-parameter group of magnifications which lead to further developments. Ames and Adams [6] demonstrated a further extension to eigenvalue problems. For certain groups Klamkin [3] and Ames and Adams [6] have characterized those classes of problems for which the method is applicable. Additional literature may be found in Ames [7].

2.1 Exact Shooting and Boundary Value Problems.

Two forms of the Emden-Fowler equation (see Ames and Adams [8])

$$\frac{d}{dt}\left[t^{\alpha}\frac{du}{dt}\right] + at^{\beta}f(u) = 0 \tag{1}$$

will be used to describe how exact shooting can be used to convert boundary value into two (sometimes more) initial value problems. The successive solution of the two problems provides a solution for the original problem.

First, consider (1) with $f(u) = \exp u$ and subject to the boundary conditions

$$u'(0) = 0, \quad u(\infty) = 0 . \tag{2}$$

Equation (1) is invariant under the transformation

$$u = U + \lambda \tag{3}$$

$$t = \mu^{-1}T$$

if $\mu = \exp[\lambda/(\beta-\alpha+2)]$, $\alpha - \beta \neq 2$. The constants λ and μ are called "constants of superposition". As a consequence if U is a solution of (1) then

$$u = U[te^{\lambda/(\beta-\alpha+2)}] + \lambda \tag{4}$$

is also a solution for any λ.

Exact shooting is characterized as follows: For the first initial value problem take -

$$U \text{ satisfies equation (1) with } U(0) = 0,\ U'(0) = 0; \qquad (5)$$

For the second initial value problem, using (4), there follows

$$u \text{ satisfies (1)}, u(0) = \lambda,\ u'(0) = 0\ . \qquad (6)$$

The value of λ is obtained from the requirement that $u(\infty) = 0$, i.e.

$$0 = u(\infty) = U(\infty) + \lambda.$$

Thus with $\lambda = -U(\infty)$ the second initial value problem can now be solved.

Existence of the solution and boundedness of U as $T \to \infty$ must be separately established (see e.g. Wong [8]) but this is not our goal here.

For the given μ equations (3) constitute a group consisting of a translation of u and a dilatation of t. The group invariant is $t \exp(su)$, $s = 1/(\beta-\alpha+2)$.

It is of interest to find all equations of the form

$$\frac{d^2u}{dt^2} = f(t,u,\frac{du}{dt}) \qquad (7)$$

which remain invariant under the one parameter group (3) with $\mu = \exp(s\lambda)$, for any s. This occurs if

$$f(\mu^{-1}T, U+\lambda, \mu P) = \mu^2 f(T,U,P) \ , \tag{8}$$

where $P = dU/dT$. Differentiating (8) with respect to λ and then setting $\lambda = 0$ (the identity) gives

$$-sT\frac{\partial f}{\partial T} + \frac{\partial f}{\partial U} + sP\frac{\partial f}{\partial P} = 2sf \ . \tag{9}$$

The general solution of this first order partial differential equation is

$$f = e^{2sU}H(Te^{sU}, Pe^{-sU}) \ ,$$

where H is arbitrary.

The foregoing procedure is illustrative of the approach used to find the class of equations left invariant under a given group.

2.2. Exact Shooting and Eigenvalue Problems.

To illustrate this procedure consider equation (1) with $\alpha = 0$, $f(u) = u^{\gamma}$, $\gamma \neq 1$ and any β. If a is thought of as a (positive) eigenvalue the problem is

$$\frac{d^2u}{dx^2} = -ax^{\beta}u^{\gamma} \tag{10}$$

with

$$u(0) = 0 \ , \ u'(0) = 1 \ , \ u(1) = 0 \ . \tag{10a}$$

First the quantity a is eliminated from the equation by a dilatation of x, or u or both. Here the transformation $X = h(a)x = a^{1/(\beta+2)}x$, $(\beta \neq -2)$ gives rise to the new problem

$$\frac{d^2u}{dX^2} = -X^{\beta}u^{\gamma} \tag{11}$$

$$u(0) = 0, \ \frac{du}{dX}(0) = a^{-1/(\beta+2)}, \ u(a^{1/(\beta+2)}) = 0 \ . \tag{11a}$$

Equation (11) is invariant under the group of magnifications

$$X = a^{\eta}\bar{X} \ , \ u = a^{\theta}\bar{u} \ , \tag{12}$$

with parameter a, if the first invariance condition

$$\theta(1-\gamma) = \eta(\beta+2) \tag{13}$$

holds.

For exact shooting the first initial value problem is

$$\left.\begin{array}{l} \bar{u}(\bar{X}) \text{ satisfies (11) with} \\ \bar{u}(0) = 0, \ \bar{u}'(0) = 1 \end{array}\right\} \tag{14}$$

and the resulting second initial value problem, from (12), is

$$\left.\begin{array}{l} u(X) \text{ satisfies (11) with} \\ u(0) = 0,\ u'(0) = a^{\theta-\eta} \end{array}\right\} . \qquad (15)$$

Comparing the second initial condition of (15) with the second condition of (11a) the second invariance condition

$$\theta - \eta = -\ 1/(\beta+2) \qquad (16)$$

is obtained. From (13) and (16) the values of θ and η are found to be

$$\theta = 1/(\beta+\gamma+1)\ ,\ \eta = (\gamma-1)/(\beta+2)(\beta+\gamma+1)\ .$$

To complete the computation of (15) the last condition of (11a) is used. The first zero $\bar{X}_1$ of $\bar{u}(\bar{X})$, whose existence must be established but is expected here, must be compatible with (12). Therefore

$$a^{\eta}\bar{X}_1 = a^{1/(\beta+2)}$$

which provides a means of calculating a.

Some results are given below:

β	0	1	2	-1/2
γ	2	2	2	3/2
$\bar{X}_1$	3.20	2.68	2.37	3.65
a	32.86	51.56	74.33	13.31

3. Monotonically Convergent Upper and Lower Bounds.

Many interacting systems (see Goel etal [9]) have mathematical models of the type

$$\frac{dy_i}{dx} = g_i(y_i)f_i(x,y_1,\cdots,y_n) \tag{17}$$

$i = 1,2,\cdots,n$, together with initial data $y_i(x_0) = \beta_i > 0$. What is desired is a computational method which preserves the (desired) positivity of the solutions and gives an automatic error estimate. One manner in which this can be accomplished is to exploit monotonicity properties of the equations, where possible, and construct antitone functional operators which are oscillatory contraction mappings. These give rise to iteratively improvable upper and lower bounds. Here the procedure is presented for a classical birth and death process. Additional examples are given in Ames and Ginsberg [10], Ames and Adams [11] and Collatz [12].

A simple birth and death (predator-prey) process is described by the dimensionless Volterra equations

$$x'(t) = ax(1-y),\ y'(t) = -cy(1-x)\ , \tag{18}$$

where a, c > 0 and x(0), y(0) are prescribed. Here it is expected that $x(t) \geq 0$, $y(t) \geq 0$. Of the several iterative forms that can be adopted for (18) the sequences

$$\left.\begin{array}{l} x_0(t) = 0;\ y_n' = -cy_n(1-x_n),\ x_{n+1}' = ax_{n+1}(1-y_n) \\ y_n(0) = y(0),\ x_{n+1}(0) = x(0),\ n \geq 0 \end{array}\right\} \tag{19}$$

are particularly attractive, as will be seen.

The integrated forms of (19) are easiest to analyze. For $n = 0$, since $x_0(t) = 0$,

$$y_0(t) = y(0)\ \exp(-ct)$$

and

$$x_1(t) = x(0)\exp\{at - ay(0)[1 - \exp(-ct)]/c\}$$

can be obtained explicitly. The iteration for $n \geq 1$ is

$$y_n(t) = y(0)\exp[-c(t - \int_0^t x_n(\tau)d\tau)] \tag{20a}$$

$$x_{n+1}(t) = x(0)\exp[a(t - \int_0^t y_n(\tau)d\tau)] \tag{20b}$$

which maintains the positivity of x and y. The iteration can be summarized as

$$x_0(t) = 0$$
$$y_n(t) = T_1 x_n$$
$$x_{n+1}(t) = T_2 y_n \ .$$

It is not difficult to show that this iteration is antitone by assuming that $\eta_{n-1}(t) < \eta_n(t)$, $\eta = x$ or y, and verifying that $\eta_n(t) > \eta_{n+1}(t)$ by induction. Thus the oscillating behavior,

$$0 \le \eta_0 < \eta_2 < \eta_4 < \cdots < \eta_{2n} < \cdots < \eta < \cdots < \eta_{2n+1} < \cdots < \eta_5 < \eta_3 < \eta_1 \ ,$$

which provides convergent upper and lower bounds for x and y, is obtained. The actual computation of (19) can be done in either differential or integral form.

The study of convergence and the development of error bounds is easily accomplished. In particular from (20b) and with elementary inequalities, on $0 < t < T$,

$$0 \le x_{2n+1} - x_{2n} < L_1 \int_0^t (y_{2n-1} - y_{2n})\, d\tau \ , \qquad (21)$$

with

$$L_1 = x(0)\ a \exp[aT] \ .$$

In a similar fashion, from (20a),

$$0 \le y_{2n-1} - y_{2n} < L_2 \int_0^t (x_{2n-1} - x_{2n-1})\, d\tau \tag{22}$$

with

$$L_2 = cy(0)\ \exp[c\int_0^T x_1 d\tau]\ .$$

Combining the results of (21) and (22) and converting the two fold integral to a single integral there results

$$\begin{aligned} 0 \le x_{2n+1} - x_{2n} &< L_1 L_2 \int_0^t \int_0^t (x_{2n-1} - x_{2n-2})\, d\tau d\tau \\ &= L_1 L_2 \int_0^t (t-\tau)(x_{2n-1} - x_{2n-2})\, d\tau \\ &< M_1 T \int_0^t (x_{2n-1} - x_{2n-2})\, d\tau \\ &< K_1 T (M_1 T)^n / n! \end{aligned}$$

where $x_1 < K_1$ on $0 < t \le T$.

A similar result holds for the sequence y_n , whereupon

$$0 \le \eta_{2n+1} - \eta_{2n} < KT(MT)^n/n! \tag{23}$$

$$0 < t \le T\ ,$$

for $M = \text{Max}\{M_1,M_2\}$, $K = \text{Max}\{K_1,K_2\}$. Uniform convergence to solutions of the original differential equations follows as a result of (23).

Theorems providing for the construction of antitone operators are not available in general but for equation (17) there is a result due to Ames and Ginsberg [10]. The result is stated here for positive solutions where the vectors y and f represent their corresponding components. Q will be the Banach space with norm $||z(x)|| = \max_i |z_i(x)|$ containing positive solutions of the initial value problem (17). Further the quantities $G_i(u) = \int \frac{du}{g_i(u)}$, $i = 1,2,\cdots,n$ and the results will be stated for $x \in I = [x_0,x_0+h]$, $h > 0$.

Suppose

1) G_i^{-1} is a positive continuous strictly antitone operator;
2) $f(x,y)$ is such that each component is positive and bounded on I and f_i is a strictly monotone increasing function with respect to all components of y.
3) Each f_i satisfies a Lipschitz condition on a closed region about the first iterate $y^{(1)}$, i.e.

$$f_i(x,\xi) - f_i(x,\eta) < K_i||\xi - \eta||,\ \xi > \eta.$$

Then

a) There exists a sequence of iterates $\{y_i^{(p)}\}$ satisfying $y_i^{(0)} = \beta_i$

$$\frac{dy_i^{(p)}}{dx} = g_i(y_i^{(p)})f_i(x,y_1^{(p-1)},y_2^{(p-1)},\cdots,y_n^{(p-1)}),$$

$$y_i^{(p)}(x_0) = \beta_i\ ,\ p = 1,2,\cdots;\ i = 1,2,\cdots,n.$$

where

$$y_i^{(p)}(x) = G_i^{-1}\ [\int_{x_0}^{x} f_i(u,y^{(p-1)}(u))du + G_i(\beta_i)].$$

b) The sequence $\{y_i^{(p)}\}$ maintains the order relations for all $x \in I$ and all $i = 1,2,\cdots,n$

$$y_i^{(1)}(x) < y_i^{(3)}(x) < \cdots < y_i^{(2p+1)}(x) < \cdots <$$
$$y_i^{(2p)}(x) < \cdots < y_i^{(4)}(x) < y_i^{(2)}(x).$$

c) The sequence $\{y_i^{(p)}\}$ has a limit $y_i(x)$ which satisfies the initial value problem (17).

Various modifications of the assumptions can be treated. These are summarized in the table below where subscripts have been dropped.

VARIATIONS OF GENERAL THEOREM FOR STRICTLY POSITIVE f(x,y) AND y

	f(x,y) strictly monotone increasing	f(x,y) strictly monotone decreasing
G^{-1} strictly antitone	oscillatory convergence -- even iterates monotonically decreasing from above and odd iterates monotonically increasing from below	one-sided convergence from above or below
G^{-1} strictly monotone increasing	one-sided convergence from above or below	oscillatory convergence, even iterates monotonically decreasing from above and odd iterates monotonically increasing from below

VARIATIONS OF GENERAL THEOREM FOR STRICTLY NEGATIVE f(x,y) AND y

	f(x,y) strictly monotone increasing	f(x,y) strictly monotone decreasing
G^{-1} strictly antitone	oscillatory convergence -- even iterates monotonically increasing from below and odd iterates monotonically decreasing from above	one-sided convergence from above or below
G^{-1} strictly monotone increasing	one-sided convergence from above or below	oscillatory convergence, even iterates monotonically increasing from below and odd iterates monotonically decreasing from above

The procedure of the section is also applicable to problems of Section 4 after some initial treatment.

4. Canonical Forms and Computational Reduction.

In many examples of kinetics there are several redundant differential equations which can be eliminated by elementary operations on the rows of the system. This gives rise to conservation laws which together with the resulting canonical form governs the system. In some realistic cases a further reduction is possible using the methods of Section 3. To illustrate this approach suppose the chemical reaction

$$\begin{aligned} A_1 + A_2 &\overset{k_1}{\rightarrow} A_3 \\ A_2 + A_3 &\overset{k_2}{\rightarrow} A_4 \\ A_5 + A_2 &\overset{k_3}{\rightarrow} A_6 \\ A_6 + A_2 &\overset{k_4}{\rightarrow} A_7 , \end{aligned} \tag{24}$$

taking place at constant volume, is modeled, using the "law of mass action", by

$$\frac{d}{dt}\begin{bmatrix} x_1 \\ x_2 \\ x_3 \\ x_4 \\ x_5 \\ x_6 \\ x_7 \end{bmatrix} = \begin{bmatrix} -k_1 & 0 & 0 & 0 \\ -k_1 & -k_2 & -k_3 & -k_4 \\ k_1 & -k_2 & 0 & 0 \\ 0 & k_2 & 0 & 0 \\ 0 & 0 & -k_3 & 0 \\ 0 & 0 & k_3 & -k_4 \\ 0 & 0 & 0 & k_4 \end{bmatrix} \begin{bmatrix} x_1 \\ x_3 \\ x_5 \\ x_6 \end{bmatrix} x_2 \tag{25}$$

In (24) $x_i \geq 0$ are the concentrations of A_i, $k_i > 0$ $i = 1,2,\cdots,7$, and $x_1(0)$, $x_2(0)$, $x_2(0)$, $x_5(0)$, $x_3(0) = x_4(0) = x_6(0) = x_7(0) = 0$ are prescribed. While A_4 is the desired product the presence of A_5 leads to undesirable byproducts. Here the quantities $\alpha = {}^{k_2}/k_1 < 1$ and $\beta = {}^{k_4}/k_3 < 1$ are used in the sequel.

By elementary transformations on the rows <u>alone</u> the canonical form (Ames [13])

$$\frac{d}{dt}\begin{bmatrix} x_1 \\ x_1 + x_3 \\ x_5 \\ x_5 + x_6 \end{bmatrix} = \begin{bmatrix} -k_1 & 0 & 0 & 0 \\ 0 & -k_2 & 0 & 0 \\ 0 & 0 & -k_3 & 0 \\ 0 & 0 & 0 & -k_4 \end{bmatrix} \begin{bmatrix} x_1 \\ x_3 \\ x_5 \\ x_6 \end{bmatrix} x_2 \tag{26}$$

is obtained, together with the algebraic relations

$$x_2(t) = 2x_1 + x_3 + 2x_5 + x_6 + [x_2(0) - 2x_1(0) - 2x_5(0)] \tag{27}$$

$$x_7(t) = x_5(0) - x_6 - x_5\,,\; x_4(t) = x_1(0) + x_3(0) - x_3(t) - x_1(t) \;. \tag{28}$$

Further reduction of (26) is readily accomplished. For example division of the second equation by the first, followed by an integration, yields

$$x_3(t) = \frac{1}{1-\alpha}\,\{(x_1(0)/x_1(t))^{1-\alpha} - 1\}x_1(t) \;. \tag{29}$$

In a similar fashion

$$x_6(t) = \frac{1}{1-\beta}\,\{(x_5(0)/x_5(t))^{1-\beta} - 1\}x_5(t) \;. \tag{30}$$

Finally, eliminating x_2 from the first and third equations results in

$$x_5(t)/x_5(0) = [x_1(t)/x_1(0)]^{\gamma},\; \gamma = {}^{k_3}/k_1 \;. \tag{31}$$

As a consequence of these preliminaries <u>only one</u>, <u>say $x_1(t)$, need be computed</u> and the remaining components follow directly from equations (27) to (31).

To compute x_1 an algorithm will be developed from the integrated form of the first equation of (26),

$$x_1(t) = x_1(0)\,\exp[-k_1\int_0^t x_2(\tau)\,d\tau] \;. \tag{32}$$

Beginning with the obvious upper bound $x_2^{(0)}(t) = x_2(0)$, for $x_2(t)$, compute recursively

$$
\begin{aligned}
&x_1^{(n+1)} = x_1(0)\ \exp[-k_1 \int_0^t x_2^{(n)}(\tau)d\tau] , \\
&x_5^{(n+1)} \text{ from (31)}, \\
&x_6^{(n+1)} \text{ from (30)}, \\
&x_3^{(n+1)} \text{ from (29)}, \\
&x_2^{(n+1)} \text{ from (27) and } x_4^{(n+1)},\ x_7^{(n+1)} \text{ from (28)}.
\end{aligned}
\tag{33}
$$

From (25) it is obvious that x_2 is monotone decreasing in t. To discover the properties of the foregoing algorithm, for $t > 0$, suppose $x_2^{(n)}(t) < x_2^{(n+1)}(t)$. Then, from (32), it follows that $x_1^{(n+1)} > x_1^{(n+2)}$ and, from (31), $x_5^{(n+1)} > x_5^{(n+2)}$. Since x_3 and x_6 are governed by similar expressions only the properties of the iterative structure of $x_3^{(n)}$ will be examined. Clearly, from (29),

$$
x_3^{(n+1)} - x_3^{(n+2)} = \frac{x_1(0)}{1-\alpha} \left[\left(\frac{x_1^{(n+1)}}{x_1(0)} \right)^{\alpha} - \frac{x_1^{(n+1)}}{x_1(0)} \right] - \left[\left(\frac{x_1^{(n+2)}}{x_1(0)} \right)^{\alpha} - \frac{x_1^{(n+2)}}{x_1(0)} \right] \tag{34}
$$

is positive when the function $f(u) = u^{\alpha} - u$, $0 < u < 1$, $0 < \alpha < 1$, is monotonically increasing. This occurs for $f'(u) > 0$, i.e. for $u < \alpha^{1/(1-\alpha)}$. Since x_1 is not produced in the system there will occur a time t_0 and a value of n, say n_0, such that

$$\frac{x_1^{(n_0)}(t_0)}{x_1(0)} < \alpha^{1/(1-\alpha)} ,$$

whereupon (34) becomes positive. Similarly, there exists t_1 and n_1 such that

$$\frac{x_5^{(n_1)}(t_1)}{x_5(0)} < \beta^{1/(1-\beta)} .$$

Thus for all $t > t^* = \max(t_0, t_1)$ and all $n > \max(n_0, n_1)$ there follows $x_3^{(n+1)} > x_3^{(n+2)}$ and $x_6^{(n+1)} > x_6^{(n+2)}$, whereupon, from (27), $x_2^{(n+1)} > x_2^{(n+2)}$. Thus, ultimately, the algorithm becomes oscillatory and provides two sided bounds.

If, on the other hand, it is assumed that $x_2^{(n)} > x_2^{(n+1)}$ then an analogous argument demonstrates that, ultimately $x_2^{(n+1)} < x_2^{(n+2)}$ for t and n sufficiently large.

5. Application of Generating Functions (Z Transforms) to Polymerization.

The Z transform (see Jury [14] for properties),

$$P(z,t) = \sum_{n=1}^{\infty} P_n(t) z^{-n}, \ |z| \geq -1 \tag{35}$$

is employed in a variety of problems to transform a discrete set (indexed by n) into a continuous set in the new variable (z). Kilkson [15] has found it especially useful for the solution of path dependent polymerization problems. Here an application is given to an irreversible condensation in a batch reactor.

Let A and B designate active end groups capable of reacting to form a link (AB) and let X designate an inert end group. The considered polymerization will be a condensation of bifunctional molecules of type $A - (BA)_{n-1} - B$ in the presence of monofunctional molecules or chain stoppers $A - (BA)_{n-1} - X$. If the reaction between A and B groups occurs irreversibly, according to second order kinetics, with a rate constant k independent of chain length then the overall decrease in concentration of end groups is given by

$$-\frac{d(A)}{dt} = -\frac{d(B)}{dt} = k(A)(B) \ . \tag{36}$$

Here t is time and (A) represents the concentration of A.

Designate a bifunctional molecule, with n AB units, by ℓ_n and its monofunctional counterpart $(AB)_{n-1} - AX$ by ℓ_{nx}. Then the condensation reaction is represented by the equations

$$\ell_m + \ell_{n-m} \xrightarrow{k} \ell_n$$
$$\ell_{mx} + \ell_{n-m} \xrightarrow{k} \ell_{nx} \,, \qquad (37)$$

that is to say that an individual species may be <u>formed</u> from any pair of bifunctional molecules whose indices add up to n and <u>destroyed</u> by reaction with any molecule, regardless of length or kind. Thus the rate equation for species ℓ_n is

$$\frac{dP_n}{d\tau} = \sum_{m=1}^{n-1} P_m P_{n-m} - P_n [2 \sum_{i=1}^{\infty} P_i + \sum_{i=1}^{\infty} P_{ix}] \qquad (38)$$

where $\tau = kt, P_n(\tau)$ = concentration of ℓ_n and $P_{nx}(\tau)$ = concentration of ℓ_{nx}.

The species ℓ_{nx} is created by indicial addition of a bifunctional molecule and a monofunctional one and destroyed by reaction with any bifunctional molecule. Thus

$$\frac{dP_{nx}}{d\tau} = \sum_{m=1}^{n-1} P_{mx} P_{n-m} - P_{nx} \sum_{i=1}^{\infty} P_i \qquad (39)$$

is the rate equation for P_{nx}.

Equations (38) and (39) define the infinite sets of nonlinear rate equations describing the time dependency of the two distributions $\{\ell_n\}$ and $\{\ell_{nx}\}$. There are solved by using the convolution property of the Z transform

$$P(z)\cdot P^*(z) = (\sum_1^\infty P_n z^{-n})(\sum_1^\infty P_n^* z^{-n})$$
$$= \sum_{n=1}^\infty z^{-n}(\sum_{m=1}^n P_m \cdot P_{n-m}^*) \tag{40}$$

To use (40) multiply equation (38) by z^{-n} and sum over n to obtain

$$\sum_{n=1}^\infty z^{-n}\frac{dP_n}{d\tau} = \sum_{n=1}^\infty [\sum_{m=1}^{n-1} P_m z^{-m}\cdot P_{n-m} z^{-(n-m)}]$$
$$- \sum_{n=1}^\infty P_n z^{-n}[2\sum_{i=1}^\infty P_i + \sum_{i=1}^\infty P_{ix}] . \tag{41}$$

From the Z transform properties in (35) and (40) equation (41) becomes

$$\frac{\partial P(z,\tau)}{\partial \tau} = [P(z,\tau)]^2 - P(z,\tau)[2P(1,\tau) + P_x(1,\tau)] . \tag{42}$$

Tranformation of (39) in a similar manner yields, for the distribution ℓ_{nx},

$$\frac{\partial P_x(z,\tau)}{\partial \tau} = P(z,\tau)P_x(z,\tau) - P_x(z,\tau)P(1,\tau) . \tag{43}$$

By application of the Z transform the two infinite sets of ordinary differential equations have been replaced by two modest, but nonlinear, partial differential equations.

Setting z = 1 in equations (42) and (43) there results

$$\frac{dP(1,\tau)}{d\tau} = -[P(1,\tau)]^2 - P(1,\tau)P_x(1,\tau) \tag{44}$$

$$\frac{dP_x(1,\tau)}{d\tau} = 0 \ , \tag{45}$$

that is the rate equations for the total number of molecules of each kind. From these it is seen that the total number of monofunctional molecules may not change but the number of bifunctional molecules decreases as a result of internal condensation and reaction with monofunctional ones.

The solution of (42) and (43) is simplified by transformation to the normalized transform $C(z,\tau)$ and $C_x(z,\tau)$ defined by means of $C(z,\tau) = \frac{P(z,\tau)}{P(1,\tau)}$ and $C_x(z,\tau) = P_x(z,\tau)/P_x(1,\tau)$. Upon taking the τ derivative of C there results

$$\frac{\partial C}{\partial \tau} = \frac{1}{P(1,\tau)} \frac{\partial P(z,\tau)}{\partial \tau} - \frac{C}{P(1,\tau)} \frac{dP(1,\tau)}{d\tau} \ . \tag{46}$$

Substituting (42) and (44) into (46) yields

$$\frac{\partial C}{\partial \tau} = P(1,\tau)C[C-1] \ . \tag{47}$$

The time dependency for the normalized transform $C_x = P_x(z,\tau)/P_x(1,\tau)$ is obtained in a similar manner to (47) as

$$\frac{\partial C_x}{\partial \tau} = P(1,\tau)C_x[C-1] \ . \tag{48}$$

Upon division of (47) by (48) and integrating it is seen that

$$\frac{C}{C_0} = \frac{C_x}{C_{x0}} \tag{49}$$

where C_0 and C_{x0} represent feed distributions. Thus the distributions are simply related for dissimilar feeds and are identical for identical feed distributions. The integration is now easily completed.

References.

1. K. Toepfer, Bemerkung zu dem ausatz von H. Blasius, grenzschichten im flussigkeiten mit kleiner reibung, Z. Math. Phys. 60, 397 (1912).

2. M. S. Klamkin, On the transformations of a class of boundary value problems into initial value problems for ordinary differential equations, SIAM Rev. 4, 43 (1962).

3. M. S. Klamkin, Transformation of boundary value into initial value problems, J. math. Analysis Applic. 32, 308 (1970).

4. T. Y. Na, Transforming boundary conditions to initial conditions for ordinary differential equations, SIAM Rev. 9, 204 (1967).

5. T. Y. Na, Further extension on transforming boundary value to initial value problems, SIAM Rev. 10, 85 (1968).

6. W. F. Ames and E. Adams, Exact shooting and eigenparameter problems, Nonlinear Analysis; Theo., Meth. Applic. 1, 75 (1976).

7. W. F. Ames, Nonlinear Partial Differential Equations in Engineering, Vol. II, Academic Press, 1972, pp. 136-142.

8. J. S. W. Wong, On the generalized Emden-Fowler equation, SIAM Rev. 17, 339 (1975).

9. N. S. Goel, S. C. Maitra and E. W. Montroll, Nonlinear Models of Interacting Populations, Academic Press, New York, 1971.

10. W. F. Ames, and M. Ginsberg, Bilateral algorithms and their applications, Computational Mechanics (Lecture Notes in Mathematics #461, J. T. Oden (Ed.)), Springer-Verlag, New York, 1975, 1-32.

11. W. F. Ames and E. Adams, Monotonically convergent two sided bounds for some invariant parabolic boundary value problems, Z. angew. Math. Mech. 56, T240 (1976).

12. L. Collatz, Functional Analysis and Numerical Mathematics, Academic Press, New York, 1966, 350-357.

13. W. F. Ames, Nonlinear Ordinary Differential Equations in Transport Processes, Academic Press, New York, 1968, 87-94.

14. E. I. Jury, Theory and Application of the Z Transform Method, Wiley & Sons, 1964.

15. H. Kilkson, Ind. Eng. Chem. Fund. 3, 1964, p. 281.

AUTOMATIC DETECTION AND TREATMENT OF OSCILLATORY AND/OR STIFF ORDINARY DIFFERENTIAL EQUATIONS

by

C.W. Gear
Department of Computer Science
University of Illinois at Urbana-Champaign
Urbana, IL 61801

Abstract

The next generation of ODE software can be expected to detect special problems and to adapt to their needs. This paper is principally concerned with the low-cost, automatic detection of oscillatory behavior, the determination of its period, and methods for its subsequent efficient integration. It also discusses stiffness detection. In the first phase, the method for oscillatory problems discussed examines the output of any integrator to determine if the output is nearly periodic. At the point this answer is positive, the second phase is entered and an automatic, nonstiff, multirevolutionary method is invoked. This requires the occasional solution of a nearly periodic initial-value problem over one period by a standard method and the re-determination of its period. Because the multirevolutionary method uses a very large step, the problem has a high probability of being stiff in this second phase. Hence, it is important to detect if stiffness is present so an appropriate stiff, multirevolutionary method can be selected. Stiffness detection uses techniques proposed by a number of authors. The same technique can be used to switch to a standard stiff method if necessary for a non-oscillatory problem, in the first phase of an oscillatory problem, or in the standard integration over one period of an oscillatory problem.

Supported in part by Department of Energy contract ENERGY/EY-76-S-02-2383.

1. Introduction

A truly automatic code for ordinary differential equations must not only handle the most general case reasonably efficiently, but must also automatically detect those classes of problems that are unreasonably expensive by general methods and switch to methods which are more efficient for those problems. This paper will consider two classes of problems, stiff and oscillatory, for which special methods can be far more efficient than general methods. This is not to say that there are not other classes of problems that are worthy of special treatment. For example, linear equations probably can be solved more efficiently if this fact is known [1], but at this time there are no methods that are sufficiently more efficient for linear problems than the general methods that it seems worth the effort to detect linear problems automatically. (Furthermore, most users can tell if a problem is linear, while it may be difficult for them to tell when it becomes stiff or oscillatory.)

Although it is common to talk about "stiff differential equations," an equation *per se* is not stiff, a particular initial value problem for that equation may be stiff, in some regions, but the sizes of these regions depend on the initial values *and* the error tolerance. For most problems the solution is initially in a transient and an accurate solution demands a stepsize sufficiently small that the truncation error of that transient is small. For such stepsizes stability is not a problem. When the transient has decayed below the error tolerance, the problem may be stiff. At this time a stiff method must be used. Many techniques and programs are available for stiff equations ([2], [3], [4], [5], [6]) so we will not repeat that material.

Until recently stiff methods have also been used in the transient region, but the fact that they are generally less efficient than nonstiff methods (both because of smaller error coefficients and the linear algebra involved) has encouraged several people to investigate automatic detection of stiffness.

The problem of highly oscillatory solutions has some parallels to the stiff problem. Again, the solution may not be nearly periodic initially, but after a transient starting phase, may tend towards a periodic solution or have a nearly periodic behavior. There are some methods that are applicable in the latter phase, for example, Mace and Thomas [9], Graff [6], Graff and Bettis [7] and Petzold [10]. However, these methods cannot be used in the transient phase so it is essential to detect the time when nearly periodic behavior has begun and to estimate the period reasonably accurately. There are, of course, problems for which the user knows that the solution is nearly periodic throughout. Satellite orbits are a case in point. (Most of the early methods were developed for these problems.) In such cases the period is known reasonably accurately so there is no detection problem. This paper is particularly concerned with those problems which may become nearly periodic in later stages and methods for detecting this behavior in order to switch to an appropriate scheme.

Methods for nearly periodic problems are generally known as multirevolutionary from their celestial orbit background. The idea of such methods is to calculate, by some conventional integrator, the change in the solution over one orbit. If the period of an orbit is T (for a moment assumed fixed), then a conventional integrator is used to compute the value of

$$d(t) = y(t + T) - y(t)$$

by integrating the initial value problem $y' = f(y)$ over one period T. If we consider the sequence of times $t = mT$, m integral, we have a sequence of values $y(mT)$ which are slowly changing if y is nearly periodic. The conventional integrator allows us to compute the first differences $d(mT)$ of this sequence at any time mT. Under appropriate "smoothness" conditions (whatever that means for a sequence) we can interpolate or extrapolate for values of $d(mT)$ from a subset of all values of d, for example from $d(kqT)$, $k = 1, 2, 3, \ldots$, where q is an integer > 1, and thus estimate $y(mT)$ by integrating only over occasional orbits.

In a satellite orbit problem it is fairly easy to define the meaning of "one period." For example, one could use a zero crossing of a particular coordinate, or even a fixed period based on a first order theory. In her thesis, Petzold considered problems for which it is difficult to find physical definitions of the period and examined a method for determining the approximate period by minimizing a function of the form

$$I(t, T) = ||y(\tau + T) - y(\tau)||_t$$

where the norm measures the values of $y(\tau + T) - y(\tau)$ approximately over the range $\tau\varepsilon(t, t + T)$. The actual norm she used was

(1) $$I(t, T) = \int_t^{t+\bar{T}} ||y(\tau + T) - y(\tau)||_2 \, d\tau$$

where $\bar{T}$ was the last estimate of the period. The use of $\bar{T}$ was for pragmatic reasons. Ignoring that detail, the value of T which minimizes $I(t, T)$ is a function of t, and $T(t)$ was said to be the period of the solution. This enabled $d(t) = y(t + T(t)) - y(t)$ to be calculated and multirevolutionary methods to be used. The variable period was handled easily by a change of independent variables to s in which the period is constant, say 1. The equation

$$t(s + 1) - t(s) = T(t)$$

was appended to the system

$$z(s + 1) - z(s) = g(s)$$

where $z(s) = y(t(s))$ and $g(s) = d(t/s)$ for integer values of s. (When T is constant, this is the analog of the old device for converting a non-autonomous system to an autonomous system by appending the differential equation $t' = 1$.)

The scheme for period calculation used by Petzold suffers from three drawbacks. The first drawback is that it is fairly expensive, involving a numerical approximation to the first two derivatives of $I(t, T)$ by quadrature which requires integration for $y(\tau)$ over two periods. In the experimental implementation, integration was repeated for every iteration of a Newton method to minimize $I(t, T)$ by solving $\partial I/\partial T = 0$. This could have been eliminated by saving all values and interpolating, but the storage cost becomes high and the quadrature/interpolation cost remains non-negligible. The second drawback is that a reasonably accurate period estimate is needed for the Newton iteration to converge. Outside the region of convergence for Newton's method a search scheme for a minimum could be used but this would be very expensive because of the computation involved in each quadrature even if all previously computed values could be saved. This makes the approach very unattractive for initial period detection when there is no starting estimate. The third drawback is that minimizing a function subject to several sources of error (including truncation errors in the integration and quadrature, and roundoff errors revealed by considerable cancellation in $\|y(\tau + T) - y(\tau)\|$) is likely to yield a fairly inaccurate answer. Since the value of $d(t) = g(s)$ is quite sensitive to small absolute changes in the period T which may be large relative to the period, the function $g(s)$ may not appear to be very smooth.

This paper discusses an alternate approach to the period identification problem. It overcomes the cost and convergence problems, and also seems to help with the sensitivity problem. This is discussed in the next section along with stiffness detection. The third section reviews multirevolutionary multistep integrators and a technique for handling variable periods based on the period identification algorithm. The fourth section discusses a numerical example while the final section discusses unsolved problems.

2. Periodic and Stiffness Detection

A fully automatic method should be able to detect problems with special properties that can be solved more efficiently, but the cost of detection should be low compared to the integration cost so that problems without those properties do not cost appreciably more. Since stiffness and nearly periodic behavior are properties that may appear at any point in the solution, the detection process must operate continuously. If it is to have a low cost, it must not take more than a few operations on available intermediate or final results of a standard integrator. This section first discusses periodic behavior detection, then stiffness detection.

We have been deliberately imprecise about the meaning of "nearly periodic," and will continue that way with the working definition in our minds of "the type of problem that can be handled efficiently by multirevolutionary methods." The types of problems that have the required properties are differential equations for which there exist functions $F(\tau, t)$ and $T(t)$ such that

$$y(t) = F(t, t)$$

is a solution of the differential equation $dy/dt = f(y, t)$, F is periodic in τ with period $T(t)$, that is,

$$F(\tau + T(t), t) = F(\tau, t),$$

for all t and τ, $\partial F/\partial t$ is very small compared to $\partial F/\partial \tau$, and $T(t)$ is slowly varying. Here, τ and t are the "fast" and "slow" times as shown in Figure 1. The "meaning" of this representation is that $P(t) = \{F(\tau, t), \tau \varepsilon [0, T(t)]\}$ is the local periodic behavior of the solution, and the change of $P(t)$ with respect to t represents the way this behavior slowly changes. (This representation is only valid for problems which are not phase locked to a periodic driving function.) $F(0, t)$ was called a <u>quasi-envelope</u> by Petzold. It is the function $z(t)$ defined earlier for a discrete set of points only.

This representation is not unique, but depends on the choice of the period $T(t)$ and the values of $F(0, t)$ over the initial period. It is convenient to consider a change of variables to (s, t) with $s = \tau/T(t)$. In the new coordinate system, $F(s,t)$ has period 1 in s and a unique quasi-envelope is defined for any fixed s in terms of $F(0, t)$.

The "period" of a nearly periodic function has not yet been defined. We could use some intuitively reasonable mathematical description, in which case we would have to seek computational algorithms for its approximation. However, the period is most easily defined in terms of the algorithm used to calculate it. It should, of course, yield the exact period for periodic functions and be close for small perturbations of periodic functions. This replaces an analysis of the accuracy of period calculation with an analysis of the efficiency of the multirevolutionary method with respect to different period definitions. This latter may be an easier task.

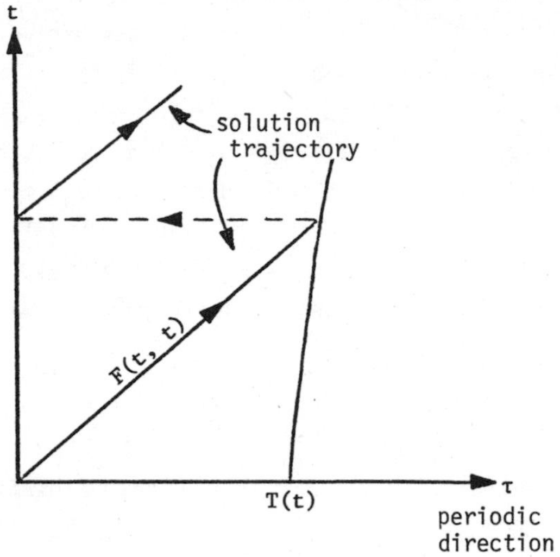

Figure 1. Nearly Periodic Solution Family

Petzold's period definition, based on minimizing the norm in eq. (1), is very expensive to apply and cannot be considered as a technique for determining if an arbitrary output of an integrator is nearly periodic. Therefore, we look for alternate definitions of the period. First, note that if the oscillation is due to a periodic driving function, we probably know its period or can examine the system which generates the driving function directly. Hence, we can restrict ourselves to autonomous systems or systems which when made autonomous are nearly periodic. (This means that the partials of the system with respect to time are small.) The solution of an autonomous system is completely determined by the specification of the value of the solution vector $\underline{y}$ at one time. That is to say, if we identify two times on the solution such that $\underline{y}(t_1) = \underline{y}(t_2)$, we know that the solution is periodic with period $t_2 - t_1$. This suggests determining the period by looking for minimum of $||\underline{y}(t_1) - \underline{y}(t_2)||$. The cost of this is not particularly low and it requires a clever adaptive program with a lot of heuristics. The form of the program is to choose an initial point t_1 and, as each new value $\underline{y}(t)$ is calculated, to see if $||\underline{y}(t_1) - \underline{y}(t)||$ has passed a local minimum. If so, interpolation is used to locate the minimum. If the value of the norm at the minimum is small (first heuristic parameter), we have possible periodic behavior. If not, we must continue to examine more $y(t)$. However, since the periodic behavior of y may not have started by t_1, we must also advance t_1 occasionally (next heuristic parameter). After some experiments, we abandoned this approach.

Another way of defining the period is to identify certain points on the solution at which a simple characterization is repeated, such as zero crossing. The solution itself may not have zero crossings and if it consists of a periodic function superimposed on a slowly growing function, it may be difficult to choose any value which is crossed periodically. However, its derivative will have periodic sign changes, so we have experimented with a definition of period based on the zero crossings of $\underline{c}^T\underline{y}'$ where $\underline{c}^T$ is the transpose of a vector of constants. The program examines the integrator output for positive-going zero crossings of $\underline{c}^T\underline{y}'$. (Currently, $\underline{c}$ is a vector of the weights provided by the user for error norm calculations.) Anything but a simple periodic solution may lead to more than one zero crossing in a single period, so the norm $||y'(t_1) - y'(t_2)||$ is also examined, where t_1 and t_2 are a pair of zero crossings. If the norm is small, the possibility of a period is considered. The procedure used is as follows:

1. Identify a positive going sign change in $\underline{c}^T\underline{y}'$.
2. Interpolate to find the t value $t_{current}$ of the zero crossing. Also compute interpolated values of $\underline{y}$, $\underline{y}'$.
3. Save these values. (Up to ten prior values are saved.)
4. Compare current value with each prior value in turn until a small $||\underline{y}'_{old} - \underline{y}'_{current}||$ is found.
5. Save $period_{new} = t_{current} - t_{old}$.
6. Compare $period_{new}$ with $period_{old}$ if one has previously been calculated. If they are relatively close, accept the new period and switch to multirevolutionary methods.
7. Examine several backward differences of recent periods. If they seem to be smoothly varying, accept new period.

The last test was found to be necessary for some problems with variable periods.

As can be seen, there are numerous heuristics, which implies that much tuning is possible. However, it is important to note that tuning effects efficiency only. If the tests for periodicity are too stringent, the standard integrator will continue too long; if they are too lenient, multirevolutionary methods will be invoked before they are efficient. However, since they will then run using a stepsize of one period, they will perform very little worse then the conventional integrator. (The only losses are additional overhead.)

The multirevolutionary method, described in the next section, is a modification of a standard integrator. It calls on a subroutine to evaluate $g(z) = z(s + 1) - z(s)$ given $z(s)$. It can suffer from stiffness in exactly the same way that an ordinary integrator can suffer from stiffness: if $H\partial g/\partial z$ is large the method may be unstable and the corrector iteration will not converge unless a Jacobian $J = \partial g/\partial z$ is used in a Newton iteration. Shampine [11] has suggested monitoring the size of the Jacobian by estimating its norm when two or more function evaluations are used in a single step. Essentially,

$$L = \max \frac{||g(z_1) - g(z_2)||}{||z_1 - z_2||}$$

is calculated, where the max is taken over all steps and z_1 and z_2 are two different values of z for which g is evaluated in one step. (In practice, it seems preferable to take an exponentially weighted max such as

$$L_{new} = \max \left(0.9\, L_{old}, \frac{||g(z_1) - g(z_2)||}{||z_1 - z_2||}\right)$$

but this is yet another tuning heuristic.)

This technique can be used in both the regular integrator used to calculate g and the multirevolutionary integrator. However, there are a number of tuning questions that pose some difficulties. One could decide to switch to a stiff method the moment that J becomes large enough to restrict the stepsize in which case there are no problems. However, that is not efficient because nonstiff methods are both considerably lower in cost per step and have a smaller error tolerance. The natural solution to consider is to continue with the nonstiff method until the estimated stepsize that could be used in a stiff method is sufficiently larger to offset the increased cost per step. This requires estimating the error in stiff methods by estimating various derivatives. This is done with suitable difference formulas but if care is not used, the derivatives estimated may appear to be such that little step increase is possible with a stiff method. The reason for this difficulty is that most codes do not directly restrict their stepsize on the basis of a Jacobian estimate and stability needs. At the most they restrict $h||J||$ so that the corrector would converge if iterated. However, higher order methods may be on the verge of instability. Once $h||J||$ is too large for stability, small errors grow rapidly. The automatic error control quickly reduces the step to keep the error near the tolerance level and a well-engineered nonstiff solver will produce a perfectly good solution, albeit slowly. However, the solution contains errors of the order of the error tolerance due to these marginally stable components, and these errors usually oscillate. When a difference formula is applied to them, large values result, and completely obscure the derivatives we want to estimate. For example, the marginal stability could introduce an error of $(-1)^n\varepsilon$ into the numerical solution y_n at the n-th step. If we form the k-th backward difference to estimate $h^k y^{(k)}$ we have a component due to this error of $(-1)^n 2^k \varepsilon$. If we now ask what stepsize αh can be used in a stiff method whose error is $C_k h^k y^{(k)}$ to achieve an error of ε, we find that $\alpha = (2^k C_k)^{-1/k}$ independent of current h. For BDF, $C_k = 1/k$ so α is always less than one, falsely indicating that the stepsize cannot be increased. To avoid this difficulty it is necessary to keep $h||J||$ small enough that components with the most negative eigenvalues are at least moderately damped.

In addition to estimating $||J||$, it is possible to estimate the largest eigenvalue (or eigenvalue pair if complex conjugate) using evaluations of g(z) from more than one step, as suggested in Gear [12]. However, experiments on that technique indicate that the eigenvalue estimates are not too reliable. The reason for wanting to know the largest eigenvalue is to know whether to use a technique efficient for real eigenvalues, such as BDF, or a technique better suited to eigenvalues close to the imaginary axis, such as BLEND [13] or implicit Runge-Kutta [3]. K. Stewart* pointed out that it is sufficient to wait until a decision to use stiff methods has been made. At that time a Jacobian must be calculated, and power methods can be used to determine the arguments of the large eigenvalues. (This poses an interesting question for the numerical linear algebraist: how to calculate cheaply the maximum argument of all eigenvalues exceeding a certain size in a matrix.) At the time the Jacobian is first calculated, it can also be checked for other properties such as bandedness and sparsity so that a decision on which linear equation scheme to use can be made.

*K.Stewart, Jet Propulsion Lab, Pasadena, CA. Private communication.

3. Variable Period Multirevolutionary Methods

In the original coordinates we have $z(t + T) - z(t) = d(t)$. For small T this says $z'(t) \cong d(t)/T$. Hence, it is not surprising that the numerical interpolation for $z(t)$ given a technique for computing $g(t) = d(t)/T$ is very similar to a numerical integration technique. In the new coordinate system, the basic structure of the program is an outer integrator which solves the equations

$$z(s + 1) - z(s) = g(z)$$
$$t(s + 1) - t(s) = T(t(s))$$

using an outer stepsize H. The method varies the order and stepsize just as an ordinary integrator does. See Petzold [10] for details. It calls a subroutine to evaluate g and T given z and t. This is done by integrating the underlying ordinary differential equation $y' = f(y)$ starting from $y(t) = z$, determining when a period has elapsed and computing $g(z) = y(t + T(t)) - y(t)$. Both methods for defining the period discussed in the pevious section have been tried. The first, minimizing $||y(t + T(t)) - y(t)||$, is now easier to implement because the left end, $y(t)$, is fixed. The only tuning difficulty is to ignore intermediate minima, and we have done this by considering only values of T starting from 0.9 of the previous period estimate. (If T changes more rapidly over H than this, either H is too large or the nearly periodic assumption is questionable.) The norm actually used has the form

$$||v||_A^2 = \sum_{j=0}^{s} \sum_{i=1}^{n} A_{si}(v_i^{(j)})^2$$

where $v_i^{(j)}$ is the j-th derivative of the i-th component, and the A_{si} are weights. It appears to be best to use $A_{si} = 0$, $s \neq 1$, and A_{1i} = weight of i-th component of error weight vector. This allows for arbitrary nonperiodic linear functions to be ignored. Higher derivatives can be included, but knowledge of them is subject to larger errors due to the inner integration.

The second method, looking for a zero crossing, has a difficulty: the function $\underline{c}^T\underline{y}'$ will not necessarily be zero at $y(t) = z$. (It can be shown that it will be zero except for roundoff error for linear problems.) This has been overcome by choosing a vector $\underline{c}$ separately for each period such that $p(t) = \underline{c}^T\underline{y}' = 0$ at the start of the period. A future zero of $p(t)$ defines the length of the period. A unique $\underline{c}$ is defined by choosing $\underline{c}$ to maximize $p'(t)$ at the start of the period subject to $||\underline{c}|| = 1$. This value is chosen because it minimizes the roundoff error effects in determining a zero of $p(t)$. The value of $\underline{c}$, apart from scaling, is

$$\underline{c} = \underline{y}'' - \frac{(\underline{y}', \underline{y}'')}{||\underline{y}'||^2}\underline{y}'$$

This requires a knowledge of y" at the initial point $y(t)$. This is available because a Runge-Kutta starter which computes the first four derivatives is used for the multistep inner integrator [14]. After $\underline{c}$ has been calculated, future positive going zero crossings of p are examined, $\underline{y}$ and its derivatives are calculated by interpolation at the zero crossing point $t + \tau$, and $||\underline{y}'(t + \tau) - \underline{y}'(t)||$ is evaluated. If it is small enough, the period is set to τ and g is calculated.

The variable period multirevolutionary integrator is based on a modified Nordsieck scheme. Each component of z is represented by the history vector

$$\underline{a} = [z, Hg, H^2g'/2, H^3g''/6, \ldots, H^kg^{(k-1)}/k!]^T$$

Petzold has shown that in this representation the predictor has the form

$$\underline{a}_{n,(0)} = A\underline{a}_{n-1}$$

where A is the Pascal triangle matrix except for the first row which is

$$[1, 1, \alpha_1(r), \alpha_2(r), \ldots, \alpha_{k-1}(r)]$$

where $r = 1/H$. She also showed that the corrector takes the form

$$\underline{a}_n = \underline{a}_{n,(0)} + \underline{\ell}\omega$$

where ω is chosen so that $\underline{a}_n$ "satisfies" the relation $z(s_n + 1) - z(s_n) = g(s_n)$ and $\underline{\ell}$ is the conventional corrector vector except in the first component which is a function of $r = 1/H$. Petzold gives these functions for generalized Adams methods. (They are polynomials in r.) The corresponding functions for BDF methods are inverse polynomials in r. They are

Order	First Coefficient of ℓ
1	-1
2	$-1/(3/2 + r)$
3	$-1/(11/6 + 2r + r^2)$
4	$-1/(25/12 + 35r/12 + 5r/2 + r^3)$
5	$-1/(137/60 + 15r/4 + 17r^2/4 + 3r^3 + r^4)$
6	$-1/(147/60 + 203r/45 + 49r^2/8 + 103r^3/12 + 7r^4/2 + r^5)$

Petzold suggests a linear combination of the generalized Adams and BDF coefficients, for example, $r \cdot \text{Adams} + (1 - r) \cdot \text{BDF}$ so that the method has the properties of BDF formula for large H ($r \to 0$), namely stiff stability, and the property of Generalized Adams for $r = 1$, namely the outer integrator is exact. Since it is not proposed to use BDF methods until stiffness has set in (when H is large), it does not seem worth considering this complication.

4. A Numerical Test

Several example problems have been constructed using the Van der Pol oscillator to give a nonlinear oscillation. Typical of these problems is the following system of four equations

$$\underline{u} = Q\underline{y}$$

$$u_1' = u_2$$

$$u_2' = -(u_1 - u_3) + 2(u_3 - (u_1 - u_3)^2)u_2$$

$$u_3' = -10^{-3}(u_3 - 1)$$

$$u_4' = 10^{-3} \sin 10^{-3}t$$

$$\underline{y}' = Q^{-1}\underline{u}'$$

All initial values were zero so $u_3 = 1 - e^{-.001t}$ and $u_4 = 1 - \cos .001t$. u_1 and u_2 are the solution and first derivative of a Van der Pol oscillator oscillation about a level u_3 and peak amplitude about $2u_3$. The period is about 2π for small u_3 and steadily increases to about 7.63 for $u_3 = 1$.

The matrix Q used was

$$1/2 \begin{bmatrix} -1 & 1 & 1 & 1 \\ 1 & -1 & 1 & 1 \\ 1 & 1 & -1 & 1 \\ 1 & 1 & 1 & -1 \end{bmatrix}$$

(It is idempotent and $Q = Q^{-1}$.) All components of y oscillate after an initial period. The periodic detector located the oscillation at about t = 156. The integration was continued to t = 10,000 with an average outer step of about 28 periods. At that stepsize the outer problem is quite stiff. The oscillatory behavior is initially close to sinusoidal and changes to the steep-edged behavior typical of the Van der Pol oscillator by t = 1000. A local error tolerance of 10^{-9} in the inner integrator required about 400 inner steps per period at first, increasing to about 1200 at the end. The outer integrator took 50 steps with local tolerance of 10^{-3} from t = 156 to t = 10,000, using 154 inner integrations over one period including those for occasional Jacobian evaluations of g by numerical differencing, an average speed up of ninefold over the standard inner integrator which would have used about 10^6 steps for the whole problem.

Plots of the phases of the solution are shown in Figures 2 to 5. In all of these figures the vertical scales for the four components have been renormalized to put y_i between 2i - 1.9 and 2i - 0.1 for i = 1 to 4. Figure 2 shows the first phase of the integration prior to detection of the oscillation.

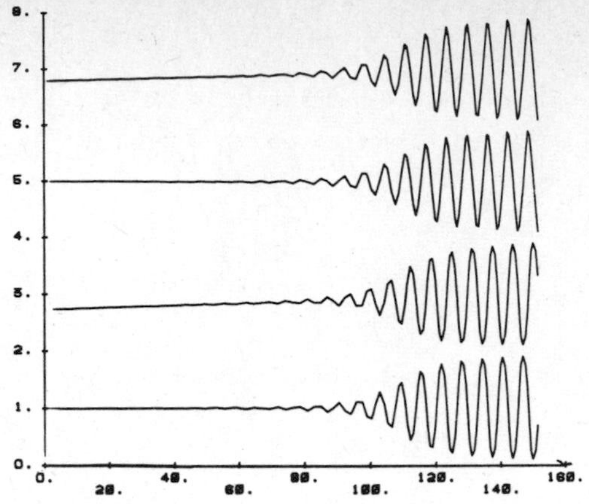

Figure 2. Initial Phase before Period Detected

(For extraneous reasons, only one integration point in ten has been plotted here, hence the jagged curves.) The amplitude of the oscillation at this point is 0.99. The shape of the oscillation at t = 156 is shown in Figure 3. This shape changes and grows in amplitude to 3.02 by t = 10,000, as is shown in Figure 4. Figure 5 shows the smooth "quasi-envelope" z found by the multirevolutionary integrator. It was generated using the 50 outer integration steps so the actual solution y is found by superimposing the oscillatory behavior of the form shown in Figures 3 and 4 and the appropriate t values in Figure 5.

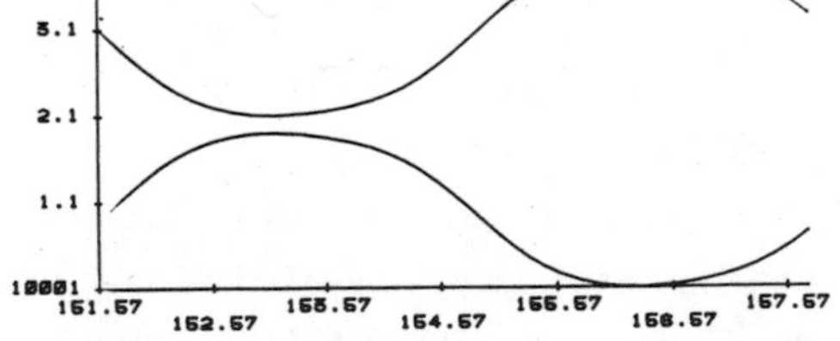

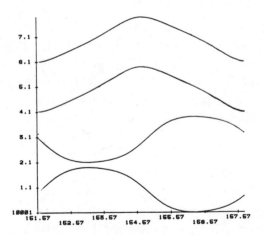

Figure 3. First Period in Multirevolutionary Integration

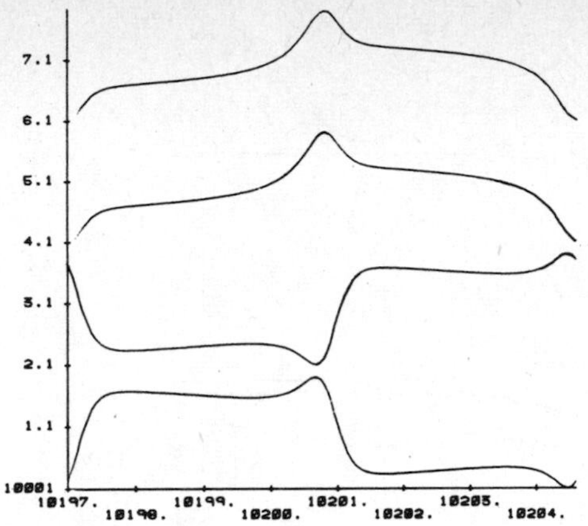

Figure 4. Fiftieth (last) Period in Multirevolutionary Integration

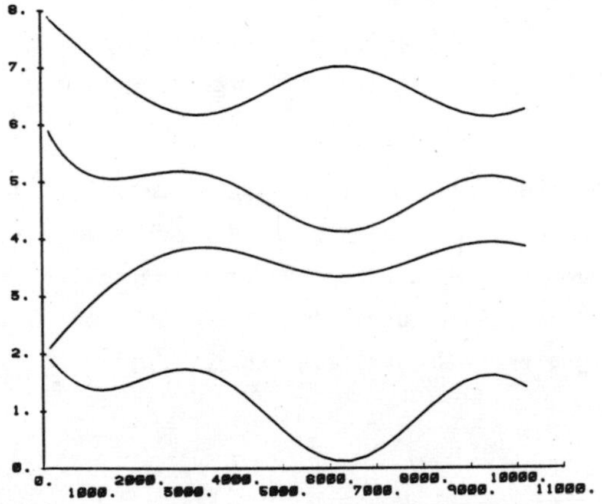

Figure 5. Quasi-envelope z

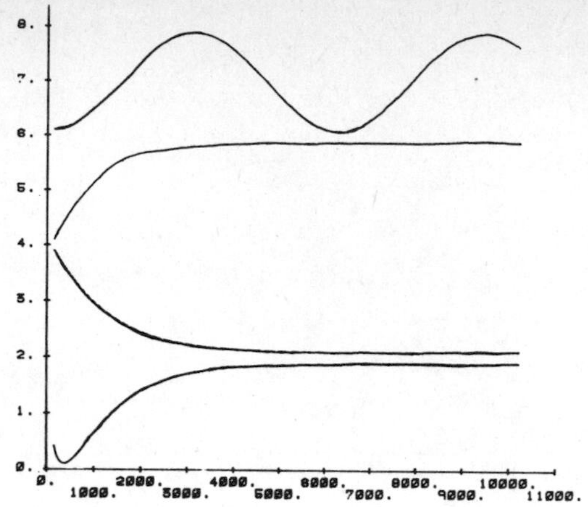

Figure 6. Quasi-envelope Corresponding to u (Qz)

It should be noted that these are not small oscillations. For example, the range of z_1 in Figure 5 (bottom line) is -1.13 to -0.6 (approx.). The oscillation changes in amplitude from 0.99 to 3.02 (peak to peak) over the interval.

In order to check the accuracy, the equivalent quasi-envelope for $\underline{u}$ was recovered from $\underline{z}$ by the transformation $\hat{\underline{u}} = Q\underline{z}$. Since u_3 and u_4 are not oscillatory, $\hat{u}_3$ and $\hat{u}_4$ should also be $1 - \exp(-10^{-3}t)$ and $1 - \cos(10^{-3}t)$, respectively. Since the cosine component is neutrally stable, any integration errors will not be damped in later steps. The relative error in the cosine component at t = 10,000 was .005 (.008 absolute).

These results were without tuning. There are a number of inefficiencies in the software that can be removed. For example, we did not give the multirevolutionary integrator the information gathered during the period detection phase about g and its differences. These can be used to allow a high order start in the multirevolutionary integrator. We believe that additional improvements are possible.

5. Further Problems

There are a number of additional problems of concern. Here we will discuss three problems: non-autonomous problems, detecting the end of periodicity, and the multiple oscillator problem. Some problems require only simple extensions; others, in particular the multiple oscillator problem, pose serious difficulties.

There are two cases of the non-autonomous problem $y' = f(t, y)$ to consider: those in which $f(t, y)$ is a slowly changing function of t and those in which the t-dependence is responsible for the oscillation--we say that the oscillation is driven. In the former case we can conceptually convert to an autonomous system by appending the usual $t' = 1$. This term is slowly varying so the solution of the enlarged problem remains "nearly periodic." In the latter case we can determine the period by examining the driving term (that is, the t-dependent terms in f) and continue to use the same method.

Some nonlinear oscilltors are such that a variable in the system increases to a point that the oscillation is quenched. Then there is a period of relaxation until it starts again. For such systems an automatic program must detect the end of the oscillation and revert to a conventional method. This is analogous to the problem of detecting a derivative discontinuity in the solution of a conventional differential equation and similar techniques can be used. When the multi-revolutionary integrator calls for an evaluation of the local period and of $g(z)$, the period detection scheme will be unable to find a period anywhere close to the prior value, or even to find one at all. After it has looked a modest distance beyond the expected value, it should report failure so that the multirevolutionary integrator can reduce its stepsize to find the "discontinuity" where the oscillator is quenched. When the stepsize has been reduced to one of only a few periods, the software can revert to a conventional integration method.

The multiple oscillator problem poses difficulties unless there is a large gap in frequencies between the two highest frequencies. In that case, the lower frequencies can be viewed as the slowly changing components of $F(\tau, t)$ and possibly the method can be used recursively for the second highest frequency, and so on. If the two highest frequencies ω_1 and ω_2 are of the same order of magnitude, the behavior will be far from nearly periodic unless $\omega_1/\omega_2 = p/q$ for small integers p and q. In that case there is a subharmonic ω_1/p of the two frequencies which can be used as the period. If not, there does not seem to be much hope unless the oscillators can be isolated and treated separately by the techniques discussed above. Suppose we can visualize the system as consisting of two oscillators $u' = p(u, y)$ and $v' = q(v, y)$ where y is a slowly varying term, and a slow part described by $y' = f(y, u, v)$. If f is linear in u and v, it is sufficient to find the behavior of the average of u and v and this can be done for each separately.

However, if f is nonlinear in u and v, we must also keep track of the relative phases of the oscillations of u and v so that each time f is evaluated on a coarse mesh, the correct relative phases can be used.

References

[1] Shampine, L.F., Linear equations in general purpose codes for stiff ODEs, Report 80-0429 Sandia Laboratories, Albuquerque, NM, February 1980.

[2] Butcher, J.C., A transformed implicit Runge-Kutta method, Report 111, Dept. Mathematics, Univ. Auckland, New Zealand, May 1977.

[3] Butcher, J.C., Burrage, K. and F.H. Chipman, STRIDE: Stable Runge-Kutta Integrator for Differential Equations, Report 150, Dept. Mathematics, Univ. Auckland, New Zealand, August 1979.

[4] Gear, C.W., The automatic integration of stiff ordinary differential equations, Proceedings IFIP Congress 1968, 1968, 187-193.

[5] Hindmarsh, A., GEAR: ordinary differential equation solver, Report UCID-30001, Rev. 3, Lawrence Livermore Laboratory, CA, 1974.

[6] Byrne, G.D. and A.C. Hindmarsh, EPISODEB: An Experimental Package for the Integration of Systems of Ordinary Differential Equations with Banded Jacobians, Report UCID-30132, Lawrence Livermore Laboratory, CA, April 1976.

[7] Graff, O.F., Methods of orbit computation with multirevolution steps, Report AMRL 1063, Applied Mechanics Research Laboratory, Univ. Texas at Austin, TX, 1973.

[8] Graff, O.F. and D.G. Bettis, Modified multirevolution integration methods for satellite orbit computation, Celestial Mechanics 11, 1975, 443-448.

[9] Mace, D. and L.H. Thomas, An extrapolation method for stepping the calculations of the orbit of an artificial satellite several revolutions ahead at a time, Astronomical Journal 65 (5), June 1960.

[10] Petzold, L.R., An efficient numerical method for highly oscillatory ordinary differential equations, Report UIUCDCS-R-78-933, Dept. Comp. Sci., Univ. Illinois at Urbana-Champaign, IL, August 1978.

[11] Shampine, L.F., Lipschitz constants and robust ODE codes, Report 79-0458, Sandia Laboratories, Albuquerque, NM, March 1979.

[12] Gear, C.W., Method and initial stepsize selection in multistep ODE solvers, Report UIUCDCS-R-80-1006, Dept. Comp. Sci., Univ. Illinois at Urbana-Champaign, IL, February 1980.

[13] Skeel, R.D. and A.K. Kong, Blended linear multistep methods, ACM Trans. Math. Software 3 (4), December 1977, 326-345.

[14] Gear, C.W., Runge-Kutta starters for multistep methods, Report UIUCDCS-R-78-938, Dept. Comp. Sci., Univ. Illinois at Urbana-Champaign, IL, September 1978, to appear ACM Transactions on Mathematical Software. 6, 263 (1980)

CHARACTERIZATION OF NON-LINEARLY STABLE IMPLICIT RUNGE-KUTTA METHODS

E. Hairer and G. Wanner

Abstract. Implicit Runge-Kutta methods, though difficult to implement, possess the strongest stability properties. This paper introduces to the theory of algebraically stable (A-contractive, B-stable) Runge-Kutta methods. These are methods for which the numerical solutions remain contractive if the (nonlinear) differential equation has contractive solutions. The proofs are sometimes omitted or sketched only , their details can be found in [13].

1. Examples of Runge-Kutta methods

Let us consider the system of differential equations

(1) $$y' = f(t,y) \ , \quad y(t_0) = y_0$$

where y and f are elements of R^n. An s-stage implicit Runge-Kutta method (IRK) is a one-step method defined by the formulas

(2) $$k_i = hf(t_0 + c_i h,\ y_0 + \sum_{j=1}^{s} a_{ij}k_j) \qquad i=1,\dots,s$$

$$y_1 = y_0 + \sum_{i=1}^{s} b_i k_i \ , \qquad c_i = \sum_{j=1}^{s} a_{ij} \ .$$

The method is said to be of <u>order</u> p, if the local truncation error satisfies

$$y_1 - y(t_0+h) = 0(h^{p+1}) \qquad \text{for} \quad h \to 0 \ .$$

A popular example of an IRK method is the <u>Backward Euler method</u>

$$y_1 = y_0 + hf(t_0+h, y_1) \ .$$

If we represent the coefficients in (2) in Butcher's notation

$$\begin{array}{c|ccc} c_1 & a_{11} & \dots & a_{1s} \\ \vdots & \vdots & & \vdots \\ c_s & a_{s1} & \dots & a_{ss} \\ \hline & b_1 & \dots & b_s \end{array}$$

the Backward Euler method is given by

$$\text{(BE)} \qquad \begin{array}{c|c} 1 & 1 \\ \hline & 1 \end{array} \qquad s = 1,\ p = 1.$$

Another example is the Trapezoidal rule

$$y_1 = y_0 + \frac{h}{2}[f(t_0,y_0) + f(t_0+h,y_1)]$$

with the coefficients

$$\text{(TR)} \qquad \begin{array}{c|cc} 0 & 0 & 0 \\ 1 & 1/2 & 1/2 \\ \hline & 1/2 & 1/2 \end{array} \qquad s = 2,\ p = 2,$$

or the classical 4-th order method of Kutta (1901)

$$\text{(RK)} \qquad \begin{array}{c|cccc} 0 & & & & \\ 1/2 & 1/2 & & & \\ 1/2 & 0 & 1/2 & & \\ 1 & 0 & 0 & 1 & \\ \hline & 1/6 & 1/3 & 1/3 & 1/6 \end{array} \qquad s = 4,\ p = 4.$$

Methods with highest possible order, based on Gaussian quadrature formulas, have been investigated by J. Butcher, 1964, [4], e.g.

$$\text{(GB(3))} \qquad \begin{array}{c|ccc} (5-\sqrt{15})/10 & 5/36 & (10-3\sqrt{15})/45 & (25-6\sqrt{15})/180 \\ 1/2 & (10+3\sqrt{15})/72 & 2/9 & (10-3\sqrt{15})/72 \\ (5+\sqrt{15})/10 & (25+6\sqrt{15})/180 & (10+3\sqrt{15})/45 & 5/36 \\ \hline & 5/18 & 4/9 & 5/18 \end{array} \qquad \begin{array}{l} s = 3, \\ p = 6. \end{array}$$

Methods with better "stability at infinity" have been introduced 1969 independently by Axelsson [1] and Ehle [11]:

$$\text{(AE(3))} \qquad \begin{array}{c|ccc} (4-\sqrt{6})/10 & (88-7\sqrt{6})/360 & (296-169\sqrt{6})/1800 & (-2+3\sqrt{6})/225 \\ (4+\sqrt{6})/10 & (296+169\sqrt{6})/1800 & (88+7\sqrt{6})/360 & (-2-3\sqrt{6})/225 \\ 1 & (16-\sqrt{6})/36 & (16+\sqrt{6})/36 & 1/9 \\ \hline & (16-\sqrt{6})/36 & (16+\sqrt{6})/36 & 1/9 \end{array} \qquad \begin{array}{l} s = 3, \\ p = 5, \end{array}$$

and, even stronger stable at infinity are the methods of "type III_C" by Chipman [6]

$$\text{(CH(3))} \qquad \begin{array}{c|ccc} 0 & 1/6 & -1/3 & 1/6 \\ 1/2 & 1/6 & 5/12 & -1/12 \\ 1 & 1/6 & 2/3 & 1/6 \\ \hline & 1/6 & 2/3 & 1/6 \end{array} \qquad s = 3,\ p = 4.$$

The actual implementation of the formula (2) is easier, if $a_{ij} = 0$ for $i < j$. We call such methods "diagonally implicit" (DIRK).

Especially interesting are methods for which in addition $a_{ii} = \gamma$ for all i ("singly diagonally IRK", SDIRK). Such methods have first been constructed by S. P. Nørsett [15]. We give as examples methods constructed independently by K. Burrage [2] and M. Crouzeix [7].

(BC(3))

$(1+\alpha)/2$	$(1+\alpha)/2$			
$1/2$	$-\alpha/2$	$(1+\alpha)/2$		
$(1-\alpha)/2$	$1+\alpha$	$-1-2\alpha$	$(1+\alpha)/2$	
	$1/(6\alpha^2)$	$1-1/(3\alpha^2)$	$1/(6\alpha^2)$	

$\alpha = (2\cos(\pi/18))/\sqrt{3}$, $s = 3$, $p = 4$.

(B(4))

.5728160625	.5728160625			
.0242889252	-.5485271373	.5728160625		
.9757110748	-.7169560624	1.1198510747	.5728160625	
.4271839375	.5450631823	-.3913115464	-.2993837609	.5728160625
	.3234580063	.1765419937	.1765419937	.3234580063

$s = 4$, $p = 4$.

2. The W-transformation

For the RK-method (2) we assume that the weights b_i are strictly positive and the c_i distinct. The corresponding quadrature formula then defines a scalar product on the space of all polynomials with degree less than s in the following way

$$(3)\qquad \langle q,r\rangle = \sum_{i=1}^{s} b_i q(c_i) r(c_i) \ .$$

Let now $p_i(t)$ $(i=0,1,\dots,s-1)$ be the polynomials of degree i orthonormal with respect to (3). If the quadrature formula is of order p, we have

$$\sum_{i=1}^{s} b_i q(c_i) r(c_i) = \int_0^1 q(t) r(t) dt$$

for polynomials q and r with $\deg(q\cdot r) \leqq p-1$. Therefore $p_i(t)$ coincides with the (shifted and normalized) Legendre polynomial

$$(4)\qquad p_i(t) = \sqrt{2i+1} \sum_{j=0}^{i} (-1)^{i+j} \binom{i}{j}\binom{i+j}{j} t^j$$

for $i = 0,1,\dots,[(p-1)/2]$.

We introduce the matrix

$$\bullet\ W = \begin{pmatrix} 1 & p_1(c_1) & \cdots & p_{s-1}(c_1) \\ \vdots & \vdots & & \vdots \\ 1 & p_1(c_s) & \cdots & p_{s-1}(c_s) \end{pmatrix}$$

which satisfies $W^T B W = I$, where $B = \mathrm{diag}(b_1,\ldots,b_s)$. For the RK-method (2) we then define

(5) $$X = W^{-1}AW = W^T BAW \ .$$

This matrix turns out to possess beautiful properties. For the above examples we obtain:

(BE) $$X = (1)$$

(TR) $$X = \begin{pmatrix} 1/2 & 0 \\ 1/2 & 0 \end{pmatrix}$$

(RK) $$X = \begin{pmatrix} 1/2 & -1/(2\sqrt{3}) & 0 & 0 \\ 1/(2\sqrt{3}) & 0 & -1/(2\sqrt{6}) & -1/(2\sqrt{2}) \\ 0 & 1/(2\sqrt{6}) & -1/4 & -\sqrt{3}/4 \\ 0 & -1/(2\sqrt{2}) & \sqrt{3}/4 & -1/4 \end{pmatrix}$$

(GB(3)) $$X = \begin{pmatrix} 1/2 & -1/(2\sqrt{3}) & 0 \\ 1/(2\sqrt{3}) & 0 & -1/(2\sqrt{15}) \\ 0 & 1/(2\sqrt{15}) & 0 \end{pmatrix}$$

(AE(3)) $$X = \begin{pmatrix} 1/2 & -1/(2\sqrt{3}) & 0 \\ 1/(2\sqrt{3}) & 0 & -1/(2\sqrt{15}) \\ 0 & 1/(2\sqrt{15}) & 1/10 \end{pmatrix}$$

(CH(3)) $$X = \begin{pmatrix} 1/2 & -1/(2\sqrt{3}) & 0 \\ 1/(2\sqrt{3}) & 0 & -1/(2\sqrt{6}) \\ 0 & 1/(2\sqrt{6}) & 1/4 \end{pmatrix}$$

(BC(3)) $$X = \begin{pmatrix} 1/2 & -1/(2\sqrt{3}) & 0 \\ 1/(2\sqrt{3}) & 0 & -1.41045 \\ 0 & 1.41045 & 2.70574 \end{pmatrix}$$

(B(4)) $$X = \begin{pmatrix} 1/2 & -1/(2\sqrt{3}) & 0 & 0 \\ 1/(2\sqrt{3}) & 0 & 0.58632 & 0 \\ 0 & -0.58632 & 1.47809 & 0.42787 \\ 0 & 0 & -0.42787 & 0.31317 \end{pmatrix}$$

One observes that the first columns of X always possess a special structure. This is, in fact, a consequence of properties of the orthogonal polynomials and the so-called simplifying assumptions

(6) $$\sum_{j=1}^{s} a_{ij}c_j^{q-1} = c_i^q/q \qquad q=1,\dots,k \text{ and all } i.$$

These simplifying assumptions express the fact that the intermediate points $g_i = y_o + \sum_{j=1}^{s} a_{ij}k_j$ are approximations to $y(t_o+c_ih)$ of order k.

<u>Theorem 1</u>. If the associated quadrature formula (with positive b_i and distinct c_i) is of order p and $k \leq [(p-1)/2]$, then (6) is equivalent with the fact that the first k columns of X are given by $(1 \leq j \leq k)$

(7) $$x_{ij} = \begin{cases} 1/2 & i = j = 1 \\ \xi_j & i = j+1 \\ -\xi_i & j = i+1 \\ 0 & \text{else} \end{cases}$$

where $\xi_i = 1/(2\sqrt{4i^2-1})$.

<u>Proof</u>. (6) means that for any polynomial q(t) of degree $\leq$ k-1

$$\sum_{j=1}^{s} a_{ij}q(c_j) = \int_0^{c_i} q(t)dt \ .$$

If we insert the polynomials $p_i(t)$ and use the integration formulas, which are valid for the Legendre polynomials (4)

$$\int_0^t p_i(t)dt = \xi_{i+1}p_{i+1}(t) - \xi_i p_{i-1}(t) \qquad i = 1,2,\dots$$

$$\int_0^t p_o(t)dt = \xi_1 p_1(t) + \frac{1}{2} p_o(t) \ ,$$

we obtain that

$$AW = W\begin{pmatrix} 1/2 & -\xi_1 & & & & * & \dots & * \\ \xi_1 & 0 & -\xi_2 & & & & & \\ & \xi_2 & 0 & & & & & \\ & & & \ddots & -\xi_{k-1} & & & \\ & & & & 0 & & & \\ & & & & \xi_k & & & \\ & & & & & * & \dots & * \end{pmatrix}$$

The statement now follows, if we multiply this equation with W^{-1} from the left. □

<u>Remark</u>. For the methods (GB(s)) of Butcher, $c_1,\dots,c_s$ are the zeros of the s-th Legendre polynomial and a_{ij} are defined by (6) with k=s. The

above proof shows that in this case all elements of X are given by (7).

The next theorem shows how the second kind of simplifying assumptions

(8) $$\sum_{i=1}^{s} b_i c_i^{q-1} a_{ij} = b_j(1 - c_j^q)/q \qquad q=1,\dots,k \quad \text{and all } j$$

influences the structure of X.

Theorem 2. Under the assumptions of Theorem 1 we have:

(8) is equivalent with the fact that the first k rows of X are given by (7).

The proof is given in [13].

These two theorems will be crucial in the characterization of non-linearly stable methods.

3. Discussion of A-stability

The oldest concept of unconditional stability has been introduced by Dahlquist [8]: Method (2) is called A-stable if, when applied to $y' = \lambda y$, $\mathrm{Re}\lambda \leq 0$, it always holds that $|y_1| \leq |y_0|$.

Introducing formulas (2) to this test equation we obtain with $h\lambda=z$

$$k = z\mathbb{1}y_0 + zAk \ ,$$

$$y_1 = y_0 + \mathbb{1}^T Bk \ ,$$

where $\mathbb{1} = (1,\dots,1)^T$, $k = (k_1,\dots,k_s)^T$. In order to use the W-transformation, we put $k = Wu\cdot y_0$ and obtain

$$(I - zX)u = zW^{-1}\mathbb{1} = ze_1 \qquad (e_1 = (1,0,\dots,0)^T)$$

$$y_1 = (1 + e_1^T u)y_0 = :R(z)y_0$$

with

(9) $$R(z) = \frac{\det(I - zX) + z\det(\tilde{I} - z\tilde{X})}{\det(I-zX)}$$

where $\tilde{I}$, $\tilde{X}$ are the (s-1,s-1)-submatrices of I, X with the first line and the first column omitted. The method is thus A-stable iff

$$|R(z)| \leq 1 \quad \text{for} \quad \mathrm{Re}z \leq 0 \ .$$

A general characterization of A-stable methods of order $\geq 2s-4$ is given in Wanner [17]. All above given examples, with the exception of (RK), are A-stable.

Another property (important for very stiff and non-linear differential equations) is "zero-stability at infinity", i.e.

$$\lim_{z\to\infty} R(z) = 0 \quad .$$

It follows from (9) that, if $\det X \neq 0$, this is equivalent to

$$\det(X) = \det(\tilde{X}) \quad .$$

Out of the above examples, the methods (BE), (AE(s)), (CH(s)) and (B(4)) possess this property.

An even stronger condition would be $\lim_{z\to\infty} zR(z) = 0$. This is satisfied by the methods (CH(s)).

4. Discussion of A-contractivity or algebraic stability

Not all A-stable methods possess also satisfactory stability properties for the general time dependent or non-linear case. Consider the following example

$$y' = (-120\exp(-0.18t) + 1.9t)y \quad , \quad 0 \leq t \leq 10 \ ,$$

whose numerical solutions from the trapezoidal rule (stepsize $h = 1$) are plotted in Fig. 1.

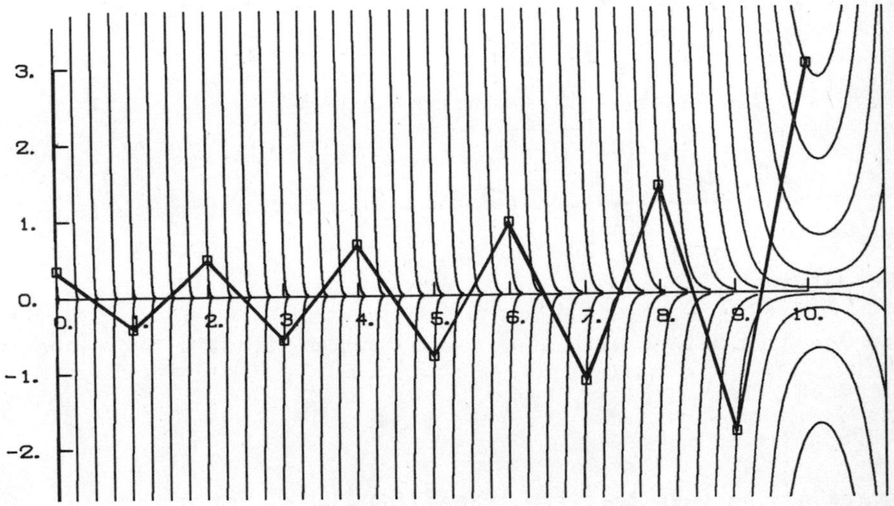

Fig. 1

It should be noted, however, that this instability is not very serious. In fact, the corresponding one-leg version of the trapezoidal rule is G-stable [9]. A more interesting example (with n=2) has been investigated by H. O. Kreiss [14] and J. Sand [16].

If the differential equation (1) is dissipative, i.e.

$$\langle f(t,y) - f(t,z),\ y - z\rangle \leqq 0 ,$$

any two solutions of (1) approach (in the euclidean norm). We want that in such cases also two numerical solutions approach. Method (2) is called A-contractive (or BN-stable), if for any dissipative differential equation two different initial values give results whose difference is bounded by the difference of the initial values. From [3] we have the following result (independently also proved in [7]):

Theorem 3. If the RK-method (2) satisfies

i) $b_i > 0$ for all i

ii) $BA + A^TB - B\mathbb{1}\mathbb{1}^TB \geqq 0$

then it is A-contractive.

Both conditions are also necessary for A-contractivity, if the c_i are distinct (see [3]) and if the method is irreducible (see [10]). Methods satisfying i) and ii) are called algebraically stable.

Because of $W^TB\mathbb{1}$ = (first column of W^TBW) = e_1 we have

$$W^T(BA + A^TB - B\mathbb{1}\mathbb{1}^TB)W = X + X^T - E$$

where

$$E = e_1e_1^T = \left(\begin{array}{c|ccc} 1 & 0 & \cdots & 0 \\ \hline 0 & & & \\ \vdots & & 0 & \\ 0 & & & \end{array}\right) .$$

Thus we obtain:

Theorem 3a. A RK-method with positive b_i and distinct c_i is A-contractive if and only if

$$X + X^T - E \geqq 0 .$$

It can readily be seen that the trapezoidal rule, for which

$$X + X^T - E = \begin{pmatrix} 0 & 1/2 \\ 1/2 & 0 \end{pmatrix},$$

is not A-contractive. Evidently, the methods (BE), (GB(s)) (for which $X + X^T - E = 0$), (AE(s)), (CH(s)), (BC(3)) and (B(4)) satisfy this property. In the following table we give a survey:

			stability			implementability		
method	stages	order	A-stable	A-contr.	zero-st. at inf.	explicit	SDIRK	SIRK
BE	1	1	yes	yes	yes	-	yes	yes
TR	2	2	yes	-	-	-	-	-
RK	4	4	-	-	-	yes	yes	yes
GB(s)	s	2s	yes	yes	-	-	-	-
AE(s)	s	2s-1	yes	yes	yes	-	-	-
CH(s)	s	2s-2	yes	yes	yes	-	-	-
BC(3)	3	4	yes	yes	-	-	yes	yes
B(4)	4	4	yes	yes	yes	-	yes	yes
SIRK6	5	6	yes	yes	-	-	-	yes
SIRK5	5	5	yes	yes	yes	-	-	yes

The following theorem, which summarizes some of the main results of [13], characterizes all algebraically stable RK-methods.

Theorem 4. Assume that the quadrature formula with positive b_i (and distinct c_i), associated to formula (2), is of order p. Then the IRK (2) is algebraically stable and of order p if and only if $A = WXW^{-1}$, where X is given by

$$X = \begin{pmatrix} 1/2 & -\xi_1 & 0 & \cdots & & & & 0 \\ \xi_1 & 0 & -\xi_2 & & & & & \vdots \\ 0 & \xi_2 & \ddots & \ddots & & & & \vdots \\ \vdots & & \ddots & 0 & -\xi_k & 0 & \cdots & 0 \\ \vdots & & & \xi_k & & & & \\ \vdots & & & 0 & & Y & & \\ \vdots & & & \vdots & & & & \\ 0 & \cdots & \cdots & 0 & & & & \end{pmatrix}$$

Here $k = [(p-1)/2]$, $\xi_j = 1/(2\sqrt{4j^2-1})$ and Y is a (s-k,s-k)-matrix satisfying $Y + Y^T \geq 0$. For p even we have in addition $x_{k+1,k+1} = y_{11} = 0$.

Proof (sketch). The order conditions for a p-th order RK-method imply (see [12])

$$\sum_{i=1}^{s} b_i \Big(\sum_{j=1}^{s} a_{ij} c_j^{q-1} - c_i^q/q \Big)^2 = 0 \quad \text{for} \quad q = 1,\dots,[(p-1)/2].$$

Since the b_i are positive, condition (6) with $k = [(p-1)/2]$ is necessary, so that by Theorem 1 the first k columns of X are given by (7). The "only if"-part now follows from the fact that $X + X^T - E \geqq 0$. For the "if"-part we observe that the special structure of X implies (Theorem 1 and Theorem 2) the validity of the simplifying assumptions (6) and (8). With their use it is not difficult to prove that the method is of order p. Algebraic stability follows from the fact that the b_i are positive and $Y + Y^T \geqq 0$. □

This theorem will be used in the next section for the construction of efficiently implementable, algebraically stable RK-methods of orders 5 and 6.

5. Singly implicit methods which are algebraically stable

The implementation of (2) requires in general the solution of an implicit system of equations of dimension $n \cdot s$. If, however, the method is DIRK, this reduces to a sequence of s equations each of dimension n only. Method (B(4)) is an example of such a method which is algebraically stable and of order 4. No algebraically stable DIRK of order 5 could be found. In fact, the following theorem is proved in [12]:

Theorem 5. If a DIRK is algebraically stable and of order p, then

$$p \leqq 4 .$$

In the fully implicit case one can take advantage of the fact that the Jacobian of the implicit system (2) is of tensor product structure $A \otimes \frac{\partial f}{\partial y}$ and simplify the computations by a transformation of A to Jordan canonical form. This is especially advantageous if this transformation is real and in particular if all eigenvalues of A are equal. We then call the method singly implicit (SIRK). The numerical work necessary for the solution of (2) is then comparable to that for SDIRK's. See Butcher [5] for more details.

Because of Theorem 5, we are interested to construct SIRK's of order greater than 4. We start from the following specialization of Theorem 4:

Corollary 6. All algebraically stable methods of order 5 with $s = 5$ are given by

$$(10) \qquad X = \begin{pmatrix} 1/2 & -1/(2\sqrt{3}) & 0 & 0 & 0 \\ 1/(2\sqrt{3}) & 0 & -1/(2\sqrt{15}) & 0 & 0 \\ 0 & 1/(2\sqrt{15}) & & & \\ 0 & 0 & & Y & \\ 0 & 0 & & & \end{pmatrix}$$

where $Y = \begin{pmatrix} y_{11} & y_{12} & y_{13} \\ y_{21} & y_{22} & y_{23} \\ y_{31} & y_{32} & y_{33} \end{pmatrix}$, and $Y + Y^T \geq 0$.

If $y_{11} = 0$, we have all methods of order 6.

If we compute the characteristic polynomial of X in (10) and compare it with $(\gamma - \lambda)^5$, we arrive at:

Theorem 7. The method given in (10) is singly implicit with quintuple eigenvalue γ iff

$$(11) \qquad \begin{aligned} \det(Y) &= 12\gamma^5 - 30\gamma^4 + 30\gamma^3 - 10\gamma^2 + 5\gamma/4 - 1/20 =: q_1 \\ y_{11}y_{22} - y_{21}y_{12} + y_{11}y_{33} - y_{31}y_{13} + y_{22}y_{33} - y_{32}y_{23} &= \\ &= 10\gamma^2 - 5\gamma/2 + 3/20 =: q_2 \\ y_{11} + y_{22} + y_{33} &= 5\gamma - 1/2 =: q_3 \\ y_{22}y_{33} - y_{32}y_{23} &= 300\gamma^4 - 300\gamma^3 + 100\gamma^2 - 25\gamma/2 + 1/2 =: q_4 \\ y_{22} + y_{33} &= -720\gamma^5 + 1800\gamma^4 - 1200\gamma^3 + 300\gamma^2 - 25\gamma + 1/2 =: q_5 . \end{aligned}$$

Inserting the conditions $\det(X) = \det(\tilde{X})$ (see section 3) or $y_{11} = 0$ we obtain:

Theorem 8. The method given by (10) and (11) is stable at infinity if

$$\gamma = \gamma_1 = .278053841136452 \qquad (\text{zero of } q_4 - 10q_1 = 0)$$

and of order six if

$$\gamma = \gamma_2 = .473268391258295 \qquad (\text{zero of } q_3 - q_5 = 0).$$

Remark. The other possible zeros of these polynomials can not produce algebraically stable methods (see Hairer - Wanner [13], Thm. 6).

Example 1. SIRK6, a singly implicit, algebraically stable RK-method of order 6 :

$$Y = \begin{pmatrix} 0 & -a & 0 \\ a & d_2 & b \\ 0 & b & d_3 \end{pmatrix} = \begin{pmatrix} 0 & -.9873820478 & 0 \\ .9873820478 & 1.5978657524 & .4441323185 \\ 0 & .4441323185 & .2684762038 \end{pmatrix}$$

The non-zero elements of Y are given by: $\gamma = \gamma_2$,

$$a^2 = q_2 - q_4 , \quad d_3 = q_1/a^2 , \quad d_2 = q_3 - d_3 , \quad b^2 = d_2 d_3 - q_4 .$$

Example 2. SIRK5, a singly implicit, algebraically stable RK-method of order 5, which is zero-stable at infinity:

$$Y = \begin{pmatrix} d_1 & -a & b \\ a & d_2 & c \\ b & -c & d_3 \end{pmatrix} = \begin{pmatrix} .3816347293 & -.2236860783 & .3403914712 \\ .2236860783 & .1000000000 & .2426100710 \\ .3403914712 & -.2426100710 & .4086344763 \end{pmatrix}$$

Here we have $\gamma = \gamma_1$, $d_2 = 0.1$ (d_2 is in fact a free parameter),

$$d_1 = q_3 - q_5 , \quad d_3 = q_5 - d_2 , \quad c^2 = q_4 - d_2 d_3 \quad (c \geqq 0)$$

and a, b are solutions of the hyperbolic equations

$$b^2 - a^2 = q_4 + d_1 q_5 - q_2 , \quad d_2 b^2 + 2cab - d_3 a^2 = d_1 q_4 - q_1 .$$

References

[1] O. Axelsson, A class of A-stable methods. BIT 9 (1969), 185 - 199.

[2] K. Burrage, Efficiently implementable algebraically stable Runge-Kutta methods. University of Auckland, Dept. of Math., Report Series No 138, Nov. 1978 (to be published in SIAM J. Num. Anal.)

[3] K. Burrage and J. C. Butcher, Stability criteria for implicit Runge-Kutta methods. SIAM J. Num. Anal. 16(1), (1979), 46 - 57.

[4] J. C. Butcher, Implicit Runge-Kutta processes. Math. Comp. 18 (1964), 50 - 64.

[5] J. C. Butcher, On the implementation of implicit Runge-Kutta methods. BIT 16 (1976), 237 - 240.

[6] F. H. Chipman, A-stable Runge-Kutta processes. BIT 11 (1971), 384 - 388.

[7] M. Crouzeix, Sur la B-stabilité des méthodes de Runge-Kutta. Numer. Math. 32 (1979), 75 - 82.

[8] G. Dahlquist, A special stability problem for linear multistep methods. BIT 3 (1963), 27 - 43.

[9] G. Dahlquist, Error analysis for a class of methods for stiff non-linear initial value problems. Numerical Analysis, Dundee 1975, Springer Lecture Notes in Math., Nr. 506, 60 - 74.

[10] G. Dahlquist and R. Jeltsch, Generalized disks of contractivity for explicit and implicit Runge-Kutta methods. Royal Institute of Technology, Stockholm, Sweden, Report TRITA-NA-7906 (1979).

[11] B. L. Ehle, On Padé approximation to the exponential function and A-stable methods for the numerical solution of initial value problems. Research Report CSRR 2010, Dept. AACS, University of Waterloo (1969).

[12] E. Hairer, Highest possible order of algebraically stable diagonally implicit Runge-Kutta methods. BIT 20 (1980).

[13] E. Hairer and G. Wanner, Algebraically stable and implementable Runge-Kutta methods of high order. SIAM J. Num. 18, 1098 (1981).

[14] H. O. Kreiss, Difference methods for stiff ordinary differential equations. SIAM J. Num. Anal. 15 (1978), 21 - 58.

[15] S. P. Nørsett, Semi explicit Runge-Kutta methods. Report No 6/74, University of Trondheim, Norway (1974).

[16] J. Sand, A note on a differential system constructed by H. O. Kreiss. Royal institute of Technology, Stockholm, Sweden, Report TRITA-NA-8004 (1980).

[17] G. Wanner, All A-stable methods of order ≧ 2m-4 . to appear in BIT.

E. Hairer
Inst. für Angewandte Mathematik
Universität Heidelberg
Im Neuenheimer Feld 293
D-6900 Heidelberg 1 , Germany

G. Wanner
Université de Genève
Section de Mathématiques
Case postale 124
CH-1211 Genève 24 , Suisse

COMPACT DEFERRED CORRECTION FORMULAS

by

Bengt Lindberg

Dept. of Computer Science

The Royal Institute of Technology

Stockholm

Sweden

Abstract

A new kind of deferred correction formulas are presented and applied to two-point boundary value problems for ordinary differential equations. The compact formulas can be considered to be generalizations of the Collatz Mehrstellenverfahren obeying certain side conditions to make them suitable for iterative deferred corrections. The ideas presented can also be applied to other types of discretization algorithms, e.g. to discretizations of elliptic boundary value problems in several variables.

Acknowledgment

This work was done at the Zentrum für interdisziplinäre Forschung, University of Bielefeld, within the program "Properties and Reactions of Isolated Atoms and Molecules".

1. Introduction

Several techniques for iteratively improving the order of accuracy of discretization algorithms have been proposed during the last decades. See Fox [1], Pereyra [11], [12], Lentini, Pereyra [7], Lindberg [8], Frank, Hertling, Ueberhuber [5], Stetter [14], Daniel, Martin [3], Keller, Pereyra [6]. For a historical survey and a treatment with unifying new theoretical results see Skeel [13].

The main idea behind these techniques is to estimate the local truncation error for a basic discretization algorithm from the supposedly smooth solution of the discrete problem and then compensate for it by solving a slightly perturbed discrete problem. The process can then be repeated.

In deferred correction as described and implemented in [3], [6], [7], [11], [12] the individual terms of the local error expansions are calculated and approximated by linear combinations of the solutions at different points.

In defect correction as described in [5], [14] polynomials interpolating the approximate solution are defined and then the local truncation error is estimated by applying the basic discretization formula to these polynomials or to a defect function calculated from the polynomials.

Lindberg [8] proves general theorems allowing for the two techniques above and also new types of local error estimation procedures. However, some types of discretization algorithms are not covered by these theorems.

All the estimation procedures above assume the existence of smooth error expansions for the basic discretization algorithms.

Skeel [13] gives theorems that cover wider classes of discretization algorithms witout the need for asymptotic error expansions for the basic discretization.

The work of this paper relies on theorem 4 of [8].

The idea behind the present paper is to use extensions of the Collatz Mehrstellenverfahren, see e.g. [2], to compute estimates of the local truncation error. Simultaneously increasingly accurate approximations to the derivatives of the solution are computed. The technique is described in detail for a scalar two-point boundary value problem in section 2 and 3. In section 4 a numerical example is given. Extentensions to other problems are sketched in section 5.

2. Algorithm

Consider the two-point boundary value problem

$$\begin{cases} y'' - f(x,y,y') = 0 \\ y(a) = A \quad y(b) = B \end{cases} \tag{1}$$

and the basic discretization

$$\begin{cases} x_n = a + nh \quad h = (b-a)/N \\ (y_{n+1} - 2y_n + y_{n-1})/h^2 - f(x_n,y_n,(y_{n+1} - y_{n-1})/2h) = 0 \\ \qquad n = 1,2,\ldots N-1 \\ y_0 = A \quad y_N = B \end{cases} \tag{2}$$

To keep this presentation at a fairly elementary level I will not introduce the formalism of [8]. To avoid confusion the following notational conventions will be used

y the exact solution of (1)

s the derivative of the exact solution of (1)

$z = (z_0, z_1, \ldots z_N)$

$t = (t_1, t_2, \ldots t_{N-1})$

$\bar{y} = (y(x_0), y(x_1), \ldots y(x_N))$

$\bar{s} = (y'(x_1), y'(x_2), \ldots y'(x_{N-1}))$

y^j an approximation to $\bar{y}$

s^j an approximatin to $\bar{s}$

The basic discretization (2) can be viewed as a special instance of

$$\phi(z,t) + r = 0$$

where ϕ is defined by

$$\begin{cases} x_n = a + nh \quad h = (b-a)/N \\ (z_{n+1} - 2z_n + z_{n-1})/h^2 - f(x_n,z_n,t_n) + r_n^1 = 0 \\ (z_{n+1} - z_{n-1})/2h - t_n + r_n^2 = 0 \qquad (3a) \\ \qquad n = 1,2,\dots N-1 \\ z_0 = A \quad z_N = B \qquad (3b) \end{cases}$$

and $r = (r_1^1, r_2^1,\dots r_{N-1}^1, r_1^2, r_2^2,\dots r_{N-1}^2)$.

We can view $\phi(z,t) = 0$ as a discretization of the equation

$$F(y,s) = 0$$

where F is defined b

$$\begin{cases} y'' - f(x,y,s) = 0 \\ y' - s = 0 \qquad (4) \\ y(a) = A \quad y(b) = B. \end{cases}$$

Note that for given r we can write (3a), (3b) as

$$\begin{cases} (z_{n+1} - 2z_n + z_{n-1})/h^2 - f(x_n,z_n,(z_{n+1}-z_{n-1})/2h + r_n^2) + r_n^1 = 0 \\ \qquad n = 1,2,\dots N-1 \\ z_0 = A \quad z_N = B \end{cases} \qquad (5)$$

This system of non-linear equations can be solved with the same technique as the system (2), e.g. Newton-iteration with a tridiagonal Jacobian matrix. From the solution z we then compute t from (3b).

To compute increasingly accurate approximations to the solution of (1) we solve the sequence of problems

$$\begin{cases} \phi(y^1,s^1) = 0 \\ \phi(y^j,s^j) - \phi(y^{j-1}, s^{j-1}) + \phi_j(y^{j-1}, s^{j-1}) = 0 \quad ; j = 2,3,\dots \end{cases} \qquad (6)$$

where ϕ_j, $j = 2,3,\ldots$ are defined such that at the exact solution $\bar{y},\bar{s}$ $\phi_j(\bar{y},\bar{s})$ is consistent with $F(y,s)$ with order of consistency $2j$, and such that at the exact solution the Frechet derivative of $\phi_j(z,t)$ with respect to z and t differ from the Frechet derivative of $\phi(z,t)$ at most by quantities that are of order 2 in h.

Then, if the original problem $F(y,s) = 0$ is sufficiently smooth we get

$$y_n^j = y(x_n) + \mathcal{O}(h^{2j}) \qquad j = 2,3,\ldots$$

$$s_n^j = y'(x_n) + \mathcal{O}(h^{2j}) \qquad j = 2,3,\ldots$$

This follows from theorem 4 of [8]. See also note 4 of the same paper.

3. Perturbation operators

To describe the perturbation operators ϕ_j, j = 2,3,... we will use the averaging operator μ and the central difference operator δ, see e.g. Björck, Dahlquist [1].

$$\delta^2 z_n = z_{n+1} - 2z_n + z_{n-1}$$

$$\mu\delta z_n = (z_{n+1} - z_{n-1})/2.$$

Note that the even powers of δ give symmetric expressions and that μ* (odd powers of δ) give anti-symmetric expressions.

Now define $\phi_j(z,t)$, j = 2,3,... by

$$\frac{1}{h^2}\,[\delta^2 z_n - h^2 f(x_n,z_n,t_n) + \sum_{k=2} a_k\,\delta^{2k} z_n + h \sum_{k=2} b_k\,\mu\delta^{2k-1} t_n$$

$$- h^2 \sum_{k=1} c_k\,\delta^{2k} f(x_n,z_n,t_n)]$$

$$\frac{1}{h}\,[\mu\delta z_n - h t_n + \sum_{k=2} A_k\,\mu\delta^{2k-1} z_n - h \sum_{k=1} B_k\,\delta^{2k} t_n$$

$$- h^2 \sum_{k=1} C_k\,\mu\delta^{2k-1} f(x_n,z_n,t_n)]$$

n = 1,2,...N-1.

For b_k, c_k, B_k, C_k all zero the same type of perturbation operators that are used in [3], [5], [6], [7], [11], [12] are obtained. We can view the expressions above as perturbations for the discretization operators δ^2 and $\mu\delta$ respectively.

The upper limits in the sums and the constants are choosen such that $\phi_j(\bar{y},\bar{s}) = O(h^{2j})$. To determine the constants we first observe that at the exact solution we have

$$s(x_n) = y'(x_n)$$

$$f(x_n,y(x_n),s(x_n)) = y''(x_n).$$

Now define

$$L_1(y) = (\delta^2 y(x_n) - h^2 y''(x_n) + \sum_{k=2} a_k \delta^{2k} y(x_n)$$
$$+ h \sum_{k=2} b_k \mu\delta^{2k-1} y'(x_n) - h^2 \sum_{k=1} c_k \delta^{2k} y''(x_n))/h^2 \tag{7a}$$

$$L_2(y) = (\mu\delta y(x_n) - hy'(x_n) + \sum_{k=2} A_k \mu\delta^{2k-1} y(x_n)$$
$$- h \sum_{k=1} B_k \delta^{2k} y'(x_n) - h^2 \sum_{k=1} C_k \mu\delta^{2k-1} y''(x_n))/h \tag{7b}$$

and observe that we need to choose the coefficients in the formulas for ϕ_j such that

$$L_1(y) = \mathcal{O}(h^{2j})$$

$$L_2(y) = \mathcal{O}(h^{2j})$$

i.e. such that the linear functionals L_1 and L_2 are accurate of order 2j.

We are here interested in formulas that are as compact as possible, i.e. formulas that achieve the wanted accuracy with as few adjacent points as possible.

With some elementary operator calculus, cf. Lindberg [9] one can easily show that with the upper summation limit m in all the sums in (7a), (7b) we get the following maximal orders p_1 and p_2 for L_1 and L_2 respectively.

m	p_1	p_2	width
1	4	6	3
2	10	12	5
3	16	18	7
4	22	24	9

Table 1

The colums width specifies the number of adjacent points needed in the formula. Remember that all the formulas use points symmetrically placed around $x = x_n$. For intermediate orders of accuracy, like 8, we can get different formulas with the same width by specifying the values of some of the coefficients in a formula for a suitable m, e.g. m = 2 for 2j = 8.

In the numerical example of section 4 the coefficients in tables 2 and 3 were used. Further coefficients together with derivations of the systems of linear equations defining the coefficients are given in [9].

j	a_2	c_1	c_2	order
2		1/12		4
3	54/1080	12/90		6
4	465/3780	780/3780	23/3780	8

Table 2

j	A_2	B_1	B_2	order
2		1/6		4
3	1/30	1/5		6
4	25/210	60/210	3/210	8

Table 3

To be able to use the same perturbation formulas at points close to the boundary the numerical solution is extended sufficiently far outside the interval (a,b) using the basic discretization formula of (2) in a forward step and backward step fashion at the right-hand side and the left-hand side respectively. See Keller, Pereyra [6] for further discussion of this extension technique.

4. Numerical example

The equation of the catenary [5]

$$y'' = \sqrt{1+(y')^2}$$

$$y(-1) = \cosh(-1) \quad y(1) = \cosh(1)$$

with the very smooth solution y = cosh(x) was solved with the technique described in the previous sections. With the perturbation operators ϕ_j, j = 2,3,4 of tables 2 and 3 the results below were obtained. The calculations were done in Algol 60 on the University of Bielefeld TR-440 with round off unit $2 \cdot 10^{-12}$.

j	h = 1/4	h = 1/8	quotient
1	$2.4\ 10^{-3}$	$5.9\ 10^{-4}$	4.1
2	$-3.1\ 10^{-5}$	$-1.9\ 10^{-6}$	16.3
3	$7.4\ 10^{-7}$	$1.1\ 10^{-8}$	67.3
4	$-2.9\ 10^{-8}$	$-2.9\ 10^{-10}$	100.0

Table 4 Maximal errors for the improved solutions

The maximal errors occured at x = 0 in all cases. It is obvious from the table that the order of the j-th approximation, j = 2,3 is 2j while for j = 4 we get approximately order 7. However, the result for j = 4, h = 1/8 is seriously afflicted by iteration and rounding errors.

5. Extensions

The technique described in this paper can equally well be applied to other problems and discretization methods.

If a non-uniform mesh is used the basic discretization and the perturbations would be different, but still compact correction formulas could be derived.

Other kinds of boundary conditions, like $A*y(a) + B*y'(a) + C = 0$ can also easily be handled.

For general linear or non-linear n-th order systems with the basic discretizations of [6] high order compact deferred correction formulas could also be derived.

Applications to Sturm-Liouville eigenvalue problems will be described in Lindberg [10].

In this section we shall present one more example. The notation will be the same as in section 2 and 3. The boundary conditions and the treatment of them will be omitted.

Consider the operator $F(y,s)$ defined by

$$\begin{cases} y'' - f(x,y) \\ & a \le x \le b \\ y' - s \end{cases}$$

with a basic discretization $\phi(z,t)$ of order 4. We use Collatz Mehrstellenverfahren combined with a first derivative approximation of order 4

$$\begin{cases} \delta^2 z_n/h^2 - f_n - \frac{1}{12}\,\delta^2 f_n \\ \\ \mu\delta z_n/h - t_n - \frac{1}{6}\,\delta^3 z_n/h. \end{cases}$$

Here we can compute perturbations ϕ_j, $j = 2,3,\ldots$ according to

$$\frac{1}{h^2}\,(\delta^2 z_n - h^2(1 + \frac{1}{12}\,\delta^2) f_n + \sum_{k=3} a_k\,\delta^{2k} z_n$$

$$+ h \sum_{k=3} b_k\,\mu\delta^{2k-1} t_n - h^2 \sum_{k=2} c_k\,\delta^{2k} f_n)$$

$$\frac{1}{h}\mu\delta z_n - ht_n - \frac{1}{6}\mu\delta^3 z_n + \sum_{k=3} A_k \mu\delta^{2k-1} z_n$$

$$- h \sum_{k=2} B_k \delta^{2k} t_n - h^2 \sum_{k=2} C_k \mu\delta^{2k-1} f_n)$$

and gain 4 orders of accuracy per iteration if the coefficients of the formulas are choosen such that

$$\phi_j(\bar{y},\bar{s}) = \mathcal{O}(h^{4j}) \qquad j = 2,3\ldots\ .$$

The coefficients are determined as in section 3 and [9].

Note that the n-th components of the Frechet derivatives $\phi'(\bar{y},\bar{s})$ applied to vectors $(\bar{u},\bar{w})$ are

$$\begin{cases} \delta^2 u((x_n)/h^2 - (1 + \frac{1}{12}\ \delta^2)\ \{\frac{\partial f}{\partial y}(x_n,y(x_n))u(x_n)\} \\ \mu\delta\ u(x_n)/h - w(x_n) - \frac{1}{6}\ \mu\delta^3\ u(x_n)/h\ . \end{cases}$$

Here

$$u = (u(x_0),u(x_1),\ldots u(x_N))$$

$$w = (w(x_1),w(x_2),\ldots w(x_{N-1}))$$

for smooth functions $u(x)$ and $w(x)$.

Further the Frechet derivative $F'(y,s)$ applied to sufficiently smooth functions (u,w) give

$$\begin{cases} u'' - \frac{\partial f}{\partial y}(x,y)x))u \\ u' - w. \end{cases}$$

From these expressions we get

$$\frac{1}{h^2}\delta^2 u(x_n) - (1 + \frac{1}{12}\ \delta^2)\ \{\frac{\partial f}{\partial y}(x_n,y(x_n))u(x_n)\} =$$

$$= (1 + \frac{1}{12}\ \delta^2)\{u''(x_n) - \frac{\partial f}{\partial y}(x_n,y(x_n))u(x_n)\} + \mathcal{O}(h^4)$$

$$(\mu\delta - \frac{1}{6}\ \mu\delta^3)u(x_n)/h - w(x_n) = u'(x_n) - w(x_n) + \mathcal{O}(h^4).$$

Hence in a certain sense $\phi'(\bar{y},\bar{s})$ approximates $F'(y,s)$ with error at most $O(h^4)$. Similarly we can show that also $\phi'_j(\bar{y},\bar{s})$ approximates $F'(y,s)$ with error at most $O(h^4)$.

With proper definitions of the table operators in theorem 4 of [8] we get

$$y_n^j = y(x_n) + O(h^{4j}) \qquad j = 2,3,\ldots$$

if y^j, $j = 1,2,\ldots$ are defined by

$$\phi(y^1,s^1) = 0$$

$$\phi(y^j,s^j) - \phi(y^{j-1},s^{j-1}) + \phi_j(y^{j-1},s^{j-1}) = 0 \qquad j = 2,3,\ldots .$$

In the same way as in section 3 we need to extend the basic solution outside the interval of interest.

To conclude we can say that for each occurence of a discretization operator in the basic discretization we substitute in the perturbation operator a more accurate discretization operator differing from the basic one only by terms that are of the same order in h as the order of accuracy of the basic method. One may consider each of the differential operators that has to be approximated individually and derive perturbations for them individually.

For example, for the two-dimensional elliptic boundary value problem

$$\frac{\partial^2 u}{\partial x^2} + \frac{\partial^2 u}{\partial y^2} + g(x,y,u,\frac{\partial u}{\partial x}, \frac{\partial u}{\partial y}) = 0$$

with a basic discretization

$$\delta x^2 u_{ij} + \delta_y^2 u_{ij} + h^2 g(x_i,y_j, u_{ij}, \mu_x \delta_x u_{ij}/h, \mu_y \delta_y u_{ij}/h) = 0$$

we can directly use the formulas derived in section 3 to define perturbation terms for the discretization operators.

$$\delta_x^2 , \delta_y^2 , \mu_x \delta_x , \mu_y \delta_y .$$

References

[1] G. Dahlquist and A. Björck, "Numerical Methods", Prentice Hall, Englewood Cliffs, 1974.

[2] L. Collatz, "The Numerical Treatment of Differential Equations", Springer-Verlag, Berlin, 1960.

[3] J. Daniel and A. Martin, "Numerov's method with deferred corrections for two-point boundary value problems", SIAM J. Num. Anal. 14 (1977), 1033-1050.

[4] L. Fox, "Numerical Solution of Two Point Boundary Value Problems", Clarendon Press, Oxford, 1957.

[5] R. Frank, J. Hertling and C.W. Ueberhuber, "An extension of the applicability of Iterated Deferred Corrections", Math. Comp. 31 (1977), 907-915.

[6] H.B. Keller and V. Pereyra, "Difference methods and deferred corrections for ordinary boundary value problems", SIAM J. Numer. Anal. 16 (1979), 241-259.

[7] M. Lentini and V. Pereyra, "A variable order finite difference method for nonlinear multipoint boundary value problems", Math. Comp. 28 (1974), 981-1004.

[8] B. Lindberg, "Error estimation and iterative improvement for discretization algorithms", To appear in BIT, also Report no UIUCDCS-R-76-820, Department of Computer Science, University of Illinois, Urbana, 1976.

[9] B. Lindberg, "Compact deferred correction formulas", Report TRITA-NA-80XX, Dept. of Numerical Analysis and Computing Science, The Royal Institute of Technology, Stockholm, Sweden, (1980).

[10] B. Lindberg, "High order approximations to eigensolutions of Sturm-Lionville problems by deferred corrections", Report TRITA-NA-80XX, Dept. of Numerical Analysis and Computing Science, The Royal Inst. of Technology, Stockholm, Sweden, (1980).

[11] V. Pereyra, "Iterated deferred corrections for non-linear operator equations", Numer. Math. 10 (1967), 316-323.

[12] V. Pereyra, "Iterated deferred corrections for nonlinear boundary value problems", Numer. Math. 11 (1968), 111-125.

[13] R.D. Skeel, "A theoretical framework for proving accuracy results for deferred corrections", Report no UIUCDCS-F-80-892, Dept. of Computer Science, Univ. of Illinois, Urbana, SIAM J. Num. 19, 171 (1982).

[14] H.J. Steter, "The defect correction principle and discretization methods", Numer. Math. 29 (1978), 425-443.

SOLVING ODES IN QUASI STEADY STATE

L. F. Shampine
Applied Mathematics Research Department
Sandia National Laboratories
Albuquerque, New Mexico 87185, U.S.A.

Abstract

Solving ordinary differential equations (ODEs) with solutions in a quasi steady state has been studied by computational chemists, applied mathematicians, and numerical analysts. Because of this, it is a very appropriate topic for this interdisciplinary workshop.

In this paper we shall first discuss what stiffness is for model problems arising in chemical kinetics. Chemists and applied mathematicians have made use of quasi steady state approximations (singular perturbation theory) to alter the problem so as to avoid stiffness. The approach is described and some difficulties noted. Numerical analysts have developed methods to solve general stiff ODEs. How they relate to the problem at hand is described and some difficulties pointed out. Finally, ideas from both approaches are combined. The new combination deals effectively with stiffness when the quasi steady state hypothesis is valid.

We are concerned with the integration of a system of N first order ordinary differential equations which in vector form are

(1) $$\frac{d\underline{y}}{dt} = \dot{\underline{y}} = \underline{f}(t,\underline{y}),\ \underline{y}(t_0) \text{ given}$$

We use underlining to denote vectors and dots to denote differentiation by the independent variable t. We shall refer to t as the "time." When no confusion seems likely, we shall suppress t in subsequent expressions. The time behavior of the concentration of chemical species reacting according to the mass action law is an important example we shall refer to continually. As formulated by Edsberg [1], the the equations are

(2) $$\dot{\underline{y}} = A\underline{p}$$

where A is a rectangular matrix with entries $A_{ij} = q_{ij} - r_{ji}$. Here the q_{ij} and r_{ji} are non-negative integers. The components p_j are the rate functions given by

(3) $$p_j = k_j \prod_i y_i^{r_{ji}}$$

where k_j is a rate constant. For some purposes terms are grouped so that

(4) $$\dot{y}_i = P_i(t,\underline{y}) - Q_i(t,\underline{y})y_i \qquad i = 1,\ldots,N.$$

The term P_i is the rate of production of y_i and the other term, the rate of loss. For (2), P_i and Q_i do not depend on t and P_i does not depend on y_i.

From the first numerical solutions of (2) on, puzzling difficulties were observed. It was recognized that the difficulties were often associated with very reactive species for which the production and loss rates were large but the time rates of change of the concentrations of the species were small -- they were nearly in equilibrium. This corresponds, at least for a short time interval, to P_i and Q_i being roughly constant and $Q_i >> 1$. We shall formulate a hypothesis about the behavior of the coefficients of a system (4).

Quasi steady state hypothesis (QSSH):

For at least one i in (4), P_i and Q_i are roughly constant and $Q_i >> 1$.

We are led to consider how numerical methods behave when applied to a set of equations

(5) $$\dot{y}_i = P_i - Q_i y_i \qquad i = 1,\ldots,N$$

with all the P_i and Q_i constant and $Q_i \neq 0$.

The forward Euler method is representative of classical methods like Runge-Kutta and Adams. A Taylor series expansion says that for any smooth function

(6) $$y_i(t_n+h) = y_i(t_n) + h\dot{y}_i(t_n) + \frac{h^2}{2}\ddot{y}_i(t_n) + \ldots$$

We wish to proceed from an approximation $y_{i,n}$ of the solution $y_i(t_n)$ to an approximation of $y_i(t_n+h)$, thus advancing the numerical solution one time step of length h. Euler's method uses the linear terms of the Taylor series. The fact that we are talking about a solution of a differential equation (1) appears when we use the relation

$$\dot{y}_i(t_n) = f_i(t_n,\underline{y}(t_n)).$$

Thus Euler's forward method is

(7) $$y_{i,n+1} = y_{i,n} + hf_i(t_n,\underline{y}_{\cdot,n}) \qquad i = 1,\ldots,N$$

It is plausible, and essentially true, that the error of (7) is the first term omitted in (6). It is called the local (truncation) error. The typical code receives from its user a desired accuracy tolerance and it attempts to adjust h so as to keep the magnitude of the local error less than this tolerance.

For the model problem (5) it is easy to understand what local error control means. First we notice that the equations have a steady state (equilibrium) solution with $\dot{y}_i \equiv 0$, namely $s_i = P_i/Q_i$. The solution of (5) with $y_i(t_n) = y_{i,n}$ is

$$y_i(t) = s_i + [y_{i,n} - s_i] \exp(-Q_i(t-t_n)) \qquad \text{for } t \geq t_n.$$

For $Q_i > 0$ there is a boundary layer (transition region) of rapid change as the solution approaches the steady state. The step size h is chosen so that

$$\left|\frac{h^2}{2}\ddot{y}_{i,n}\right| \leq \text{given constant}$$

at each step. Here

$$\ddot{y}_{i,n} = [y_{i,n} - s_i]\, Q_i^2.$$

If the numerical solution $y_{i,n}$ is far from the steady state, we see that

(8) $$h \sim \frac{1}{|Q_i|}.$$

As the steady state is approached, the step size permissable becomes arbitrarily large. In both cases this is just what we would expect from the behavior of the solution.

What about the stability of the method? The formula is

$$y_{i,n+1} = y_{i,n} + h[P_i - Q_i y_{i,n}].$$

A perturbation of $y_{i,n}$ by $\delta_{i,n}$ leads to a perturbation of $y_{i,n+1}$ by $[1 - hQ_i]\ \delta_{i,n}$. Numerical stability of the formula requires that perturbations not be amplified, hence that $|1 - hQ_i| \leq 1$. This means a condition like (8) for all t! This is extremely frustrating because we must use small step sizes for $Q_i \gg 1$ even when the solution is barely changing.

The stability restriction of Euler's forward method is typical of classical numerical methods and is one manifestation of stiffness. If the QSSH is valid, the model analysis says we shall suffer from stiffness. We may suffer from stiffness if the QSSH is not valid, but in the context of chemical kinetics it seems from experience that stiffness is usually accompanied by the QSSH.

One way to avoid stiffness is to change the problem. With the model problem it is pointless to continue integrating a solution component nearly in steady state. If one replaces the differential equation by an algebraic equation expressing equilibrium, he arrives at the QSS approximation (QSSA). In detail, if the QSSH holds for equation i

$$\dot{y}_i = P_i(t,\underline{y}) - Q_i(t,\underline{y})y_i,$$

it is replaced by

$$0 = P_i(t,\underline{y}) - Q_i(t,\underline{y})y_i.$$

The package [2] is an example of this for chemical kinetics. More generally, applied mathematicians have encountered many physical problems for which there is a natural parameter $\varepsilon > 0$ and the differential equation (1) can be written in partitioned form as

(9) $$\dot{\underline{u}} = \underline{g}(\underline{u},\underline{v}),$$
$$\varepsilon\dot{\underline{v}} = \underline{G}(\underline{u},\underline{v}),$$

and the solution $\underline{y}(t;\varepsilon) = (\underline{u}(t;\varepsilon),\ \underline{v}(t;\varepsilon))^T$. It is natural to consider the "reduced" problem

(10) $$\dot{\underline{u}} = \underline{g}(\underline{u},\underline{v}),$$
$$\underline{0} = \underline{G}(\underline{u},\underline{v}),$$

with solution $\underline{y}(t;0) = (\underline{u}(t;0),\ \underline{v}(t;0))^T$. The relation between the problems for

$\varepsilon > 0$ and $\varepsilon = 0$ can be quite complex. We shall be interested in those problems for which $\underline{y}(t;\varepsilon) \to \underline{y}(t;0)$ on some time interval of interest as $\varepsilon \to 0$. The solution $\underline{y}(t;0)$ is the "outer" approximation for $\underline{y}(t;\varepsilon)$ in a singular perturbation analysis. Of course the solution of the reduced problem cannot, in general, satisfy all the initial conditions of the full problem so it cannot be expected to be a good approximation in an initial boundary layer.

The QSSA is an outer approximation where we can identify ε as a characteristic magnitude for those Q_i with equation i satisfying the QSSH. Robertson [3] argues that the chemical kinetic equations (2) have the form (9) where the variables $\underline{v}$ are those species participating in "fast" reactions.

In the package [2] the user designates which species react "fast," and they are assumed in equilibrium for all time.

The QSSA has been widely used, but a number of serious objections have been raised. For one thing, how is the partition of fast and slow variables to be made? The reaction rates may span a wide range, but not fall into two distinct groups. A partition may change for nonlinear problems. Consider the famous van der Pol equation

$$\ddot{x} - \mu(1-x^2)\,\dot{x} + x = 0.$$

Written as a system with $y_1 = x$, $y_2 = \dot{x}$ we have

$$\dot{y}_1 = y_2$$
$$\dot{y}_2 = -y_1 + \mu(1-y_1^2)y_2$$

We are interested in relaxation oscillations with the parameter $\mu \gg 1$. Plots of the solution with $\mu = 1000$ can be found in [4,p.326]. This problem has the form (4), but the sign of Q_2 is wrong when $1-y_1^2 < 0$. All solution curves approach rapidly a steady oscillation which has a maximum of about 2. The QSSH is valid and the solution y_1 slowly decreases to about 1, at which point the character changes and there is an almost discontinuous drop to about -2. Once again the QSSH is valid and there is a slow increase of y_1 to about -1, followed there by an almost discontinuous jump to about +2. It is clear that application of the QSSA is not straightforward.

It is not apparent how to estimate the error of the QSSA even when it is applicable. It is still less clear how to handle boundary layers. The chemical kinetic equations allow linear conservation laws, e.g. mass balance. These are preserved exactly by the typical numerical methods but not by the QSSA.

Because of all these factors, using the QSSA sometimes leads to results which are qualitatively incorrect. Workers have devised variants designed to respond to some of the difficulties and research continues. As we shall see, numerical analysts in the meantime produced a reliable alternative.

Stiff differential equations arise in the most varied contexts, see e.g. the survey [5]. Numerical analysts have tried to respond to the need to solve such problems by developing formulas which do not suffer the stability limitations of the forward Euler. The backward Euler formula is an example. It is derived similar to the forward formula with the derivative taken at the other end of the step:

(11) $$y_{i,n+1} = y_{i,n} + hf_i(\underline{y}_{\cdot,n+1}).$$

If $y_{i,n}$ is perturbed by $\delta_{i,n}$, the result $y_{i,n+1}$ for the model problem (5) is perturbed to

$$y_{i,n+1} + \frac{\delta_{i,n}}{1 + h\,Q_i}.$$

Thus the formula is stable for all $Q_i > 0$, and there is no restriction on the size of h.

The backward Euler and similar formulas do away with the stability limitation, more or less successfully, but a price must be paid. As (11) exemplifies, they are all implicit, i.e. a system of algebraic equations must be solved at each step for the new approximation $\underline{y}_{\cdot,n+1}$. The generic form is

(12) $$\underline{z} = \psi + h\gamma\underline{f}(\underline{z}) \text{ or } \underline{F}(\underline{z}) = \underline{z} - h\gamma\underline{f}(\underline{z}) - \underline{\psi} = \underline{0}.$$

Here $\underline{z}$ represents the new solution approximation, $\underline{\psi}$ comes from previously computed quantities, γ is a constant determined by the formula, and h and $\underline{f}$ have their usual meanings. Let $\underline{z}^*$ denote a solution of the algebraic equations (12).

At first implicit formulas were evaluated by simple or functional iteration

(13) $$\underline{z}^{(m+1)} = \underline{\psi} + h\gamma\underline{f}(\underline{z}^{(m)}).$$

It is possible using past data to predict efficiently a starting approximation $\underline{z}^{(0)}$. For <u>non</u>-stiff problems the iteration is very efficient. If we apply it to the model problem (5), we easily find that

$$(z_i^{(m+1)} - z_i^*) = -h\gamma Q_i(z_i^{(m)} - z_i^*).$$

This means that we must have $|h\gamma Q_i| < 1$ to get convergence and the rate $R_1 = |h\gamma Q_i|$. We have not gained a thing, because once again we have a restriction on h like (8). The answer is to go to a Newton iteration scheme. If we linearize $\underline{f}(\underline{z})$ about some iterate $\underline{z}^{(m)}$, we are led to Newton's iteration

$$\underline{z}^{(m+1)} = \underline{\psi} + h\gamma\underline{f}(\underline{z}^{(m)}) + \mathcal{J}(\underline{z}^{(m)})(\underline{z}^{(m+1)} - \underline{z}^{(m)}).$$

Here $\mathcal{J}$ is the Jacobian matrix $\left(\frac{\partial f_i}{\partial y_j}\right)$. It is too expensive to reevaluate $\mathcal{J}$ at each iterate, so codes form an approximation J and use it as long as convergence is adequate. The resulting iteration is actually carried out in the form

(14) $$M(\underline{z}^{(m+1)} - \underline{z}^{(m)}) = -\underline{F}(\underline{z}^{(m)})$$

where

(15) $$M = I - h\gamma J.$$

Let

(16) $$H(\underline{z}^*) = I - M^{-1}(I - h\gamma\mathcal{J}(\underline{z}^*)).$$

With some smoothness assumptions, if $\|H(\underline{z}^*)\|_\infty \leq r < 1$, then the iteration (14) converges for all $\underline{z}^{(0)}$ sufficiently close to $\underline{z}^*$ and the (linear) rate of convergence is at least r. Here

$$\|H\|_\infty = \max_i \sum_{j=1}^{N} |H_{ij}|.$$

Notice that if J is sufficiently close to $\mathcal{J}(\underline{z}^*)$ and (15) is used, the iteration converges and does so rapidly.

The procedure described is closely related to the singular perturbation approach. The algebraic equations (12) for the differential equations (9) assume the partitioned form

$$\underline{F}(\underline{z}) = \begin{pmatrix} \underline{z}_1 - h\gamma\underline{g}(\underline{z}_1,\underline{z}_2) - \underline{\psi}_1 \\ \\ \underline{z}_2 - \frac{h\gamma}{\varepsilon}\underline{G}(\underline{z}_1,\underline{z}_2) - \underline{\psi}_2 \end{pmatrix} = \underline{0}$$

Scaling the second set of equations by $\varepsilon/h\gamma$ does not affect the solution, but reveals what is going on:

$$\begin{pmatrix} \underline{z}_1 - h\gamma\underline{g}(\underline{z}_1,\underline{z}_2) - \underline{\psi}_1 \\ \frac{\varepsilon}{h\gamma}\underline{z}_2 - \underline{G}(\underline{z}_1,\underline{z}_2) - \frac{\varepsilon}{h\gamma}\underline{\psi}_2 \end{pmatrix} = \underline{0}$$

If ε were 0, these would be the algebraic equations arising from application of the integration formula to the reduced problem (10). Thus the general approach does much the same thing as the singular perturbation approach _when the latter is valid_, but it is generally applicable and reliable.

The general approach to stiff ODEs described is quite effective, but the task is far more difficult and expensive than the solution of non-stiff problems. For one thing, it is necessary to approximate $\mathcal{J}(\underline{z}^*)$. This is often very expensive and/or a lot of trouble. For chemical kinetics it is neither. Recall that $\dot{\underline{y}} = A\underline{p}$ where A is constant. Because

$$\frac{\partial p_i}{\partial y_j} = r_{ij} \frac{p_i}{y_i}$$

it is easy and inexpensive to form the Jacobian. The storage required for the solution of the differential equation goes from a multiple of N for non-stiff problems to a multiple of N^2 because of the Jacobian. This can be very serious, even prohibitive.

The costs of the linear algebra, which have no analog for non-stiff problems, can be quite important. The main cost is that of decomposing M. Repeated solution of the linear systems in (14) can also be a significant amount of work.

The costs mentioned lead to a less satisfactory adaptation of step size h and method (as reflected in γ) than in the non-stiff case. This is because changes of h and/or γ may force a factorization of a new M and possibly the formation of a new J.

For the reasons sketched the methods for stiff problems are much more expensive per step than the methods for non-stiff problems. Despite this, fantastic savings are possible for stiff problems because the step size can be so much larger. Naturally a great deal of research is being devoted to reducing still more the costs of these procedures.

The general approach furnishes a reliable, effective solution procedure, but takes no advantage of special structure. It is quite plausible that with equations in the production-loss form (4), one can do a much better job when the QSSH is valid. We shall suggest here ways to do this.

There are several ways to motivate the first two variations on a theme that we propose. For brevity we take a rather formal approach. First we note that the iteration matrix M does not have to arise directly from an approximation to $I - h\gamma\mathcal{J}$. It is only necessary that it be a reasonably good approximation. Second we note that most of the disadvantages listed for the general approach would disappear if M were a diagonal matrix. Thus formally we suggest a variant which uses the diagonal of the matrix arising from Newton's method:

V1: Take M in (14) to be $\mathrm{diag}\{1 - h\gamma\frac{\partial f_i}{\partial y_i}(\underline{z}^{(m)})\}$.

The iteration for solving (12) is then

$$\left(1 - h\gamma\frac{\partial f_i}{\partial y_i}(\underline{z}^{(m)})\right)(z_i^{(m+)} - z_i^{(m)}) = -F_i(\underline{z}^{(m)}).$$

This form is to be used mainly for theoretical purposes. A more practical variant is:

V2: Take M in (14) to be $\mathrm{diag}\{1 - h\gamma J_{ii}\}$.

Thus we are simply using the diagonal of the usual iteration matrix (15).

We shall state a convergence result, but first let us consider why it might suffice to use a diagonal matrix. We are interested in

(4) $$\dot{y}_i = P_i(t,\underline{y}) - Q_i(t,\underline{y})y_i = f_i(t,\underline{y}) \qquad i = 1,\ldots,N.$$

The QSSH is that for some i, P_i and Q_i are roughly constant, which we take to mean that their partial derivatives are all "small." Also $Q_i >> 1$. This means that

QSSH: $$\frac{\partial f_i}{\partial y_i} << -1, \quad \left|\frac{\partial f_i}{\partial y_i}\right| >> \left|\frac{\partial f_i}{\partial y_j}\right| \quad j \neq i.$$

That is, the diagonal element of the Jacobian dominates strongly the off-diagonal elements in a row corresponding to an equation for which the QSSH is valid.

The form (4) suggests another iteration which has been noticed by a number of authors. Namely, to solve (12), which is for (4)

$$F_i(\underline{z}) = z_i - \psi_i - h\gamma[P_i(t,\underline{z}) - Q_i(t,\underline{z})z_i] = 0,$$

use the iteration

V3: $$(1+h\gamma Q_i(t,\underline{z}^{(m)}))(z_i^{(m+1)} - z_i^{(m)}) = -F_i(\underline{z}^{(m)}).$$

If the QSSH is valid, V3 is essentially the same as V1, but perhaps more convenient. The relationship is especially close for the chemical kinetic equations (2). In general

$$\frac{\partial f_i}{\partial y_i} = -\frac{\partial P_i}{\partial y_i} + \frac{\partial Q_i}{\partial y_i} y_i + Q_i.$$

For (2) we group so that $\partial P_i/\partial y_i = 0$. The QSSH assumes that $\partial Q_i/\partial y_i$ is "small." Moreover Robertson [3] has argued that the very reactive species for which the QSSH should be valid should also have y_i "small."

If we apply the convergence result already stated to the M of V1, we have the sufficient condition for convergence at rate r

$$\|H\|_\infty = \max_i \frac{\sum_{j\neq i} h\gamma|\mathscr{J}_{ij}|}{|1-h\gamma\mathscr{J}_{ii}|} < r,$$

where we write

$$\mathscr{J}_{ij} = \frac{\partial f_i(\underline{z}^*)}{\partial y_j}$$

Simpler sufficient conditions are that for all $i = 1,\ldots,N$

(17a) if $\mathscr{J}_{ii} > 0$, $h\gamma \sum_j |\mathscr{J}_{ij}| < r$

(17b) if $\mathscr{J}_{ii} < 0$, $\sum_{j\neq i} |\mathscr{J}_{ij}| < r|\mathscr{J}_{ii}|$.

The first condition is the same kind of condition as arises from simple iteration, but the second is quite noteworthy because it does not restrict h <u>at all</u>. If the QSSH is valid for equation i, condition (17b) obviously holds, and the step size is not restricted. Notice that the magnitudes of the entries of the Jacobian in such a row do not matter, just that the diagonal entry be negative and dominate. In particular, for a singularly perturbed equation

$$\varepsilon \dot{y}_i = f_i(t,\underline{y}),$$

the size of ε does not matter. If (17b) holds, $\varepsilon \to 0$ does not make the algebraic problem harder to solve by the iteration V1. It is easy to establish similar results for V2 and V3.

Two different ways of using these results suggest themselves. Suppose we are using a general purpose code with $M = I - h\gamma J$. We can test (17) to see if the diagonal of M would suffice to yield the desired rate of convergence r with the step size h. If it will, we use V2. This avoids a matrix decomposition and the cost of repeatedly solving linear systems, and we can adjust the step size and formula as we please. Thus if the QSSH should at any time be valid, we can take full advantage of it. If the QSSH is not valid, or is not responsible for the stiffness, the problem is solved in the usual, reliable way.

Alternatively one could ask for the problem in the form (4) and always use the iteration V3 with a suitable formula. The author is pursuing this matter with a code based on both explicit and implicit Runge-Kutta methods. For simplicity suppose we have an efficient, highly stable, implicit Runge-Kutta formula. Because V3 is essentially simple iteration for non-stiff problems, we can expect to do about as well with it as is possible for non-stiff problems. In addition, we solve efficiently a class of stiff problems. It is important to realize that it is the nature of the stiffness and not how stiff a problem is that counts in such a code. We regard such a code as an extension of classical Runge-Kutta codes. Young and Boris [6,7] have been concerned with reactive flow problems. Storage is so critical there that they believe the general purpose codes cannot be used. Furthermore, based on a combination of physical insight and experience, they believe that for many important problems, the QSSH is valid and that it is the source of the stiffness. They use a refined QSSA which overcomes some of the objections raised, but not all. The code outlined here removes all the objections Young and Boris raise in connection with general purpose methods and, of course, does not experience the difficulties of the QSSA. For the problems treated by Young and Boris, the approach presented here appears to be extremely attractive.

References

[1] Edsberg, L., Integration package for chemical kinetics, in R. A. Willoughby, ed., Stiff Differential Systems, Plenum Press, New York, 1974.

[2] Clark, R. L. and G. F. Groner, A CSMP/360 precompiler for kinetic chemical equations, Simulation, 19(1972) 127-132.

[3] Robertson, H. H., Numerical integration of systems of stiff ordinary differential differential equations with special structure, in G. Hall and J. M. Watt, eds., Modern Numerical Methods for Ordinary Differential Equations, Clarendon Press, Oxford, 1976.

[4] Villadsen, J., and M. L. Michelsen, Solution of Differential Equation Models by Polynomial Approximation, Prentice-Hall, Englewood Cliffs, NJ, 1978.

[5] Bjurel, G., et al., Survey of stiff ordinary differential equations, Rept. NA 70.11, Dept. Inf. Proc. Comp. Sci., Royal Inst. Tech., Stockholm, 1970.

[6] Young, T. R., and J. P. Boris, A numerical technique for solving stiff ordinary differential equations associated with the chemical kinetics of reactive-flow problems, J. Phys. Chem., 81 (1977) 2424-2427.

[7] Young, T. R., CHEMEQ - a subroutine for solving stiff ordinary differential equations, Rept. NRL 4091, Naval Res. Lab., Washington, D.C., 1980.

A SINGULAR PERTURBATIONS APPROACH TO REDUCED-ORDER MODELING AND DECOUPLING FOR LARGE SCALE LINEAR SYSTEMS

Robert E. O'Malley, Jr.
Program in Applied Mathematics
University of Arizona
Tucson, AZ 85721

Singular perturbation concepts have been very helpful in control and systems theory, both in eliminating difficulties associated with stiffness and in reducing dimensionality (cf. Kokotovic et al. (1976) and O'Malley (1978)). In this short report, we wish to emphasize the second aspect, especially in the context of large-scale problems. Such an approach has been applied to power system modeling (cf. Kokotovic et al. (1979)) and to aircraft maneuvering and structural dynamics (cf. Anderson (1978) and Anderson and Hallauer (1980)).

Let us first consider initial value problems on a bounded interval, $0 \leq t \leq T$, for linear systems of the form

$$\dot{x} = Ax + Bu , \tag{1}$$

where A and B are constant matrices and the control $u(t)$ is considered, for simplicity in the present discussion, to be given.

A low-order example is provided by a model of the longitudinal dynamics of an F-8 aircraft (cf. Etkin (1972) and Teneketzis and Sandell (1977)). It involves four states x_i and one control u, the elevator deflection. Two of the four states, velocity variation and flight path angle, are described through physical intuition as being "primarily (or predominantly) slow" variables compared to the other "primarily fast" states, angle of attack and pitch rate. It is a natural temptation in such a situation to seek to approximate the solution away from $t = 0$ by neglecting the dynamics of the primarily fast variables and to integrate only the resulting initial value problem for the slow states. (This has often been proposed in the engineering literature, cf., e.g., Harvey and Pope (1976)). The trouble with this approach is that the physical coordinates are not actually provided separately as slow and fast modes, in contrast to the situation for the traditional singularly perturbed system $\dot{\alpha} = f(\alpha,\beta,t)$, $\varepsilon\dot{\beta} = g(\alpha,\beta,t)$ where β can respond on a much faster time-scale than α as $\varepsilon \to 0$.

The analytical solution of the linear system (1) is commonly expressed using variation of parameters and the matrix exponential solution e^{At} of the homogeneous system. Moler and van Loan's 1978 article "Nineteen dubious ways to compute the exponential of a matrix" illustrates the difficulties involved in computationally exploiting specific representations for this fundamental matrix. The difficulties are compounded when the dimensions of A are large. We, nonetheless, recall that the columns of e^{At} are of the form $v_i(t)e^{\lambda_i t}$ where v_i is a vector polynomial in t in the span of the eigenspace corresponding to the eigenvalue λ_i of A. We shall seek an approximation away from $t = 0$ which avoids doing a complete eigenanalysis of A. Our approximation will involve neglecting modes corresponding to large stable eigenvalues of A. This corresponds to the reduced order model in singular perturbations (cf. O'Malley (1974)) and to the smooth approximate solution of the stiff equations literature (cf. Dahlquist (1969) and Oden (1971)).

For the F-8 aircraft model, we have the two "slow" (i.e., small in magnitude) eigenvalues $s_{1,2} = (-0.75 \pm i7.6) \times 10^{-2}$ compared to the two "fast" (i.e., large) eigenvalues $f_{1,2} = -0.94 \pm i3.0$. The smallness of the parameter $\mu = |s_2/f_1| = 0.024$ indicates a large time-scale separation between slow and fast modes, and the smallness of $\varepsilon = -(1/\text{Re } f_1)/T = 1.06/T$ for a long enough time interval $0 \leq t \leq T$ provides fast-mode stability and a rapid transition to pseudo-steady state behavior. We shall determine a second-order nonstiff model for the solution away from $t = 0$. A basic question is what initial values should be used for the second-order problem. (The need to "project" the initial vector occurs in many related situations (cf. O'Malley and Flaherty (1980), Lentini and Keller (1980), and de Hoog and Weiss (1979))). Our approximation, based on asymptotic methods, will improve as the parameters ε and μ both tend toward zero. In practice, however, there is often a need to use such reduced-order methods even when these parameters are only moderately small. In contrast, our related suggestion of "equilibrating" the primarily fast states will produce an error which persists for all $t > 0$ (cf. Figure 1).

In the applications of interest, very high dimensional systems are common. We shall seek approximate nonstiff models of (differential) order $n_1 << n$, the dimension of x, where n_1 is the number of "slow", or small in magnitude, eigenvalues of A compared to its other (large) eigenvalues. Systems, for which the eigenvalues of the state matrix A can be so divided, will be called <u>two-time-scale</u>. Finer subdivisions of eigenvalues into slow, fast, and very fast categories could be treated by repeating our procedure. For our initial value problems, we'll also assume <u>fast-mode-stability</u>, i.e. that the $n_2 = n - n_1$ large eigenvalues of A also have large negative real parts. We are thereby eliminating consideration of systems with large, purely imaginary

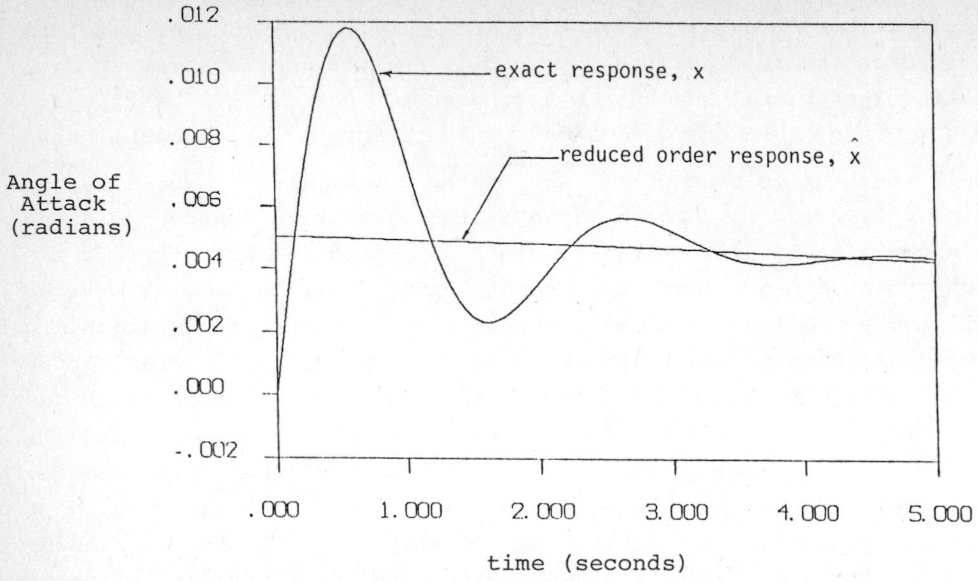

FIGURE 1. F-8 aircraft model: Angle of attack (a fast variable) vs. time.

eigenvalues. Their rapidly oscillatory solutions require techniques (cf. Miranker and Wahba (1976) or Petzold (1978)) much different from the boundary-layer method we wish to develop.

One moderate-size problem we've solved is a recently popular contest problem in the control literature (cf. Sain (1977) and De Hoff and Hall (1979)). It is a sixteenth order model for a turbofan engine. Indeed, it is one of thirty six different linearizations used to simulate the engine. The sixteen eigenvalues of A are all stable and have magnitudes ranging from 0.65 to 577. Their distribution suggests using either a third or a fifth order reduced model. We obtained very good approximations in both cases for $t > 10\varepsilon$, with ε being determined in terms of the decay rate of the smallest large eigenvalue (cf. O'Malley and Anderson (1980)).

To proceed, let us transform the constant coefficient problem

$$\dot{x} = Ax + Bu \tag{1}$$

into a new problem

$$\dot{y} = \tilde{A}y + \tilde{B}u \ , \tag{2}$$

where the fast and slow modes of the homogeneous problem are decoupled.

Specifically, let

$$y = \mathcal{T}x \ , \tag{3}$$

where

$$\tilde{A} = \mathcal{T}A\mathcal{T}^{-1} = \begin{pmatrix} \tilde{A}_{11} & 0 \\ 0 & \tilde{A}_{22} \end{pmatrix}$$

and

$$\tilde{B} = \mathcal{T}B$$

for

$$\mathcal{T} = \mathcal{T}_1\mathcal{T}_2 = \begin{pmatrix} I_{n_1} & 0 \\ L & I_{n_2} \end{pmatrix} \begin{pmatrix} I_{n_1} & K \\ 0 & I_{n_2} \end{pmatrix} . \tag{4}$$

Here, $\tilde{A}_{11}$ shall have the n_1 small eigenvalues of A and $\tilde{A}_{22}$ its n_2 (relatively) large eigenvalues (with large negative real parts as well). We note that

$$\mathcal{T}^{-1} = \mathcal{T}_2^{-1}\mathcal{T}_1^{-1} = \begin{pmatrix} I_{n_1} & -K \\ -L & I_{n_2} + LK \end{pmatrix} , \tag{5}$$

since, e.g., $\mathcal{T}_1^{-1} = \begin{pmatrix} I_{n_1} & 0 \\ -L & I_{n_2} \end{pmatrix}$. Partitioning $A = \begin{pmatrix} A_{11} & A_{12} \\ A_{21} & A_{22} \end{pmatrix}$ compatibly,

$$\mathcal{T}_1 A \mathcal{T}_1^{-1} = \begin{pmatrix} \tilde{A}_{11} & A_{12} \\ 0 & \tilde{A}_{22} \end{pmatrix} \equiv \begin{pmatrix} A_{11} - A_{12}L & A_{12} \\ 0 & A_{22} + LA_{12} \end{pmatrix} \tag{6}$$

provided the $n_2 \times n_1$ matrix L satisfies the algebraic Riccati equation

$$LA_{11} - A_{22}L - LA_{12}L + A_{21} = 0 \ . \tag{7}$$

(The connection to invariant imbedding is signaled by the entrance of a Riccati equation.) This matrix quadratic equation will have many solutions, but only one which decouples the fast modes of the homogeneous problem, as in (6). Indeed, if the spectrum of $\tilde{A}_{11}$ coincides with the slow eigenvalues λ_{s_i}, $i = 1,\ldots,n_1$, of A, i.e. if

$$\lambda(\tilde{A}_{11}) = \left\{\lambda_{s_1},\ldots,\lambda_{s_{n_1}}\right\} , \tag{8}$$

it is easy to show that

$$L = -M_{21}M_{11}^{-1} , \tag{9}$$

where the $n \times n_1$ matrix $\begin{bmatrix} M_{11} \\ M_{21} \end{bmatrix}$ spans the n_1 dimensional slow eigenspace of A (we rearrange the rows of A, if necessary, so that M_{11} is invertible) (cf. Anderson (1978), Medanic (1979), and van Dooren (1980) (which develops numerically stable computational methods)). Rather than compute this eigenspace, we prefer to iterate in a rearranged version of the Riccati equation (7) and obtain L as the limit of iterates

$$L_{i+1} = (A_{22} + L_i A_{12})^{-1}(L_i A_{11} + A_{21}) \ . \qquad (10)$$

Not unexpectedly, this procedure relates closely to Stewart (1975)'s method of finding the dominant eigenspace of A. The iteration can be shown to be robust with respect to its initialization, and its rate of convergence is proportional to the time-scale separation parameter μ, i.e. the ratio of the largest slow eigenvalue in magnitude to the smallest fast eigenvalue in magnitude.

Using the L so obtained, we will next block diagonalize the $\tilde{A}$ of (2) provided the $n_1 \times n_2$ matrix K satisfies the linear (Liapunov) equation

$$K\tilde{A}_{22} - \tilde{A}_{11}K + A_{12} = 0 \ . \qquad (11)$$

Since $\tilde{A}_{11}$ and $\tilde{A}_{22}$ have no eigenvalues in common, the linear system has a unique solution (cf., e.g., Bellman (1970)) which can be found as the limit of the iteration

$$K_{j+1} = (\tilde{A}_{11}K_j - A_{12})\tilde{A}_{22}^{-1} \ . \qquad (12)$$

Again, the procedure has a rate of convergence proportional to μ.

Having now determined the transforming matrix T of (4), there remains a decoupled system

$$\dot{y}_1 = \tilde{A}_{11}y_1 + \tilde{B}_1 u \qquad (13)$$

and

$$\dot{y}_2 = \tilde{A}_{22}y_2 + \tilde{B}_2 u \ , \qquad (14)$$

where $y = Tx = \begin{pmatrix} y_1 \\ y_2 \end{pmatrix}$ for $y_i = \tau_i x$ and $\tilde{B}_i = \tau_i B$ for $\tau_1 = (I_{n_1} + KL \ K)$ and and $\tau_2 = (L \ I_{n_2})$. We note that this decoupling procedure has used the the two-time-scale hypothesis, but not the hypothesis requiring fast-mode stability.

Since we've assumed that the eigenvalues of $\tilde{A}_{22}$ lie far into the left half plane, we can expect y_2 to decay rapidly to its steady-state

$$y_{2s} = -\tilde{A}_{22}^{-1}\tilde{B}_2 u \tag{15}$$

(presuming u doesn't vary rapidly). A good approximation to y away from t = 0 will then follow by simply integrating the nonstiff system of differential equations (3) of order n_1 with initial vector $y_1(0) = \tau_1 x(0)$. Returning to the original variables, we will obtain the slow-mode approximation

$$x_s(t) = \mathcal{T}^{-1}\begin{pmatrix} y_1(t) \\ y_{2s}(t) \end{pmatrix} . \tag{16}$$

It is generally a poor approximation near t = 0 because the "boundary layer correction" $y_{2f} \equiv y_2 - y_{2s}$ is neglected. If a better approximation is needed there, one can integrate the linear system $\dot{y}_{2f} = \tilde{A}_{22}y_{2f} + \dot{y}_{2s}$ with $y_{2f}(0) = \tau_2 x(0) - y_{2s}(0)$. Because the eigenvalues of $\tilde{A}_{22}$ are large and (very) stable, we will need to use a short stepsize only on a short initial interval, since y_{2f} should rapidly decay to zero. This would fail to be true only if y_{2s} were not slowly-varying (i.e., $\dot{y}_{2s}$ were not small) for t > 0.

A more substantial problem results for initial value problems for the time-varying system

$$\dot{x} = A(t)x + B(t)u(t) \tag{17}$$

on $0 \leq t \leq T$. Assuming A(0) is two-time scale with n_1 slow modes, we can seek a transformation $\mathcal{T}$, as in (4), but now with time-varying decoupling matrices L(t) and K(t) determined so that the transformed problem for

$$y = \mathcal{T}(t)x \tag{18}$$

remains two-time-scale and fast-mode stable with the same fixed number n_1 of (relatively) small eigenvalues throughout $0 \leq t \leq T$.

We must cautiously note that eigenvalue stability does not imply stability for time-varying systems (cf., e.g., Coppel (1978)), but this is nearly true for certain singularly perturbed systems satisfying appropriate stability hypotheses (cf. the statement of Tikhonov's theorem in Wasow (1965) or Vasil'eva and Butuzov (1973)). Kreiss (1978, 1979) exhibits counterexamples, however, for systems with non-smooth coefficients.

Using our transformation (4), L(t) will block-triangularize the system matrix A(t) provided it now satisfies the matrix Riccati differential equation

$$\dot{L} = -LA_{11}(t) + A_{22}(t)L + LA_{12}(t)L - A_{21}(t) . \qquad (19)$$

If $\dot{L}(0) = 0$, the transformation $T_1(0)$ will be a similarity transformation, and we can ask (as before) that the eigenvalues of $\tilde{A}_{11}(0) = A_{11}(0) - A_{12}(0)L(0)$ coincide with the n_1 slow eigenvalues of A(0). The initial matrix L(0) can then be obtained, as before, by iterating in the algebraic Riccati equation $\dot{L}(0) = 0$. We will then obtain L(t) on $0 \leq t \leq T$ by integrating the resulting initial value problem. We would stop the integration and abandon our procedure if we encountered a finite escape time, or if the transformed problem ceased to be two-time-scale and fast-mode-stable within our t interval. (Existence criteria for the symmetric matrix Riccati equation commonly encountered in control theory are known, but analogous criteria for the non-square problem do not seem to be available.) The possibility for intermittent reinitialization should be investigated, corresponding to the reorthonormalizations of Scott and Watts (1977).

Knowing an appropriate L(t), the matrix K(t) will produce a block-diagonalization of the state matrix provided it satisfies the linear differential system

$$\dot{K} = \tilde{A}_{11}(t)K - K\tilde{A}_{22}(t) - A_{12}(t) . \qquad (20)$$

Since fast-mode stability implies that the eigenvalues of $\tilde{A}_{22} = A_{22} + LA_{12}$ have large negative real parts, singular perturbation concepts (cf., e.g., O'Malley (1974)) suggest that the solution of the initial value problem for K, like that for $\varepsilon\dot{z} = z$, will blow up for $t > 0$. Instead, it is natural to integrate a terminal value problem for K and expect that an asymptotically valid approximation will result by using its "pseudo-steady state" approximation, i.e. the unique solution of the algebraic system

$$K_s(t) = (\tilde{A}_{11}(t)K_s(t) - A_{12}(t))\tilde{A}_{22}^{-1}(t) . \qquad (21)$$

The need for nonuniform convergence at terminal time will be eliminated by picking $\dot{K}(T) = 0$, i.e. $K(T) = K_s(T)$.

Since the homogeneous variational equation for L(t), $\dot{\ell} = -\tilde{A}_{11}\ell + \ell\tilde{A}_{22}$, is also "singularly perturbed", but opposite in stability to the system for K, we might also attempt to approximate L(t) for $t > 0$ as a smooth solution $\tilde{L}(t)$ of the algebraic Riccati equation $\dot{\tilde{L}}(t) = 0$. This would certainly need $\tilde{L}(t)$ to be slowly-varying with the eigenvalues of $A_{22} + \tilde{L}A_{12}$ remaining large, and far into the left half-plane, compared to those of $A_{11} - A_{12}\tilde{L}$. Otherwise, we need to actually integrate the full initial value problem for L(t).

Now note that the fast-mode stability of $\tilde{A}_{22}(t)$ implies that the slow-mode or pseudo-steady state approximation

$$y_{2s}(t) = -\tilde{A}_{22}^{-1}(t)\tilde{B}(t)u(t) \tag{22}$$

should nicely approximate y_2 for $t > 0$, provided y_{2s} remains slowly-varying. Thus, we are finally left with the need to integrate the non-stiff reduced-order system (13) for $y_1(t)$. By inverting our transformation, as in (16), we find a slow-mode approximate $x_s(t)$ for $t > 0$.

Note that all initial value problems for x(t) would be solved in terms of an $n \times n$ fundamental matrix X(t) for the homogeneous system. In the two-time scale situation, our problem splits into four separate problems for the $n_2 \times n_1$ matrix L(t), for the $n_1 \times n_2$ matrix K(t), for the $n_2 \times n_1$ fundamental matrix for $y_2(t)$, and the $n_1 \times n_1$ fundamental matrix for $y_1(t)$. With the fast-mode stability assumption, pseudo-steady state approximations can be used to eliminate the differential systems for K, y_2, and sometimes L. Thus, a substantial order reduction is achieved for the $t > 0$ approximation.

Substantially more detail regarding time-varying problems is contained in O'Malley and Anderson (1980). Various related problems might be treated similarly. If, for example, a two-time-scale system produced a matrix $\tilde{A}_{22}$ with only moderate-sized eigenvalues, one might consider y_1 to be approximately constant on finite T intervals, so that only a differential system for y_2 need be integrated. Likewise, if the eigenvalues of $\tilde{A}_{22}$ are large with both large negative and large positive real parts, one might seek reduced-order approximations for appropriate two-point problems. Finally, extensions of these ideas to nonlinear problems must be sought.

Acknowledgments

The author would like to recognize the ongoing assistance and collaboration of Leonard Anderson. This work was supported in part by the Office of Naval Research under Contract No. N00014-76-C-0326, and while visiting Stanford University, by the Air Force Office of Scientific Research, the Army Research Office, the Office of Naval Research, and the National Science Foundation.

References

1. L. Anderson, "Decoupling of two-time-scale linear systems," Proceedings, 1978 Joint Automatic Control Conference, vol. 4, 153-164.

2. L. R. Anderson and W. L. Hallauer, Jr., "A method of order reduction for structural dynamics," Proceedings, 21st Structures, Structural Dynamics, and Materials Conference, 1980.

3. R. Bellman, Introduction to Matrix Analysis, second edition, McGraw-Hill, New York, 1970.

4. W. A. Coppel, Dichotomies in Stability Theory, Lecture Notes in Math. 629, Springer-Verlag, Berlin, 1978.

5. G. Dahlquist, "A numerical method for some ordinary differential equations with large Lipschitz constants," Information Processing 68, A. J. H. Morell, editor, North-Holland, Amsterdam, 1969, 183-186.

6. R. L. deHoff and W. E. Hall, Jr., "Optimal control of turbine engines," J. Dynamic Systems, Measurement, and Control 101 (1979), 117-126.

7. F. de Hoog and R. Weiss, "The numerical solution of boundary value problems with an essential singularity," SIAM J. Numerical Analysis 16 (1979), 637-669.

8. B. Etkin, Dynamics of Atmospheric Flight, Wiley, New York, 1972.

9. C. A. Harvey and R. E. Pope, "Synthesis techniques for insensitive aircraft control systems," Proceedings, 1976 IEEE Decision and Control Conference, 990-1001.

10. P. V. Kokotovic, J. B. Cruz, Jr., J. V. Medanic, and W. R. Perkins, editors, Multimodeling and Control of Large Scale Systems, Report DC-28, Coordinated Science Laboratory, University of Illinois, Urbana, 1979.

11. P. V. Kokotovic, P. Sannuti, and R. E. O'Malley, Jr., "Singular perturbations and order reduction in control theory—an overview," Automatica 12 (1976), 123-132.

12. H.-O. Kreiss, "Difference methods for stiff ordinary differential equations," SIAM J. Numerical Analysis 15 (1978), 21-58.

13. H.-O. Kreiss, "Problems with different time scales for ordinary differential equations," SIAM J. Numerical Analysis 16 (1979), 980-998.

14. M. Lentini and H. B. Keller, "Boundary value problems on semi-infinite intervals and their numerical solution," SIAM J. Numerical Analysis, 17, 577 (1980).

15. J. Medanic, "Geometric properties and invariant manifolds of the Riccati equation," Technical Report, Coordinated Science Laboratory, University of Illinois-Urbana, 1979.

16. W. L. Miranker and G. Wahba, "An averaging method for the stiff highly oscillatory problem," Math. Computation 30 (1976), 383-399.

17. C. Moler and C. Van Loan, "Nineteen dubious ways to compute the exponential of a matrix," SIAM Review 20 (1978), 801-836.

18. L. Oden, "An experimental and theoretical analysis of the SAPS method for stiff ordinary differential equations," technical report, Department of Information Processing, Royal Institute of Technology, Stockholm, 1971.

19. R. E. O'Malley, Jr., Introduction to Singular Perturbations, Academic Press, New York, 1974.

20. R. E. O'Malley, Jr., "Singular perturbations and optimal control," Lecture Notes in Math. 680 (1978), Springer-Verlag, Berlin, 170-218.

21. R. E. O'Malley, Jr. and L. R. Anderson, "Decoupling and order reduction for linear time-varying two-time-scale systems," Optim. Contr. 3, 133 (1982).

22. R. E. O'Malley, Jr. and J. E. Flaherty, "Analytical and numerical methods for nonlinear singular singularly perturbed initial value problems," SIAM J. Applied Math. 38 (1980), 225-248.

23. L. R. Petzold, "An efficient numerical method for highly oscillatory ordinary differential equations," technical report 78-933, Department of Computer Science, University of Illinois, Urbana, 1978.

24. M. K. Sain, "The theme problem," Proceedings, International Forum on Alternatives for Multivariable Control, 1977, 1-12.

25. M. R. Scott and W. A. Watts, "Computational solution of linear two-point boundary value problems via orthogonormalization," SIAM J. Numerical Analysis 14 (1977), 40-70.

26. G. W. Stewart, "Methods of simultaneous iteration for calculating eigenvectors of matrices," Topics in Numerical Analysis II, J. J. H. Miller, editor, Academic Press, London, 1975, 185-196.

27. D. Teneketzis and N. R. Sandell, Jr., "Linear regulator design for stochastic systems by a multiple time-scales method," IEEE Trans. Automatic Control 22 (1977), 615-621..

28. P. Van Dooren, "Updating the QZ-algorithm for the computation of deflating subspaces," internal report, Department of Computer Science, Stanford University, 1980.

29. A. B. Vasil'eva and V. F. Butuzov, Asymptotic Expansions of Solutions of Singularly Perturbed Equations, Nauka, Moscow, 1973.

30. W. R. Wasow, Asymptotic Expansions for Ordinary Differential Equations, Wiley, New York, 1965.

GLOBAL CODES FOR BVODEs AND THEIR COMPARISON

by

Robert D. Russell

§I. Introduction

In this paper we briefly trace the development of some current software for solving boundary value problems for ordinary differential equations (BVODEs). General features of this software are considered, and a finite element code is discussed in some detail with a numerical example. We then discuss the task of comparing this code, COLSYS, to the finite differences code PASVA3, which is briefly described in Pereyra's article in this proceedings. These are the two principal codes for solving general BVODEs which obtain global solutions, as opposed to those based on initial value techniques which use the corresponding initial value software. A more complete discussion of such a comparison will appear elsewhere [Pereyra-Russell 1980].

Our purpose in presenting some comparison here is three-fold. First, it will give a reasonable idea of some of the features which have been incorporated in the modern software. Showing the direction in which this code development has gone should be helpful for users wishing to better understand the codes and for those who want to consider incorporating some of these features into their own codes. Second, the numerical examples will demonstrate the performance of these codes, at least on a small set of problems. Third, we shall show some of the difficulties and dangers involved in comparing numerical software.

In the conclusion, some aspects of comparing BVODE software are summarized. General recommendations are made relating to the advantages of this software from a user's point of view.

II. BVODE Software Developments

In the last twenty years, the personal involvement of numerical analysts in scientific computation has increased markedly. This is evidenced by the rapid development of mathematical software. Reliable codes were quick to appear in many areas, e.g. numerical quadrature, solution of (scalar) nonlinear equations, and solution of initial value problems for ordinary differential equations (IVODEs). For other areas, such as solution of BVODEs, this development has been slower in coming.

Many of the reasons for this delay can be seen by contrasting the cases of IVODE and BVODE code development. The latter has only seen the production of portable, robust software the last 5-10 years, one reason being that the BVODE theory is generally much more difficult and less understood. Experience at handling important specialized problems, such as in the articles in this proceedings by Gear and Shampine for IVODEs, has not been gained for BVODEs where concerns are much more basic (such as which methods are appropriate for which types of problems). A related difficulty is that BVODEs can arise in many forms, and it is not straightforward to choose what class of problems a code should be able to handle and to see what other types of problems can be artificially converted to this form once the decision has been made [Ascher-Russell 1980].

Not only are BVODEs of diverse forms, but the approaches for solving them are varied. Many of the codes for solving them incorporate the IVODE software, viz. the "shooting" type codes [Gladwell 1970, Scott-Watts 1977] and the multiple shooting codes which combine these initial value techniques with a global viewpoint [Bulirsch et al.1980]. The other basic type of codes uses global methods [Lentini-Pereyra 1977, Ascher et al. 1980], where one calculates a solution on a predetermined mesh over the entire region of interest.

All of this combines to make the task of evaluating BVODE software a difficult one indeed. The "complexity" of the area-viewed here as the extent to which it relies of necessity on other areas of numerical analysis - is high since numerical linear algebra, approximation theory, optimization theory, numerical solution of IVODEs, numerical quadrature, and rounding error analysis all play necessary roles at various points. Nevertheless, some form of comparison of codes in this and similarly complex areas is necessary to give potential users guidance concerning which code(s) will best suite their individual needs.

III. COLSYS

The two popular global methods for solving BVODEs are finite difference and finite element methods. The collocation method using a spline (piecewise polynomial) solution, whereby this solution is determined by satisfying the differential equation exactly at certain points, can be shown to be computationally competitive with the other finite element methods and with finite differences [Ascher et al. 1978]. This is the method used by COLSYS.

An early use of collocation with a polynomial solution was in chemical engineering applications [Finlayson 1972]. The fairly general theory for this case [Vainniko 1966] was extended to the case of spline solutions in [Russell-Shampine 1972]. It was apparently first observed by Finlayson that the use of Gauss points for collocation produces a high rate of convergence [Carey-Finlayson 1975], and the analysis of this method [de Boor-Swartz 1973] showed in fact that superconvergence occurs at the mesh points. Various extensions of this theory have been done, and the most general one in [Cerruti 1974] is the theoretical basis for COLSYS.

The capabilities of COLSYS are treated in detail elsewhere [Ascher et al. 1980], so we only give a description for a simple example and one numerical example. Consider the BVODE

$$N(y) = y''(x) - f(x,y,y') = 0, \quad a \le x \le b$$

$$g_1(y(a),\ y'(a)) = \alpha, \quad g_2(y(b),\ y'(b)) = \beta \ .$$

Given a mesh $\pi : a = x_1 < x_2 < \ldots < x_{J+1} = b$, COLSYS seeks a spline function $s(x)$ such that $s(x) \in C^{(1)}[a,b]$, $s(x)$ satisfies the boundary conditions, and $s(x)$ is a polynomial of degree $k+1$ (for some pre-selected integer $k > 0$) which satisfies the differential equation at the k Gauss points $\{x_{ij}\}_{j=1}^{k}$ in (x_i, x_{i+1}) $(1 \le i \le J)$. The B-splines are used as the basis functions for representing $s(x)$, and its coefficients are determined by solving the resulting banded systems of equations with Gaussian elimination and partial pivoting. For nonlinear BVODEs this necessitates linearizing the differential equation and a modified Newton method is the nonlinear iteration strategy. New meshes are chosen adaptively (to conform to the solution behaviour) and error estimation is done to determine when the user's desired accuracy has been achieved. The fairly sophisticated automatic mesh selection and nonlinear iteration strategies are necessary features for solving difficult problems. The ability to directly apply collocation with nonuniform meshes to high order equations is one advantage of the method over finite differences with variable meshes, where conversion to a first order system is generally necessary (but see Pereyra's article in this proceedings for a different approach).

To demonstrate COLSYS, consider the fairly straightforward example of the radial Schrödinger equation with harmonic oscillator potential (Schoombie-Botha 1980],

$$y''(x) = (x^2 + \frac{2}{x}2 - \lambda)y(x) \qquad 0 < x < \infty$$

$$y(0) = y(\infty) = 0 \ .$$

By adding the equations

$$\lambda' = 0$$

$$D' = y^2 + (y')^2$$

$$D(0) = 0 \ , \ D(\infty) = 1$$

the BVODE is in suitable form for COLSYS using a finite value L to approximate ∞. The first two eigenvalues $\lambda = 5,9$ were easily computed and we give results for the second. With initial guess $y(x) = \sin x$, $\lambda = 9.5$, and $D(x) = x/L$, an initial mesh of 10 equally spaced subintervals and $k = 4$, and requested accuracy 10^{-5} for y, y' and λ, the problem was solved for $L = 5$ and $L = 7$. Initial convergence was achieved after 12 and 13 modified Newton steps and it was resolved with 20 and 20, 40 subintervals, respectively - generating each meshes by doubling the subintervals of the previous meshes - after which the accuracy was achieved. The estimate of λ is superconvergent, so in fact the actual accuracy is 6 and 10 digits respectively. This error in the first case is from using L too small.

The problem was also solved by mapping to (0,1), so that the BVODE becomes

$$y'' = \left(\frac{x^2}{(1-x)^6} \quad \frac{2}{x^2(1-x)^2} - \frac{\lambda}{(1-x)^4}\right)y + \frac{2y'}{(1-x)} \qquad 0 < x < 1$$

$$y(0) = y(1) = 0 \ .$$

Adding the λ, D equations as before (except with $D(1) = 1$), COLSYS is used without any further modifications. With initial guess $\lambda = 9.5$, $D = x/1$ and $y = \frac{\sin(6.3x)}{6.3}$, the mesh sequence was 10, 10, 20, 40, and results were comparable to those for the second case before ($L = 7$). The solution y at every fourth point of the last mesh is given below.

x	0.	.256	.446	.562	.648	.718	.767
y	.2(-35)	.4314(-1)	.1404(0)	.1005(0)	-.9027(-1)	-.1633(0)	.6592(-1)
				.820	.876	.937	1.00
				-.2391(-2)	-.5659(-8)	-.2565(-17)	0.0

(The author is grateful to Jan Christiansen for running this example.)

IV. Comparison

One of the first steps in evaluating the performance of software is the selection of a set of performance criteria. Obviously, the relative importance given to the various criteria is a critical factor in forming conclusions in a comparison. Moreover, a given method can be implemented any number of ways, so conclusions drawn about the codes do not necessarily apply for alternate implementations of the same methods. In a high complexity area (in the sense used previously) there is the increased difficulty of insuring that one's performance criteria are consistent with the design criteria of the codes (i.e., with what the codes are intended to do), or if not that this is explicitly stated.

In this section these difficulties are demonstrated in more detail for some numerical examples run with COLSYS and PASVA3. The problems are for illustration and not claimed to be representative of BVODEs and for simplicity they are generally artificial problems. We classify them in 5 basic types - where solutions have "spikes" (SPK), turning points (TP), boundary layers (BL), oscillations (OS), or are smooth (SM). More information concerning the solutions is given in the references.

SPK [Russell-Shampine 1972]

$$y'' + (3\cot(x) + 2\tan(x))y' + .7y = 0, \quad y(30°) = 0, \quad y(60°) = 5$$

OS1,2 $$y'' + \frac{2y'}{x} + \frac{y}{x^4} = 0$$

$$y(\alpha) = 0 \qquad y(1) = \sin(1) \qquad \alpha = \frac{1}{3\pi}, \frac{1}{9\pi}$$

$$(y(x) = \sin\frac{1}{x})$$

TP1 [Ascher et al. 1979]

$$\varepsilon y'' + xy' = -\varepsilon\pi^2\cos(\pi x) - \pi x\sin(\pi x), \quad y(-1) = -2, \quad y(1) = 0$$

TP2 [Lentini-Pereyra 1977]

$$y'' = \frac{3\varepsilon y}{(\varepsilon+x^2)^2}, \quad y(-1) = -y(.1) = \frac{-.1}{\varepsilon+.01}$$

<u>SM1</u> [Russell-Shampine 1972]

$$y'' = e^{y},\ y(0) = y(1) = 0$$

<u>SM2</u> [Ascher-Russell 1980]

$$y'' = -\frac{1}{16}\sin y(x) - (x+1)\,y(x-1) + x \qquad 0 \le x \le 2$$

$$y(2) = -\frac{1}{2},\ y(x) = x - \frac{1}{2},\ -1 \le x \le 0$$

<u>SM3</u> [Ascher et al. 1980]

$$(x^3 y'')'' = 1,\quad y(1) = y''(1) = y(2) = y''(2) = 0$$

<u>BL1</u> [Ascher et. al. 1979]

$$y'' = \mu \sinh(\mu y),\ y(0) = 0,\ y(1) = 1$$

<u>BL2</u> [White 1979]

$$y'' + y' - \varepsilon y = \frac{1}{\varepsilon}\sin x,\quad y(0) = 0,\quad y(\pi) = 0$$

<u>BL3</u> [Lentini-Pereyra 1977]

$$\varepsilon y'' + y = 0,\ y(-1) = 1,\ y(1) = 2$$

Numerical results for these problems were produced on a Burroughs 6700 in single precision (~ 13 decimal digits). More complete results will appear in [Pereyra-Russell 1980]; here the examples are only used to show the basic difficulties in comparing the codes.

In comparing numerical software, the tendency is often to emphasize the more objective evaluation criteria such as CPU time, storage, and portability. Unfortunately, even these have an element of uncertainty:

1a. CPU time. Relative running times are to some extent machine dependent, e.g., because s.p. (single precision) is sufficient on some machines and d.p. (double precision) necessary on others or because large storage, which is often required with these codes, can affect CPU times for machines differently. Storage can also significantly affect the "user's time" (e.g., if overnight runs are necessary with one code and not with another), and this would be a more important measure of time for some people.

1b. Storage requirements. This depends upon whether s.p. or d.p. is used. More fundamentally, however, there is the question of how one even measures a code's storage needs on a given problem. COLSYS, for example,can have very different performances depending upon whether it is given "unlimited" storage or a restrictive upper limit. If chosen appropriately, the upper limit still allows COLSYS to successfully solve the problem more efficiently.

1c. Portability. Both programs have been run extensively on most large scale machines, including IBM, CDC, AMDAHL, and Burroughs. Still, portability cannot be assumed if not tested, e.g. the standard FORTRAN version of COLSYS in [Ascher et al.1980] required minor modifications before running on the Burroughs 6700.

One desire when comparing codes is to be able to make subjective statements about their ease of use and robustness:

2a. Ease of use. Measurement is a particularly difficult problem for criteria like this one. In learning how to run COLSYS and PASVA3, students have become adept moderately quickly, although difficulties occur for some time, particularly in specifying Jacobians. Ease of use is strongly related to the other criteria and to what one is trying to achieve, as we discuss later.

2b. Robustness. The ability to efficiently solve problems for a large class, recognize when one cannot, and exist gracefully is important for most codes. COLSYS and PASVA4 are generally reliable in solving the ten problems given here. A code can be justly criticized if it tries too hard on some problems and a more complete investigation of COLSYS and PASVA3 for robustness should contain additional types of problems, e.g., highly nonlinear problems or ones with no solutions.

We now consider evaluation criteria arising for BVODEs and show by example some difficulties in measuring them. Again, the purpose in giving them here is for illustration, and they should not be assumed representative and generalizable. The results are too brief to include all the necessary considerations for such a purpose. For example, one problem in comparing two codes which perform in different ways is to insure that for a given problem the input information is similar for both.

The following notation will be used: $4.8(-5) = 4.8 \times 10^{-5}$, TOL = requested tolerance (on all components), C = COLSYS, P = PASVA3, E(j) = measured or estimated error in the jth derivative (at the mesh points for P and globally for C), EN(j) = measured error in the jth derivative at the mesh points for C, N = (# of subintervals for P) and (# of subintervals times # of collocation points per subinterval for C) needed to achieve TOL, and T = estimated CPU time in seconds.

3a. Form of solution. This is one of the best examples of different design criteria. COLSYS produces a spline solution (with superconvergence at mesh points) and PASVA3 produces a discrete solution (where interpolation could be used to give a globally defined solution). The conclusions that one drew from the run

	Tol		N	E(0)	T
SM1	10^{-4}	P	20	4.5(-5)	.72
		C	2x4	2.5(-7)	1.9

could be strongly affected by the form of the solution in which one is interested.

3b. User feedback. In contrast to some other BVODE software, PASVA3 and COLSYS are similar in being able to provide at any stage the current mesh and solution and and account of how the nonlinear iteration is proceeding.

3c. Error estimation. The codes put heavy emphasis on estimation of the global error. Both can be somewhat unreliable at very low (1-2 digit) accuracies. Their philosophies are different for higher accuracy. PASVA3 generally provides an error estimate $\tilde{e}$ with 1-2 digits of agreement with the exact error e, COLSYS a cruder error estimate $\tilde{e}$ satisfying $\frac{1}{10} < \frac{|e-\tilde{e}|}{|e|} < 10$ in each subinterval.

3d. Stopping criteria. In this respect the codes have different purposes. If u and $\tilde{u}$ are the exact and approximate solutions, respectively, then PASVA3 tries to satisfy $\|u-\tilde{u}\| = \|e\| = \|\tilde{e}\| < TOL$ and COLSYS tries to satisfy $\|u-\tilde{u}\|_i \leq TOL(1 + \|\tilde{u}\|_i)$ for each subinterval where $\|\tilde{u}\|_i$ is the magnitude of $\tilde{u}$ in the subinterval. As the example below shows, the result is that the codes can produce uncomparable results for a given TOL.

	TOL		N	E(0)	E(1)	T
BL3 ($\varepsilon=10^{-3}$)	10^{-4}	P	280	5.0(-9)	5.6(-6)	40.03
		C	24*4	2.0(-6)	1.5(-2)	13.0

Solving TP2 from $\varepsilon = 10^{-5}$ to 10^{-6} with continuation (the solution for $\varepsilon = 10^{-5}$ is used as the initial approximation for $\varepsilon = 10^{-6}$) gives

TOL		N	E(0)	E(1)	EN(1)	EN(2)	T
10^{-4}	P	350	7.1(-9)	2.3(-5)			140.
	C	20*4	4.0(-3)	1.2(1)	6.5(-8)	6.3(-5)	12.3

Clearly, different stopping criteria and forms of solutions make comparison very difficult.

3e. Requested precision. COLSYS compares more favorably with PASVA3 for low order solution derivatives than for high order ones. Since COLSYS has the option of requesting that different tolerances be achieved for individual components and not necessarily for them all, as in most BVODE codes) using this option can considerably affect results for a given problem. Moreover, the amount of precision desired is yet another parameter which can affect the outcome of a comparison. For COLSYS, TOL affects the mesh selection strategy and thus the efficiency, as the surprising example below shows.

	ε		TOL	N	E(0)	T
BL3	10^{-2}	C	10^{-2}	18*4	1.5(-7)	9.4
		C	10^{-4}	20*4	1.3(-6)	8.2

3f. Use of code parameters. In addition to TOL, the codes have other parameters which can improve performance. For both, external continuation, where the code is called at each step, is easy to do. For TP2, solving for $\varepsilon = 10^{-5}$, 10^{-6}, and 10^{-7} with and without continuation produced the following results:

TOL		T without continuation			T with continuation		
10^{-4}	P	46.8	89.7	176*	46.5	81.22	140.
	C	14.0	23.7	39.3	16.2	12.3	10.8

*TOL not quite achieved because of a storage restriction.

(The results correspond to very different accuracies for the two codes.)

There are other features of the codes which we have not considered, viz, for PASVA3 an automatic continuation option and for COLSYS use of fixed points in the mesh or no mesh selection (only halving each subinterval at each step), varying the methods order by changing k , and use of a nonlinear iteration option for particularly sensitive problems. Inclusion of these could affect one's ease of use considerably.

3g. Form of BVODE. As discussed in [Ascher-Russell 1980], there are artificial techniques for converting many types of problems into the forms required by BVODE software. For example, problems that are nonlinear, of high order, have integral constraints, eigenvalues, simple delays, switching points, nonseparated boundary conditions, interfaces, singularities, or conditions at infinity can often be solved. Still, efficiency is generally affected by the extent to which a code has the ability to handle a problem's original form directly. We illustrate this with two examples, the first where COLSYS handles directly the fourth order system and the second where PASVA3 handles directly the non-separated boundary conditions arising after one transformation.

		TOL		N	E(0)	E(1)	T
1)	SM3	10^{-5}	P	12	6.1(-8)	5.7(-6)	1.92
			C	2*5	3.6(-8)	2.3(-4)	1.0

		TOL		N	E(0)	E(1)	T
2)	SM2	10^{-4}	P	11	4.0(-6)	8.9(-6)	1.45
			C	2*3*	6.2(-7)	2.0(-5)	3.85

*Twice as many differential equations as for PASVA3.

3h. Difficulty of problem. This can affect the relative performance of the codes. The examples below show that for the oscillatory solution sin(1/x) COLSYS' relative performance is better for the more difficult case OS2.

	TOL		N	E(0)	E(1)	T
OS1	10^{-5}	P	120	1.1(-9)	9.9(-6)	15.9
		C	4*30	5.3(-8)	2.3(-5)	9.4
OS2	10^{-5}	P	304	2.3(-9)	1.2(-6)	128.
		C	4*72	4.4(-8)	3.3(-5)	20.8

In additon to the above possible performance criteria for comparing codes, other factors which complicate the process include insuring that drivers are error free and efficient or checking sensitivity with respect to change in initial mesh or solution. The initial mesh is an extremely important factor, so important that on many problems one could make either code's performance look better, depending upon which numerical results are taken. Phenomena similar to the one given below occur surprisingly often.

		Initial Mesh	N	E(0)	T
PK	P	15	257	4.3(-7)	41.5
	C	5*3	80*3	1.5(-6)	24.2
	P	45	125	6.2(-8)	10.8
	C	15*3	60*3	1.2(-5)	11.3

§V. Conclusions.

Applying a method in an area of high complexity such as for BVODE software involves many decisions concerning what implementation features to include. It thus becomes almost impossible to do a meaningful comparison of methods. Except in a quite limited sense, this is even true of comparing codes, since they are designed to perform very different tasks. Indeed, we have even considered two "similar" codes and have ignored the initial value type codes whose design criteria are still more different! Care must always be taken to distinguish "limitations" of methods from "limitations" because of these design criteria.

For example, one could argue that the global codes are impractical for solving scattering problems when only the unknown boundary conditions at infinity are desired, because storage requirements are exhorbitant. However, these codes could be modified to generate only a few of the matrix equations, perform the Gaussian elimination on these, generate a few more equations, and continue such that at any one time only the equations needed for latter elimination steps are saved. Upon completing the forward elimination, the desired boundary conditions are then easily recovered. This demonstrates the artificiality of much of the tendency to distinguish between initial value and global methods (see also [Keller-White 1975]).

The problems which arose in attempting to compare PASVA3 and COLSYS have been largely not dealt with, and more attention must be given finding ways to deal with these difficulties. This is not to say that people do not often make their own superficial comparisons and decide that one code is "better" than another, e.g., as has been our experience with students using the codes. An appropriate quote in [Forsythe et al 1977] is "it is an order of magnitude easier to write two good subroutines than it is to decide which one is best".

If one is going to do a software comparison, we recommend keeping a record of all computer runs, including all parameter values required to insure the results are reproducible. This record can give some idea of each code's ease of use and also allow re-evaluation of the results if one later decides to change the performance evaluation criteria. Also, care should be taken to insure that the results reported in the comparison are representative. Note that this is distinct from fine tuning a code on a set of problems, which one could argue is reasonable because their set is representative of the problems of interest in an area. Satisfying the above, one could draw various conclusions about the current state of codes in an area and the performance of general features, and we shall do this for BVODE's elsewhere [Pereyra-Russell 1980].

Writing computer programs to solve BVODEs can be an unpleasant, difficult, and expensive proposition, and on these grounds alone one should consider using tested numerical software when it is available and applicable for one's problem. Moreover, even if one already has a code, there are significant advantages to using two. Each provides its own insights when solving a problem; simple programming errors are detected more quickly by comparing results; the situation of a method converging to an extraneous solution (see [Doedel 1980] and the article by Dahlquist in these proceedings) is more easily recognized; and considerable confidence can be attached to a solution obtained by two different methods. A criticism of much BVODE software could be not so much that persons are disappointed with it when used for the class of problems for which it is intended, but that more effort should be put into modifying codes to deal with needs of special problem types (such as occur in scattering theory). Significant effort in the near future will probably be determined by the degree of cooperation between these users and code designers.

Acknowledgements: I am extremely grateful to Victor Pereyra for providing financial aid, student support, and many helpful discussions during my visit at Universidad Central in Caracas, Venezuela, where this work was largely a joint effort.

§VI. References

1. U. Ascher, J. Christiansen and R.D. Russell, A collocation solver for mixed order systems of boundary value problems. Math. Comp. 33 (1978), 659-679.
2. U. Ascher, J. Christiansen and R.D. Russell, COLSYS - A collocation code for boundary value problems, 1979, in codes for Boundary Value Problems in Ordinary Differential Equations. Lecture Notes in Computer Science 76,

Springer-Verlag.

3. U. Ascher, J. Christiansen and R.D. Russell, Collocation software for boundary value ODE's, 1980, ACMT, Math. 7, 209 (1981).
4. U. Ascher and R.D. Russell, Reformulation of boundary value problems into "standard" form, 1980, SIAM Rev. 23, 238 (1981).
5. C. de Boor and B. Swartz, Collocation at Gaussian points, SIAM J. Numer. Anal. 10 (1973), 582-606.
6. R. Bulirsch, J. Stoer and P. Dauflhand, Numerical Solution of Nonlinear Two-Point Boundary Value Problems, 1980, Num. Math., Handbook Series Approximation, in preparation.
7. G.F. Carey and B.A. Finlayson, Orthogonal collocation on finite elements, Chem. Engr. Sci. 30 (1975), 587-596.
8. J.H. Cerutti, Collocation for systems of ordinary differential equations, Comp. Sci. Tech. Rep. #230 (1974), University of Wisconsin, Madison.
9. E. Doedel, The numerical computation of branches of periodic solutions, 1980, submitted for publication.
10. B.A. Finlayson, The Method of Weighted Residuals and Variational Principles, Academic Press, N.Y., 1972.
11. G.E. Forsythe, M.A. Malcolm and C.B. Moler, Computer Methods for Mathematical Computations, 1977, Prentice-Hall, Englewood Cliffs, N.J.
12. I. Gladwell, A survey of subroutines for solving boundary value problems in ordinary differential equations, 1979, in Proc. of Conf. on Computational Techniques for ODEs, University of Manchester.
13. H.B. Keller and A.B. White, Jr., Difference methods for boundary-value problems in ordinary differential equations, SINUM 12 (1975), 791-801.
14. M. Lentini and V. Pereyra, An adaptive finite difference solver for non-linear two point boundary problems with mild boundary layers, SIAM J. Numer. Anal. 14 (1977), 91-111.
15. V. Pereyra and R.D. Russell, Manuscript in preparation, 1980.
16. R.D. Russell and L.F. Shampine, A collocation method for boundary value problems, Numer. Math. 19 (1972), 1-28.
17. S.W. Schoombie and J.F. Botha, Error estimates for the solution of the radical Schrödinger equation by the Rayleigh-Ritz finite element method, 1980, to appear in J. Comp. Phys.
18. M.L. Scott and H.A. Watts, Computational solutions of linear two-point boundary value problems via orthonormalization, SIAM J. Numer. Anal. 14 (1977), 40-70.
19. A.B. White, Jr., On selection of equidistributing meshes for two-point boundary-value problems, SINUM 16 (1979), 472-502.
20. G.M. Vainniko, On convergence of the collocation method for nonlinear differential equations, USSR Comp. Math. & Math. Phys. 6 (1966), 35-42.

GLOBAL ERROR ESTIMATION
IN ORDINARY INITIAL VALUE PROBLEMS

H. J. Stetter
Technical University of Vienna
A-1040 Vienna

(for W.F. Ames)

A student, for her dissertation,
had to solve a differential equation.
But the numbers she 'd gotten
from the computer were rotten,
so she failed at her examination.

When her boyfriend hears it he swears:
" - - - - - - - - - - - - - - - - - -!
What a stupid state of affairs:
Such a code should generate
a *global error estimate*!
But it seems that nobody cares."

It is surprising indeed that the many users of software for the solution of initial value problems in ordinary differential equations have not protested against the state of affairs: None of the codes available in the IMSL-library provides any information about the error of the approximate solution values which it generates; in the NAG-library, one code (DO2BD) was introduced at Mark 7 (December 78) which computes along with the approximate solution vector an estimate of its error and outputs it on request; and a recent list of available software products for initial value problems in ordinary differential equations contains only one further code (GERK; see [3]) which performs this task.

Even worse, many users may believe that they have such information because the codes ask them to specify a value for an "accuracy parameter"; it is called TOL in the NAG-routines and in DVERK, EPS in a number of other routines. Although the documentation of these codes does not pretend that this parameter will be a bound on some norm of the errors of the approximate solution values generated along the interval of integration, the wording is sufficiently vague that a non-expert user may be fooled. (E.g., in DVOGER one reads: EPS is used to specify the maximum error criterion. The stepsize and/or the order is adjusted so that the single step error estimates divided by YMAX(I) are less than EPS in the Euclidean norm. ... The global error will depend upon both EPS and the number of steps taken.) The situation becomes even more misleading through the fact that the user is often asked to indicate a choice of weights in relation to that quantity TOL or EPS (absolute error, mixed absolute-relative error, etc.).

Actually this parameter is to be viewed as nothing but a knob which permits to "turn" the accuracy of the computation higher or lower. The reaction of the error of the computed solution to the setting of this parameter may be vaguely proportional (cf. equ. (7)), at least when the overall tendency over a larger range of values of TOL is considered (say, from 10^{-8} to 10^{-2}). However, there may be serious deviations from such a behavior even with trivial differential equations, when the error values obtained for two close values of TOL (say, 10^{-3} and 10^{-4}) are compared. In any case, the value of TOL does not give any information about the size of the error.

Therefore, it seems very important that the user may obtain information on the accuracy of the computed solution values (and their derivatives perhaps), at least a posteriori. The following aspects of this information seem important:

The size of a norm of the error vector at an output point would not be satisfactory since it will normally be dominated by the error of one particular component. If the error in an other component were more crucial, it would have to be obtained through sophisticated scaling. Also the sign of some error component may often be important.

Furthermore, an error vector should be available at each output point. Although "statistical" information about the behavior of each error component along the interval of integration will also be of interest, it may not suffice in cases where the accuracy of the computed solution varies considerably over that interval.

Essentially, what we would like to know is the number of correct digits in each component of the approximate solution. Hence we will be satisfied with the order of magnitude of the error; a gross overestimation of the error will be as misleading as a gross underestimation. Thus we will not attempt to obtain strict error bounds because they will generally be unrealistic (except when a very high effort is spent in their computation, see, e.g. [1]).

We claim that good software for ordinary initial value problems should provide information on the error, of the kind just specified, as an option. In the remainder of this paper we will consider in which fashion and to which extent this may be achieved.

The "Secondary Problem"

Obviously, we request a solution to the following problem:

Given: a) An initial value problem for a system of s ordinary differential equations

$$\begin{aligned} y'(t) &= f(t,y(t)), \qquad & t \in [0,T], \\ y(0) &= y_0, \qquad & y(t) \in \mathbb{R}^s. \end{aligned} \tag{1}$$

(We denote the solution of this "primary problem" by y(t).)

b) An approximate solution to (1), i.e. a "grid" $\{t_n, n = 1(1)N\} \subset [0,T]$ and a sequence of values $\eta_n \in \mathbb{R}^s$, $n = 1(1)N$, such that

$$\eta_n \approx y(t_n), \qquad n = 1(1)N.$$

Find: A numerical approximation $\{\varepsilon_n\}$ to the global error $\{e_n\}$ of (2), i.e. a sequence of values $\varepsilon_n \in \mathbb{R}^s$, $n = 1(1)N$, such that

$$\varepsilon_n \approx e_n := \eta_n - y(t_n), \qquad n = 1(1)N. \tag{2}$$

The requested relative accuracy in this "secondary problem" is rather low; according to the previous discussion, we will be satisfied with obtaining the sign and the leading digit of e_n correctly, even if the accuracy requirement in the primary computation has been high.

Normally it will be desirable to obtain the ε_n in parallel with the η_n: If the $\|\varepsilon_n\|$ become intolerably large the computation may have to be terminated, and - more important - it should be possible to utilize all information which has been generated in the computation of η_n but which is not saved for the computation of ε_n.

One fact should be noted from the begining: If ε_n is a reasonable approximation to e_n then

$$\bar{\eta}_n := \eta_n - \varepsilon_n \tag{3}$$

is a better approximation to $y(t_n)$ than η_n. Thus the old dilemma of numerical computation reappears: Should we use $\bar{\eta}_n$ at the price of having no error estimate for this better value. (One will tend to assume that $\|\bar{\eta}_n - y(t_n)\|$ is <u>less</u> than $\|\varepsilon_n\|$ anyway!). We will not dwell further on this question; in fact, one important approach to finding ε_n is via the (a posteriori) computation of a presumably better solution $\bar{\eta}_n$.

The equivalence of the computation of an error estimate ε_n and a more accurate approximate solution $\bar{\eta}_n$ suggests that the necessary computational effort may not be negligeable. Nevertheless, cost considerations should not be critical: First of all, without reliable information on the global error we may have to spend considerably more in the primary computation in order to achieve some security regarding accuracy. Furthermore, the error estimation option may be switched off during production runs. As a rule of thumb, the computational effort in the secondary problem should not exceed what was spent in the primary problem.

There are two basically different <u>approaches</u> to the estimation of the global error in a finite-difference solution of (1):

- Compute <u>two</u> primary approximations from a sequence of approximations with an asymptotic behavior.
- Compute the approximate <u>defect</u> of the primary solution, compute the approximate <u>effect</u> of this perturbation of (1).

<u>Asymptotic Estimates</u>

I) <u>Richardson extrapolation</u>: For a finite-difference method of "order" p, we expect

$$\eta_{n,h} = y(t) + e(t)h^p + O(h^{p+1}), \quad t = nh \text{ fixed}, \qquad (4)$$

if the constant stepsize h has been used; here e(t) is a vector-valued function independent of h. The stepsize 2h will produce (for even n)

$$\eta_{n/2,2h} = y(t) + e(t)(2h)^p + O(h^{p+1}). \qquad (5)$$

(4) and (5) imply

$$e_{n,h} \approx e(t)h^p \approx \frac{1}{2^p - 1}[\eta_{n/2,2h} - \eta_{n,h}] =: \varepsilon_n . \qquad (6)$$

With a tolerance-controlled variable step code one has to preserve coherence between the grids $\mathbb{G}_h$ and $\mathbb{G}_{2h}$ ([4], p.73): $\mathbb{G}_h$ must be formed by halving each step in $\mathbb{G}_{2h}$, see fig. 1. Therefore the auxiliary solution $\eta_{n/2,2h}$ must be computed first, under stepsize control. $\eta_{n,h}$, the approximate solution proper, is then computed on the predetermined grid $\mathbb{G}_h$, without stepsize control. The computation of the $\eta_{n/2,2h}$ and $\eta_{n,h}$ can be advanced concurrently so that ε_n is available at the same time as $\eta_{n,h}$. Obviously, the extra effort is somewhat less than 50%.

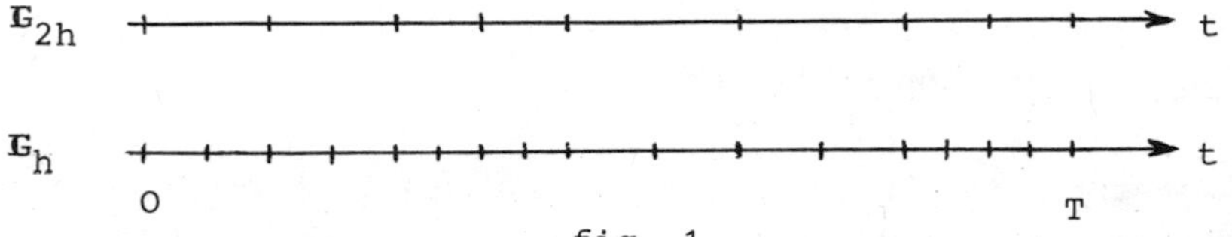

fig. 1

The approach is valid for fixed order one-step methods only, it cannot be used with multistep or variable order methods. This error estimation procedure has been suggested by Shampine-Watts ([3]) for Runge-Kutta codes, and considerable evidence for its reliability and efficiency has been compiled. The procedure will fail if (4) is not valid, i.e. if the problem (1) to be solved is not sufficiently smooth; it will be unreliable if the steps h are large (relative to variations in the solution), i.e. at low accuracies.

II) Tolerance extrapolation: In a tolerance-controlled code for (1), let $\eta_{TOL}(t)$ and $e_{TOL}(t)$ be the solution value and its error obtained with tolerance TOL. Then we expect, for $r > 0$,

$$e_{r\times TOL}(t) \approx r\, e_{TOL}(t), \qquad t \in [0,T]. \tag{7}$$

If we rely on this "tolerance proportionality" we may compute $\eta_{TOL}(t)$ and $\eta_{r\times TOL}(t)$ $(r > 1)$ and form

$$e_{TOL}(t) \approx \frac{1}{r-1}\,[\eta_{r\times TOL}(t) - \eta_{TOL}(t)] =: \varepsilon_{TOL}(t). \tag{8}$$

Here, both computations have to run under stepsize control; thus, for $r = 2^p$, the evaluation of (8) is slightly more expensive than that of (6).

However, even if (7) holds rather tightly there is a further difficulty: t will not be a common point of the two grids $\mathbb{G}_{r\times TOL}$ and $\mathbb{G}_{TOL}$, hence one of the η-values in (8) will have to be formed by interpolation. This has to be done quite carefully if it is to preserve (7).

None of the present codes satisfies (7) well enough to make (8) a reliable procedure for error estimation; see [8]. On the other hand, if no other information is provided by the code, (8) is the only way by which the user may obtain <u>some</u> indication of the error in the solution values.

Defect Integration Estimates

We interpret the computed approximate solution values η_n as exact values of the solution $\tilde{y}(t)$ of a perturbed problem (1):

$$\begin{aligned} \tilde{y}'(t) &= f(t,y(t)) + u(t), \qquad t \in [0,T], \\ \tilde{y}(0) &= y. \end{aligned} \tag{9}$$

The effect $e_n = \tilde{y}(t) - y(t)$ of the perturbation u(t) may then be approximately determined at the gridpoints.

If we request that u is a step function with a constant value $u_n \in \mathbb{R}^s$ in each step (t_{n-1},t_n), these <u>local defects</u> u_n are defined by the following problems (see fig. 2):

fig. 2

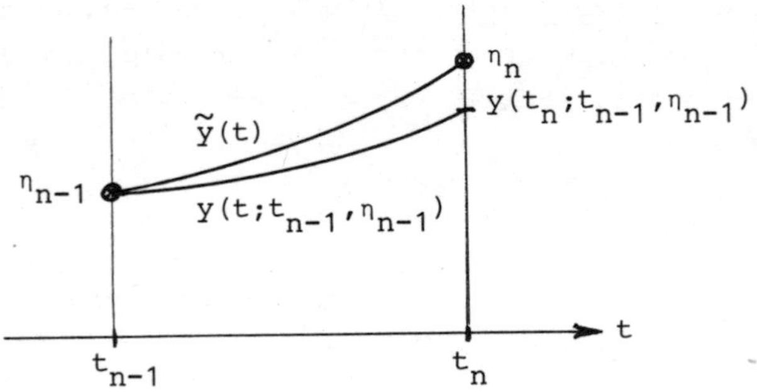

$$\begin{aligned} \tilde{y}'(t) &= f(t,\tilde{y}(t)) + u_n, \qquad t \in [t_{n-1},t_n], \\ \tilde{y}(t_{n-1}) &= \eta_{n-1}, \quad \tilde{y}(t_n) = \eta_n. \end{aligned} \tag{10}$$

Under suitable technical assumptions and for sufficiently small h_n, (10) has a unique solution u_n. We may call it the backward error of our solution, in agreement with the usage of this term in other contexts.

The defect u_n is closely related to the local error per unit step L_n:

$$L_n := \frac{\eta_n - y(t_n;t_{n-1},\eta_{n-1})}{h_n} = u_n(1+O(h_n)), \qquad (11)$$

see fig. 2. Furthermore, if $\bar{\eta}_n$ is an approximation to $y(t_n;t_{n-1},\eta_{n-1})$ of higher order than η_n, i.e. if

$$\|\bar{\eta}_n - y(t_n;t_{n-1},\eta_{n-1})\| \leq O(h_n)\ \|\eta_n - y(t_n;t_{n-1},\eta_{n-1})\| \qquad (12)$$

then

$$L_n = \frac{\eta_n - \bar{\eta}_n}{h_n}(1+O(h_n)). \qquad (13)$$

Therefore the computation of a first order correct estimate of u_n is equivalent to the computation of a more accurate local solution.

Numerical evaluation of the defect: There are numerous possibilities of obtaining an approximate value of the defect u_n defined by (10):

(i) An asymptotically correct estimate $\tilde{l}_n$ of the local error $l_n := \eta_n - y(t_n;t_{n-1},\eta_{n-1}) = h_n L_n$ may have been formed during the primary computation for control purposes; we may then use (11). Note that the local error estimates arising in codes with "local extrapolation" are not asymptotically correct.

(ii) A valid local error estimate may be formed by some further computation; see, e.g., [6] in the case of variable order, variable step Adams predictor-corrector codes.

(iii) By suitable interpolation of the η_n, we may obtain a function $\tilde{y}(t)$ which yields an approximation to u_n upon substitution into the differential equation (10). The substitution of the η_n into a linear multistep formula of sufficiently high order may be an equivalent approach.

(iv) Repetition of the step $t_{n-1} \to t_n$, with two steps of length $h_n/2$, and subsequent Richardson extrapolation yields a value $\bar{\eta}_n$ which satisfies (12) and may thus be used in (13). Naturally, a value $\bar{\eta}_n$ satisfying (12) may also be computed in various other ways. (This approach is normally too expensive to be realistic.)

In the design of the evaluation procedure for u_n, one should attempt to make only use of information which has been used or generated in the computation of η_n. Avoiding the use of previous information is essential during the starting (or restarting) phase of the code, in intervals of strong variation, etc. In [6], this objective was fully achieved. Strict adherence to this objective, on the other hand, will be difficult for high order Runge-Kutta codes.

In BDF-codes for stiff equations, asymptotically correct local error estimates based on the predictor corrector comparison are widely used for control purposes. It is not obvious whether these estimates are sufficiently accurate for our purposes in seriously stiff situations.

Computation of the global error estimate: From the local defects, an approximation of the global error may be obtained via "defect correction". The principle may be explained thus (see, e.g., [5]):

Let $\eta(t)$ be a suitable interpolation of the η_n-values from the primary computation. We have found that the numerical solution of

$$y'(t) - f(t,y(t)) = 0 \tag{14}$$

by our given discretization method yielded

$$\eta'(t) - f(t,\eta(t)) \approx \tilde{u}(t). \tag{15}$$

Hence we conclude that the numerical solution of

$$\bar{y}'(t) - f(t,\bar{y}(t)) = -\tilde{u}(t) \tag{16}$$

by the same discretization method will yield

$$\bar{\eta}'(t) - f(t,\bar{\eta}(t)) \approx 0. \tag{17}$$

Therefore the values $\bar{\eta}_n$ arising from this second computation will be better approximations of $y(t_n)$ than η_n and the global error e_n of the η_n may be found approximately as (cf. (4))

$$\varepsilon_n := \bar{\eta}_n - \eta_n.$$

In this form, we would have to bear the cost of the defect computation plus that of another run of the primary computation (minus its control activities). In view of the low relative accuracy requirements for ε_n, we would prefer to use a simpler discretization method for the secondary computation. There are two possibilities of doing this without any essential loss in accuracy:

(i) Simplified defect correction: Assume, e.g., that we wish to use Euler's method for the secondary integration. We realize that our approximate solution values η_n could have been obtained (formally) as ζ_n from

$$\frac{\zeta_n - \zeta_{n-1}}{h_n} - f(t_{n-1},\zeta_{n-1}) = \left[\frac{\eta_n - \eta_{n-1}}{h_n} - f(t_{n-1},\eta_{n-1})\right] =: \delta_n, \quad (18)$$

which is Euler's method for (14), with an appropriate perturbation δ_n. Hence the $\bar{\eta}_n$ of (17) will be obtainable from

$$\frac{\bar{\eta}_n - \bar{\eta}_{n-1}}{h_n} - f(t_{n-1},\bar{\eta}_{n-1}) = \delta_n - \tilde{u}_n, \quad (19)$$

which is Euler's method for (16), with the same perturbation.

(ii) Linearization: From (14) and (15) we see that $e(t) = \eta(t) - y(t)$ will approximately satisfy

$$e'(t) - f_y(t,\eta(t))e(t) = \tilde{u}(t). \quad (20)$$

Hence we may find our global error estimates $\varepsilon_n \approx e(t_n)$ by a numerical solution of (20) which need not employ the original discretization method.

A successful implementation of (19) for a primary code based on a variable order, variable step Adams predictor corrector code has been reported in [7]. This approach should work well for all "non stiff codes" where the steps chosen in the primary computation will yield stability also for the Euler method. The evaluation of δ_n must be carefully coded to avoid an unnecessary loss of accuracy.

The use of (20) is sensible only when the Jacobians $J_n := f_y(t_n,\eta_n)$ have been formed in the primary computation in any case, i.e. essentially in "stiff codes". Actually, the primary code normally does not store J_n but a factorized form of $I - h_n\beta J_n$, $0 < \beta < 1$. This suggests the use of the scheme

$$\eta_n = \eta_{n-1} + h_n[(1-\beta)f_{n-1} + \beta f_n] \quad (21)$$

for the secondary integration in one of the following two ways:

a) (21) is used in place of Euler's method in (18)/(19) and $I - h_n\beta J_n$ is employed in the Newton process for solving $\bar{\eta}_n - h_n\beta f(t_n,\bar{\eta}_n) = \ldots$

b) (21) is used for the integration of (20) so that $I - h_n\beta J_n$ occurs directly. In this case, we also need $I + h_n(1-\beta)J_{n-1}$ which we may write as

$$I + h_n(1-\beta)J_{n-1} = -\frac{1-\beta}{\beta}\,[I - h_n\beta J_{n-1}] + \frac{1}{\beta}\,I \tag{22}$$

according to Prothero & Robinson ([2]) who have implemented this global error estimation approach with a code based on BDF-procedures.

Note that a) and b) are not equivalent and may make different requests on the accuracy and the updating of the J_n.

Also, (21) is A-stable for $\beta \geq 1/2$ only. For BDF-procedures with $k > 3$, values of $\beta < 1/2$ occur. Here one will have to use the original procedure also for the secondary integration which should be hardly more expensive than the use of (21).

Output at non-grid points

In a number of codes for initial value problems in ordinary differential equations, the generated grid does not conform with the user-specified output abscissae; instead, the approximate solution values at the output points are formed by interpolation.

If this is not done sufficiently carefully, it will introduce an extra error which may constitute a major part of the global error but which is naturally not accounted for by the estimation procedures.

This situation has been analyzed in [6] in the context of Adams PC-codes.

Conclusions

The results reported in [2], [3], [7], and some other publications indicate that sufficiently reliable and inexpensive global error estimates may be provided by "black-box" library routines for ordinary initial value problems. The automatic generation of such global error estimates would considerably enhance the safety and efficiency of present ODE-codes: The user could "switch on" the global error estimation option to determine the appropriate value of the tolerance parameter for his problem in a preliminary run; he would then use the code most efficiently in subsequent production runs.

There is no serious reason why global error estimation should not become a matter of course in library routines within the next few years.

References

[1] U. Marcowitz: Fehlerabschätzung bei Anfangswertaufgaben für Systeme gewöhnlicher Differentialgleichungen, Num. Math. 24 (1975) 249-275.

[2] A. Prothero: Estimating the Accuracy of Numerical Solutions to Ordinary Differential Equations, Proc. Conf. on Comput. Techniques for O.D.E., Manchester, 1978.

[3] L.F. Shampine, H.A. Watts: Global Error Estimation for Ordinary Differential Equations, TOMS 2 (1976) 172-186.

[4] H.J. Stetter: Analysis of Discretization Mehtods for Ordinary Differential Equations, Springer-Verlag, Berlin-Heidelberg-New York, 1973.

[5] H.J. Stetter: The Defect Correction Principle and Discretization Methods, Num. Math. 29 (1978) 425-443.

[6] H.J. Stetter: Interpolation and Error Estimation in Adams PC-Codes, SINUM 16 (1979) 311-323.

[7] H.J. Stetter: Global Error Estimation in Adams PC-codes, TOMS 5, (1979) 415-430.

[8] H.J. Stetter: Tolerance Proportionality in ODE-codes, SIGNUM meeting on Numerical Ordinary Differential Equations, Urbana, 1979.

LOWER BOUNDS FOR THE ACCURACY OF LINEAR MULTISTEP METHODS

Rolf Jeltsch and Olavi Nevanlinna

1. Introduction

We consider linear multistep methods (2) for solving initial value problems

(1) $\quad y' = f(t,y), \quad y(0)$ given

where one has large differences in the time constants. In the classical theory [1], [7] one requests stability (= zero stability) and accuracy of the method for small stepsizes h. It has however long been observed that in the presence of large differences in the time constants one has to consider stability for large steps too, see e.g. Gear [5]. This leads to the definition of the stability region S. For measuring the accuracy of the method one uses classically error order and error constant which describe the first term of the asymptotic expansion of the error as h tends to zero. Hence these are concepts for small values of h. If one applies methods with "large" stability regions to the above mentioned stiff systems stability allows to use a stepsize h for which the asymptotic error expansion is no longer adequate. In this case the concepts of error order and error constants are no longer appropriate. We shall measure the accuracy for fixed positive h by the L_1-norm of the Peano-kernel of the error functional. This allows to compare the accuracy of methods of _different error order_. The main result which we announce in this paper is, loosely speaking, the following. For methods of order higher than 2 the "accuracy" decreases as the "size" of the stability region increases. This can be considered as a refinement of the celebrated Dahlquist barrier [2] which says that the order of an A-stable method cannot exceed 2.

In section 2.1, 2.2 a short review of the classical theory is given. In section 2.3 the stability region and its relevance for stiff initial value problems is discussed. In section 2.4 we show that the concepts of error order and error constant are not appropriate in some situations. In section 3.1 we introduce Peano-kernels K_q for measuring the accuracy of a method for any h. In section 3.2 we give our main result: lower bounds for the size of the Peano-kernel. These bounds depend on the radius R of the largest disk $D_R = \{ \mu \in \mathbb{C} \mid |\mu+R| \leq R\}$ included in the stability region. In the last section we motivate why one requests $D_R \subset S$ and indicate how one could use the presented results to give estimates for the work which is needed to solve an initial value problem within a certain tolerance.

2. The methods and their application to stiff problems

2.1 Linear Multistep Methods

Let $h > 0$ be the stepsize and $t_n = nh$ for $n = 0,1,\ldots$. Then we compute recursively y_{n+k} using

$$(2) \qquad \sum_{i=0}^{k} \alpha_i \, y_{n+i} = h \sum_{i=0}^{k} \beta_i \, f(t_{n+i}, y_{n+i}) \;, \text{ for } n = 0,1,2\ldots \;.$$

Here we assume that the starting values already have been found and that $\alpha_k \neq 0$. y_n is an approximation to $y(t_n)$. If $\beta_k = 0$ then the method is called explicit, since (2) becomes a linear equation in y_{n+k}. However if $\beta_k \neq 0$ then (2) is in general a nonlinear equation and the method is called implicit. For non-stiff differential equations (2) is solved using an iteration scheme while for stiff equations one resorts to a Newton-like procedure.

Formulas of the form (2) have been derived in various ways, see e.g. Henrici [7] and Lambert [10]. For example one obtains the Adams-Moulton formulas by replacing the integrand $f(t,y(t))$ in the integral form of (1)

$$y(t_{n+k}) - y(t_{n+k} - h) = \int_{t_{n+k-1}}^{t_{n+k}} f(\tau,y(\tau)) \, d\tau$$

by the polynomial $P(\tau)$, which interpolates $f(\tau,y(\tau))$ at $\tau = t_n, t_{n+1}, \ldots, t_{n+k}$. If $k = 1$ one obtains the trapezoidal rule

$$(3) \qquad y_{n+1} - y_n = \frac{h}{2} \left(f(t_n,y_n) + f(t_{n+1},y_{n+1})\right) \;.$$

The socalled backward differentiation formulas (BDF) are derived by requesting that the interpolation polynomial $Q(t)$ with

$$Q(t_{n+i}) = y_{n+i} \quad \text{for } i = 0,1,\ldots,k$$

satisfies the differential equation (1) at t_{n+k}.

Hence

$$Q'(t_{n+k}) = f(t_{n+k},Q(t_{n+k}))$$

This leads to formulas of the form

$$\sum_{i=0}^{k} \alpha_i \, y_{n+i} = hf(t_{n+k},y_{n+k}) ,$$

for example one has

$$k = 1: \quad y_{n+1} - y_n = hf(t_{n+1},y_{n+1}) , \quad \text{implicit Euler}$$

(4) $$k = 2: \quad y_{n+2} - \tfrac{4}{3}y_{n+1} + \tfrac{1}{3}y_n = \tfrac{2h}{3} f(t_{n+2},y_{n+2}).$$

2.2 Accuracy

In order to measure the error $y(t_n) - y_n$, we normalize the coefficients α_i, β_i such that

(5) $$\sum_{i=0}^{k} \beta_i = 1$$

and introduce the "local error"

(6) $$[L_h y](t) := \sum_{i=0}^{k} (\alpha_i \, y(t+ih) - h\beta_i \, y'(t+ih))$$

The method is said to have <u>error order</u> p if the Taylor series expansion of (6) in powers of h has the form

(7) $$[L_h y](t) = c_{p+1}h^{p+1}y^{(p+1)}(t) + c_{p+2}h^{p+2}y^{(p+2)}(t) + O(h^{p+3}) , \quad c_{p+1} \neq 0$$

for sufficiently smooth $y(t)$. c_{p+1} is called the <u>error constant</u>. At this point it is convenient to introduce the characteristic polynomials

(8) $$\rho(\zeta) := \sum_{i=0}^{k} \alpha_i \zeta^i , \qquad \sigma(\zeta) := \sum_{i=0}^{k} \beta_i \zeta^i .$$

A linear multistep method is said to be stable (= zero-stable) if the roots of $\rho(\zeta)$ have a modulus not exceeding one and those of modulus one are simple.

From a classical theorem by Dahlquist [1],[7] it follows that for appropriately chosen starting values one has

$$y(t_n) - y_n = O(h^p)$$

as h tends to zero while t_n is kept fixed, provided the method is zero-stable. Moreover zero-stability is a necessary condition for convergence. While Adams methods are stable for all k, it is known that BDF are stable for $k \leq 6$ only.

2.3 Stability regions

It is very instructive if one applies (2) to the linear test equation

(9) $$y' = \lambda y \ , \quad y(0) = 1 \ , \quad t \in [0,\infty), \quad \lambda \in \mathbb{C}.$$

Here the exact solution is

(10) $$y(t_n) = (e^{\mu})^n \quad \text{where} \quad \mu := h\lambda \ .$$

The numerical solution of a convergent method (2) when applied to (9) has the form

(11) $$y_n = d_1\zeta_1{}^n + \sum_{j>1} d_j(n)\zeta_j{}^n \ ,$$

where ζ_j are the roots of the characteristic equation

(12) $$\rho(\zeta) - \mu\sigma(\zeta) \equiv 0 \ ,$$

$d_j(n)$ is a polynomial in n of degree one less than the multiplicity of ζ_j and the coefficients of d_j are dependent on the starting values $y_0, y_1, \ldots, y_{k-1}$ only. The set

$$S = \left\{ \mu \in \mathbb{C} \;\middle|\; \begin{array}{l} \text{If } \Phi(\zeta,\mu) = 0 \text{ then either } |\zeta| < 1 \\ \text{or } \zeta \text{ simple root with } |\zeta| = 1 \end{array} \right\}$$

is called the stability region of the method since $\{y_n\}_{n=0,1,2,\ldots}$ is bounded if and only if $h\lambda = \mu \in S$. Stability regions of explicit methods are always bounded, see Fig. 1a. Clearly a method is zero-stable if and only if $0 \in S$.

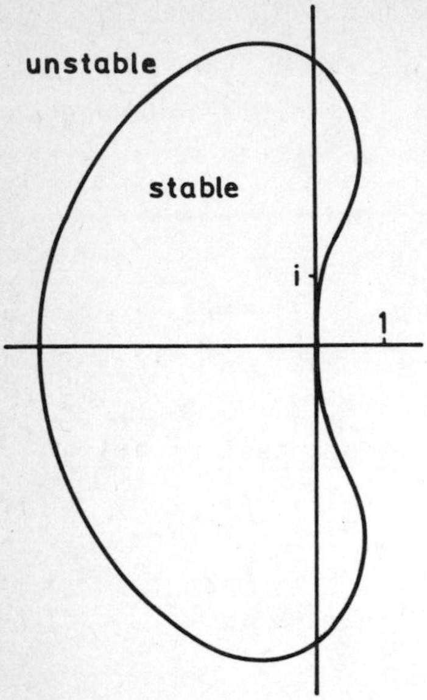

Fig. 1a: Typical stability region of an explicit method

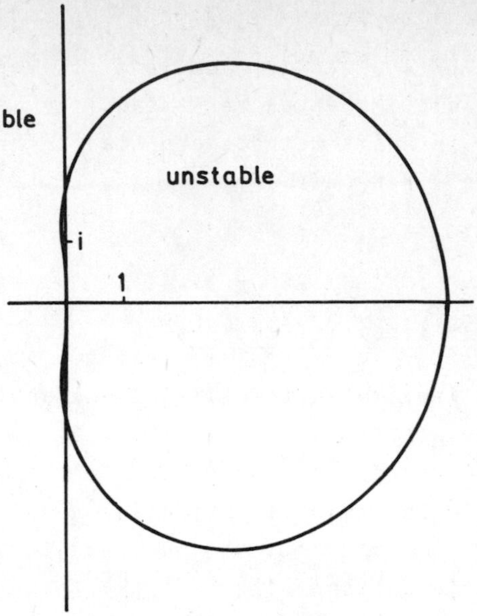

Fig. 1b: Stability region of BDF method with k = 3

It is very informative to observe that the roots ζ_j depend on $\mu = h\lambda$ and that a convergent method has exactly one root ζ_1, the principal root, which approximates the exact solution e^{μ} for μ's close to $\mu = 0$. The other roots ζ_j, $j \geq 2$, do not contribute to the accuracy of a method.

Let us consider for a moment the linear autonomous system of differential equations

(13) $\quad y' = T \operatorname{diag}(\lambda_1,\lambda_2,\ldots,\lambda_m)T^{-1}y$.

Applying (2) to (13) yields numerical values y_n. Using the transformation

(14) $\quad y_n = T^{-1}z_n$

one easily finds that the j-th component $z_n^{[j]}$ of z_n satisfies the recurrence relation one obtains when applying (2) to the scalar test equation $z' = \lambda_j z$. Thus $\{y_n\}_{n=0,1,2,...}$ is bounded for all possible starting values if and only if

(15) $h\lambda_j \in S$ for $j = 1,2,...,m$.

If the eigenvalues λ_j of the Jacobian in (13) satisfy

(16) $\mathrm{Re}\lambda_1 \leq \mathrm{Re}\lambda_2 \leq ... << ... \leq \mathrm{Re}\lambda_m \leq 0$

then (13) is called a stiff system of ordinary differential equations. Assume now that a method with a bounded stability region like the one in Fig. 1a is applied to (13), (16). Hence by (15) h must be kept small so that $h\lambda_1 \in S$ even so the exact solution of $z' = \lambda_1 z$ drops to zero very fast compared to the exact solution of $z' = \lambda_m z$. In fact after a few integration steps the contribution to y_n of the first component $z_n^{[1]}$ is negligible. Still one is not allowed to enlarge h because of (15). For this reason one has tried to find methods where S contains large parts of the left half plane $\mathbb{C}^- = \{\mu \in \mathbb{C} | \mathrm{Re}\mu < 0\}$.

According to Dahlquist [2] a linear multistep method is <u>A-stable</u> if $\mathbb{C}^- \subset S$. He showed that A-stable linear multistep methods have at most error order 2 and among those with error order 2 the trapezoidal rule (3) has the smallest error constant. This result is of a very negative nature. To avoid this strong restriction on the error order Gear [4] introduced the concept of stiff stability.

2.4 Stiff stability and an appraisal of the concepts of error order and error constants

According to Gear [4] a method is <u>stiffly stable</u> if it is <u>"accurate"</u> in a neighborhood of $\mu = 0$ and $\{\mu \in \mathbb{C} \mid \mathrm{Re}\mu \leq -D\} \subset S$ for some $D > 0$. D is called the stiff stability parameter. Jeltsch [8] has given a rigorous difinition for stiff stability and showed the following positive result. To any natural number k and any $D > 0$ there exists a stiffly stable linear multistep method with the error order $p = k$ and D as stiff stability parameter. The proof is constructive in the sense that for given k and D a method was explicitely provided with the required properties. However, these methods are not accurate, since one can show that $c_{k+1} = O(D^{-k})$ as D tends to 0. Thus the high order of the method is offset by the large error constant. This may indicate that in addition to a high error order one should request that the error constant is kept small too. This can indeed be done when $p = k = 3$. Genin [6] has given a two parameter family of stiffly stable methods where

$$(17) \qquad D = \frac{1}{-36c_4\,(b_2 - 144c_4^2)} \quad .$$

Here b_2 is a positive parameter and the error constant c_4 is a negative parameter. Thus for any $c_4 < 0$ and $D > 0$ there exists a linear 3-step method of order 3 with the prescribed error constant c_4 and stiff stability parameter D. Hence this third order method can be made almost A-stable and the error constant can be made as small as one wishes. These methods are however bad since one can show that $c_{p+2} = c_5 = O(D^{-1})$ as D tends to 0. These results indicate that it is not enough to describe the accuracy of a method by just considering the order and the error constant. In order to compare the accuracy of methods of different order for a fixed h we shall introduce in the next section the Peano-kernel of a method. The size of these Peano-kernels is then related to the size of the stability region. We report here on some results of a forthcoming paper [9].

3. Lower bounds for the accuracy due to stability

3.1 The Peano-kernel of a linear multistep method

Assume that the linear multistep method has order $p \geq 1$ and $y(t) \in C^q$ where $q \leq p+1$. Then by the Peano-kernel theorem one has

$$(18) \qquad L_h y\,(t) = h^q \int_{-k}^{0} K_q(s) y^{(q)}(t - hs)ds$$

where the Peano-kernel is given by

$$K_q(s) = L_1 \frac{s_+^{q-1}}{(q-1)!} \quad ,$$

$$s_+ = \begin{cases} s & \text{for } s > 0 \\ 0 & \text{for } s \leq 0 \end{cases}$$

Clearly $K_q(s) \in C^{q-3}$ and $\frac{d^{q-2}}{ds^{q-2}} K_q(s)$ can have jumps only at $-k, -k+1, \dots, -1, 0$. $K_q(s)$ is a polynomial of degree $q-1$ in each of the intervals $(-i, -i+1)$, $i = k, k-1, \dots, 1$ and identically 0 if $s < -k$ or $s > 0$.

As an example we give $K_2(s)$ of the BDF (4)

$$(19) \qquad K_2(s) = \begin{cases} 0 & \text{if } s < -2,\ s > 0 \\ s + \frac{4}{3} & \text{if } -2 < s < -1 \\ -\frac{s}{3} & \text{if } -1 < s < 0 \end{cases}$$

Let M be the smallest constant such that for all $y \in C^q$ one has

$$(20) \qquad |[L_h y](t)| \leq h^q M \sup_{s \in [t, t+kh]} \left| y^{(q)}(s) \right| .$$

By (18) we see that

$$(21) \qquad M = \int_{-k}^{0} |K_q(s)| ds =: \|K_q\|_1 .$$

The estimate (20) for the local error is the best possible which is valid for all functions in C^q. Hence it will be realistic for some functions, but rather pessimistic for very smooth ones. Observe that the estimate (20) consists of three factors. M depends on the method only. In particular it is independent of the stepsize h. The last factor measures the smoothness of the solution of the underlying problem and is almost independent on h and the method. Observe that

$$c_{p+1} = \int_{-k}^{0} K_{p+1}(s) ds$$

and thus

$$(22) \qquad \|K_{p+1}\|_1 = |c_{p+1}|$$

if and only if $K_{p+1}(s)$ does not change sign.

3.2 Lower bounds for $\|K_q\|_1$

Let us introduce the following disks

$$D_R := \{\mu \in \mathbb{C} \mid |\mu + R| \leq R\} .$$

In [9] the following theorems are proved:

Theorem 1 Let $k \in \mathbb{N}$. There exists a constant C_k such that the following holds. If

$$(23) \qquad D_R \subset S$$

and $q \leq p+1$ then

$$(24) \qquad \|K_q\|_1 \geq C_k \left(\frac{R}{3}\right)^{p-2} .$$

C_k can be specified more precisely if $q = p+1$ for one has

Theorem 2 Let $k \in \mathbb{N}$ and $k+1 \geq p \geq \max\{k,2\}$. Assume that

$$(25) \qquad \begin{cases} D_R \subset S & \text{if } p = k+1 \\ D_R \cup \{\infty\} \subset S & \text{if } p = k . \end{cases}$$

Then

$$(26) \qquad -c_{p+1} \geq \frac{1}{12} \left(\frac{R}{6}\right)^{p-2} .$$

Observe that $c_{p+1} > 0$ if $p \leq k$. These results show very clearly that A-stability cannot be approximated with "accurate" methods of order $p > 2$. Applying (26) to Genin's example we see that even so one can make D in (17) as small as one wishes for any fixed $c_4 < 0$, the radius R will remain bounded by $-72c_4$.

In Jeltsch, Nevanlinna [9] similar results for explicit methods are given. In addition results of the same flavor are given where the size of the Peano-kernel is related to stability intervals on the real axis.

4. Contractive integration and final remarks

The stability region S gives a complete description of the stability of the numerical solution when solving the linear system (13). In a nonlinear problem one could linearize and request that the eigenvalues λ_i of $\frac{\partial f(t,y)}{\partial y}$ satisfy (15). However it is well known that this does not imply boundedness of the numerical solution.

In order to treat nonlinear problems we request here that $f : [0,\infty) \times \mathbb{C}^s \to \mathbb{C}^s$ satisfies for some $K > 0$ the monotonicity condition

$$(27) \qquad \mathrm{Re} < f(t,y) - f(t,z), y-z > \; \leq \; -\frac{1}{K} \| f(t,y) - f(t,z) \|^2$$

$$\text{for all } y,z \in \mathbb{C}^s,\ t \in [0,\infty) .$$

Here $< , >$ is an inner product in $\mathbb{C}^s$ and $\| \ \|$ the corresponding norm. Assume that the numerical solution $\{y_n\}_{n=0,1,\ldots}$ was found by a linear multistep method with $D_R \subset S$. Using Theorem 3.1 Nevanlinna [11] leads to the following error estimate. If

$$(28) \qquad Kh_o < 2R$$

then there exists $C = C(Kh_o)$ such that for all $h \leq h_o$ and $n \geq k$ we have

$$(29) \qquad \|y(t_n)-y_n\| \leq C\{ \max_{i=0,\ldots,k-1} \|y(t_i)-y_i\| + \sum_{j=0}^{n-k} \|(L_h y)(t_j)\| \} .$$

Let us first illustrate the condition (27). If $f(t,y) = \lambda y$ then (27) is equivalent to $\lambda \in D_{K/2}$. One can easily show that (27) implies that K is a Lipschitz constant of $f(t,y)$. The numerical range of the Jacobian of f and hence its spectrum lies in $D_{K/2}$. Moreover let $y(t)$ and $z(t)$ be two solutions of (1) then (27) implies that $\|y(t) - z(t)\|$ does not grow. Dahlquist [3] showed that if a linear multistep in its one-leg formulation is used to solve (2) with (27), then one has in an appropriate space and norm, that the distance between two numerical solutions does not grow, provided

$$(30) \qquad Kh \leq 2R .$$

Thus the discretisation has the same property of contractivity in an appropriate space as the solution of the continuous problem. If $Kh < 2R$ then this result can be transformed to linear multistep methods and one gets a different proof for the error bound (29).

The main result in section 3 is a step towards the goal to prove a lower bound for the amount of work which is needed to solve a stiff initial value problem with a given error tolerance. In order that (29) holds, one wants to integrate contractively, i.e. (28) has to hold. This gives an upper bound for h which increases linearly in R. Further one wants to carry out each integration step such that the local error

$|(L_h y)(t)|$ does not exceed a given error tolerance δ. Assuming that one has equality in (20) one could use the results of Theorem 1 and 2 to find another upper "bound" for h. This bound decreases with a negative power of R. Thus there is an optimal R and h. Unfortunately in general even almost equality in (20) does not hold. Moreover it is not clear how conservative the estimates in Theorem 1 and 2 are.

Theorem 1 and 2 show that there is no sense in trying to find linear multistep methods with $D_R \subset S_{,R}$ large with a high order and a small error constant.

References

[1] G. Dahlquist, Convergence and stability in the numerical integration of ordinary differential equations, Math. Scand. 4, 1956, 33-53.

[2] G. Dahlquist, Stability and error bounds in the numerical integration of ordinary differential equations, Trans. Roy. Inst. Tech., Stockholm, Nr. 130, 1959.

[3] G. Dahlquist, G-stability is equivalent to A-stability, BIT 18, 1978, 384-401.

[4] C.W. Gear, The automatic integration of stiff ordinary differential equations, Information processing 68, ed. A.J.H. Morell, North Holland Publishing Co., 1969, 187-193.

[5] C.W. Gear, Numerical initial value problems in ordinary differential equations, Prentice Hall, Englewood Cliffs, N.J., 1971.

[6] Y. Genin, A new approach to the synthesis of stiffly stable linear multistep formulas, IEEE Trans. on C.T., 20, 1973, 352-360.

[7] P. Henrici, Discrete variable methods in ordinary differential equations, Wiley, New York, 1962.

[8] R. Jeltsch, Stiff stability and its relation to A_0- and A(0)-stability, SIAM J. Numer. Anal. 13, 1976, 8-17.

[9] R. Jeltsch, O. Nevanlinna, Stability and accuracy of time discretisations for initial value problems, Numer. Math. 37, 61 (1981).

[10] J.D. Lambert, Computational methods in ordinary differential equations, Wiley, London, 1973.

[11] O. Nevanlinna, On the numerical integration of nonlinear initial value problems by linear multistep methods, BIT 17, 1977, 58-71.

Rolf Jeltsch
Institut für Geometrie und
praktische Mathematik
Institute of Technology
RWTH Aachen
Templergraben 55
D-5100 Aachen
Fed. Rep. of Germany

Olavi Nevanlinna
Institute of Mathematics
Helsinki University of
Technology
SF-02150 Otaniemi
Finland

ASYMPTOTIC ERROR EXPANSIONS

AND DISCRETE NEWTON METHODS

FOR ELLIPTIC BOUNDARY VALUE PROBLEMS

Klaus Böhmer

0. Introduction

Since this interdisciplinary workshop combines numerical analysts and physical chemists I want to start out with a chemical example. From this motivating example we proceed to the essential goal in this paper: We begin in 2. with a low order numerical method which we improve via discrete Newton methods. We give the basic error asymptotic results and then describe discrete Newton methods in 3. and give some numbers in 4.. Since we want to avoid formalism we restrict our discussion to the simple equation (2.1). More general problems are discussed in Böhmer [4,5,6].

1. A chemical example

Let a monomolecular reaction with the temperature T, the thermal conductivity k, the heat Q, and the velocity V of the reaction take place in a bounded (open) domain $\Omega \subset \mathbb{R}^2$. Then, with the Laplacian Δ, the following equation holds, see Ames [2],

$$k \, \Delta T = -QV \text{ in } \Omega. \tag{1.1}$$

Replacing V in (1.1) and using the Arrhenius relation

$$V = c\nu \exp(-E/RT), \tag{1.2}$$

where c is the concentration, E the energy of activity, R the universal gas constant and ν a scaling factor, we find

$$\begin{cases} k \, \Delta T = - c \, \nu \, Q \exp(-E/RT) \text{ in } \Omega \\ \quad T = g \text{ on } \delta\Omega. \end{cases} \tag{1.3}$$

So, nonlinear elliptic boundary value problems do occur in chemistry.

2. Error asymptotics for elliptic boundary value problems

We discuss an elliptic problem in the form

$$(2.1)\quad Fz := \begin{cases} \Delta z(x,y)+f(x,y,z(x,y))=0 & \text{in } \Omega \subset \mathbb{R}^2 \\ z(x,y)-g(x,y)=0 & \text{on } \partial\Omega . \end{cases}$$

To approximate the exact solution z we introduce gridlines Γ_ℓ and grid points Γ^h as

$\Gamma_\ell := \{(x,y)\,|\,x \in \mathbb{R}\,,\ y=nh \ \text{ or } \ x=mh,\ y \in \mathbb{R}\,,\ m,n \ \mathbb{N}\}$ and

$\Gamma^h := \{(x,y)\,|\,x=mh,\ y=nh,\ m,n \in \mathbb{N}\}$.

In grid points $(x,y)\in\Omega\cap\Gamma^h$ with $(x\pm h,y),(x,y\pm h)\in\Omega\cap\Gamma^h$, the so-called regular grid points, we replace $\Delta z(x,y)$ by the well known symmetric divided differences, thus defining an approximation ζ^h for the exact solution z of (2.1) by

$$(2.2)\quad \begin{cases} (\varphi^h F)\zeta^h(x,y) := \\ \frac{1}{h^2}\,\{\zeta^h(x+h,y)+\zeta^h(x-h,y)+\zeta^h(x,y+h)+\zeta^h(x,y-h)-4\zeta^h(x,y)\} \\ +\ f(x,y,\zeta^h(x,y))=0 \quad \text{in regular grid points } (x,y). \end{cases}$$

For an irregular grid point $(x,y) \in \Gamma^h \cap \Omega$, so at least one of the $(x\pm h,y)$, $(x,y\pm h)$ is no longer in Ω, we have to replace the corresponding $\zeta^h(x\pm h,y)$ or $\zeta^h(x,y\pm h)$ by a provisional value obtained by the following construction, which we describe along the gridline y=const = nh with $(x+h,y)\notin\Omega$: Let $\partial\Omega$ be smooth enough and such that for small enough h there is exactly one intersection point x^* of $\partial\Omega$ with the gridline y=nh within the interval $[x,x+h]$ and none within the interval $[x-(k-1)h,x]$. Then let p_k be the polynomial of degree k interpolating the k+1 points

$$(x^*,\ g(x^*,y)),\ (x,\zeta^h(x,y)),\ldots,(x-(k-1)h,\zeta^h(x-(k-1)h,y)).$$

Now we replace $\zeta^h(x+h,y)$ in (2.2) by $p_k\,(x+h)$. The above geometric condition has to be satisfied for all irregular grid points.

This method has a long history which is indicated in the following table

Author		Year	k	Order of Convergence
Gerschgorin	[10]	1930	0	1
Collatz	[7]	1933	1	2
Shortley-Weller	[15]	1938	2	2
Mikeladse	[11]	1941	2	2
Kreiss, Pereyra-Proskurowski-Widlund	[14]	1978	$k \leq 6$	2

Wasow [16] has shown that for $k \leq 1$ there is no error asymptotic, whereas for $k \leq 6$ we have (for a **proof** see [5]):

Theorem 1: *Let* $k \leq 6$ *and in* (2.1) $\partial\Omega$ *satisfy the geometric conditions described above and let* f *be smooth enough and* $\partial f/\partial z \leq c < 0$. *Then* $\varphi^h F$ *is stable and the unique solutions* z *and* ζ^h *satisfy*

$$(2.3) \quad \| \zeta^h - (z + h^2 e_2 + h^4 e_4) \|_2 = O(h^{k-1/2}),$$

where $\| \ \|_2$ *indicates the Euclidean norm for the "vectors"* $\zeta^h(x,y)$, $z(x,h), \ldots$, *with* (x,y) *ranging over all regular and irregular grid points.*

Comparing (2.3) with the Gerschgorin and Collatz results we see that $O(h^{k-1/2})$ in (2.3) does not provide orders 1 and 2 which we expect. So Pereyra-Proskurowski-Widlund [14] , who had proved Theorem 1 for f in (2.1) independent of z, conjectured for this case, that $O(h^{k-1/2})$ may be replaced by $O(h^{k+1})$. This conjecture is partly contained in

Theorem 2: *Let* $k \leq 4$ *and in* (2.1) *let* $\partial\Omega$ *satisfy the geometric conditions and* f *be linear in* z *and satisfy* $\partial f/\partial z \leq 0$. *Then we have in the maximum norm* $\|\cdot\|_\infty$ *that*

$$(2.4) \quad \|\zeta^h - (z + h^2 e_2 + h^4 e_4)\|_\infty = O(h^{k+1}).$$

This result proves the conjecture stated above for $k\leq 4$. The fact that only $k\leq 6$ are possible is not a really practically relevant restriction, since large values of k would require strong conditions on the geometry of $\partial\Omega$ and small values of h. This would cause large and widely banded matrices.

Since the proof for Theorem 1 is much too lengthy to be indicated here, see [5], we concentrate our discussion to shetch (detailled arguments see [6]) the

Proof of Theorem 2: It is based on results due to Bramble-Hubbard [17,18]. Let $P=(x,y)$ and $\overline{P}=(\overline{x},\overline{y})$ be neighbouring gridpoints. If $\sigma(P,P)$ and $\sigma(P,\overline{P})$ indicate the nonzero coefficients of $\zeta''(x,y)=\zeta^h(P)$ and $\zeta^h(x\pm h,y)$, $\zeta^h(x,y\pm h)=\zeta^h(\overline{P})$ in (2.2), respectively, and their modification in irregular points, we need the conditions

$$(2.5)\quad \begin{cases} \sigma(P,P)<0 \text{ for all grid points } P \text{ in } \Omega; \\ \left.\begin{array}{l} \sigma(P,\overline{P})\geq 0 \text{ for } P\neq\overline{P} \\ |\sigma(P,P)|\geq \sum_{\overline{P}\neq P} \sigma(P,\overline{P}) \end{array}\right\} \text{for all regular grid points } P \text{ in } \Omega \text{ and all grid points } \overline{P} \text{ in } \Omega; \\ \text{there exists a } \delta,\ 0<\delta<1, \text{ such that for all irregular grid points } P,\ \delta|\sigma(P,P)|\geq \sum_{\overline{P}\neq P} |\sigma(P,\overline{P})|. \end{cases}$$

If the set of grid points in Ω is "connected" [17,18] and if we have

$$(2.6)\quad \begin{array}{ll} |(\varphi^h F)\zeta^h(P)|\leq C\cdot h^n & \text{for all regular grid points } P \text{ in } \Omega, \\ |(\varphi^h F)\zeta^h(P)|\leq C\sigma(P,P)h^n & \text{for irregular grid points } P \text{ in } \Omega \end{array}$$

then, by [17,18], $\|\zeta^h(P)-z(P)\|_\infty \leq C^* h^n$.

In a sequence of Lemmas, see [6], one shows that the coefficients in (2.2) and its modifications in irregular points satisfy (2.5) and that the consistency conditions (2.6) hold.

By computing an asymptotic expansion for

$(\varphi^h F)z(P)=h^2 g_2(z,f)(P)+h^4 g_4(z,f)(P)+\ldots$

and equating $Fe_2=g_2(z,f)$ in Ω and $e_2=0$ on $\partial\Omega$, one finds e_2. Then one proceeds in the wellknown inductive way to compute e_4 a.s.o., hence, to obtain (2.4). □

3. Discrete Newton methods

Let E, E_o be Banach spaces and $F: C \to E_o$, $C \subseteq E$, be differentiable in C. Then a modified Newton method is defined by

$$(3.1) \quad \begin{cases} \tilde{z}_o \in C, \quad \text{compute } \tilde{z}_\ell \text{ from} \\ F'(\tilde{z}_o)\,(\tilde{z}_\ell - \tilde{z}_{\ell-1}) = -F\tilde{z}_{\ell-1}\,. \end{cases}$$

Since the equations in (3.1) usually are not solvable directly we use discretizations. Now we may state as a general principle, that practically every discretization method, e.g. the φ^h defined in section 2, has the following property, see Böhmer [3] : With $\mathcal{L}(.,.)$, the set of linear bounded operators relating two spaces, let

$$3.2) \qquad F_o w := Lw+d, \quad L \in \mathcal{L}(E,E_o),\; d \in E_o$$

then

$$(3.3) \quad \begin{cases} (\varphi^h F_o)\omega^h = (\varphi^h L)\omega^h + \Omega_h d \quad \text{with} \\ \varphi^h L \in \mathcal{L}(E^h, E^h{}_o) \quad \text{and} \quad \Omega_h \in \mathcal{L}(E_o, E^h{}_o); \end{cases}$$

here E^h and $E^h{}_o$ are discrete spaces, corresponding to E and E_o respectively. The application of (3.1) to (2.1) yields, with $w_\ell := z_\ell - z_{\ell-1}$ the equation

$$(3.4) \quad \begin{cases} \Delta w_\ell(x,y) + \frac{\partial f}{\partial z}(x,y,\tilde{z}_o(x,y))\, w_\ell(x,y) = \\ -(\Delta z_{\ell-1}(x,y) - f(x,y,\tilde{z}_{\ell-1}(x,y))) \text{ in } \Omega\,, \\ w_\ell(x,y) = -(\tilde{z}_{\ell-1}(x,y) - g(x,y)) \text{ in } \partial\Omega\,. \end{cases}$$

We discretize (3.4) with the method in 2. to obtain

$$(3.5) \quad \begin{cases} \frac{1}{h^2}\{\omega^h_\ell(x+h,y)+\omega^h_\ell(x-h,y)+\omega^h_\ell(x,y+h)+\omega^h_\ell(x,y-h)-4\omega^h_\ell(x,y)\} \\ \frac{\partial f}{\partial z}(x,y,\tilde{z}_o(x,y)).\,\omega^h_\ell(x,y) = - \{\Delta\tilde{z}_{\ell-1}(x,y)-f(x,y,\tilde{z}_{\ell-1}(x,y))\} \\ \text{in regular grid points.} \end{cases}$$

In irregular grid points the corresponding modifications have to be used. The immediate question arising here is, how to obtain $\tilde{z}_o$ and $\tilde{z}_{\ell-1}$, respectively. For the discretization of (3.4) in (3.5) we need to know $F\tilde{z}_{\ell-1}(x,y)$ an $\frac{\partial f}{\partial z}(x,y,\tilde{z}_o(x,y))$ only in grid points. Therefore it is enough to use univariate interpolation in x and y directions, respectively, to extend the ζ^h-values from 2. to obtain $\tilde{z}_o = \zeta^h$ and $\Delta\tilde{z}_o$ in grid points (x,y). This may be achieved by either using piecewise polynomials of degree $\geq k$ or good enough symmetric or unsymmetric (near the boundary) divided difference formulas for $\Delta\tilde{z}_{\ell-1}$, see e.g. Böhmer [3,4,5], so

$$z_o := T\zeta^h \quad , \quad \zeta_1^h := \zeta^h + \omega_1^h$$

or in general

$$(3.6) \qquad z_{\ell-1} := T\zeta^h_{\ell-1} \quad , \quad \zeta^h_\ell := \zeta^h_{\ell-1} + \omega^h_\ell \ .$$

In (3.6) we have used instead of the $\tilde{z}_{\ell-1}$ in (3.1), (3.4) the functions $z_{\ell-1}$, since usually $z_{\ell-1} \neq \tilde{z}_{\ell-1}$. For the ζ^h_ℓ, defined by (3.5), (3.6), where we replace in (3.5) $\tilde{z}_o$ and $\tilde{z}_{\ell-1}$ by z_o and $z_{\ell-1}$ respectively, we have the following

Theorem 3: Let ζ^h_ℓ be defined by (3.5), (3.6) and let f and $\partial\Omega$ be smooth enough.

$$(3.7) \qquad \begin{cases} \|\zeta_1^h - z\|_\delta = O(h^{k-1-3/\delta}) \\ \|\zeta_2^h - z\|_\delta = O(h^{k-3-3/\delta}) \end{cases}$$

where $\|\cdot\|_\delta$ is defined as in (2.3) for $\delta = 2,\infty$. For nonlinear problems we have $\delta=2$ and $k \leq 6$, whereas for linear problems $\delta=\infty$ and $k \leq 4$ ist available.

Numerical experience seems to indicate that the reduction from $O(h^{k+1})$ in (2.3) to $O(h^{k-1})$ and $O(h^{k-3})$ in (3.7) does not happen in some cases.

Proof of Theorem 3 (Sketch): For the smoothness properties, discussed here, (3.4) is essentially the equation defining e_2, which is, by the discretization (3.5), reproduced of order $O(h^{k+1-2})$ for the $\|\cdot\|_\infty$. The reduction by -2 in the exponent reflects the fact, that in the transition from z_o to Fz_o two powers of h are lost. A similar argument holds for ζ_2^h, proving (3.7). A detailed proof for different cases is given in [3,6]. ◻

The main advantage of the discrete Newton methods compared with other acceleration techniques, e.g. Pereyra's [12,13] deferred corrections or Frank's and Ueberhuber's [8,9] iterated defect corrections is given by the following fact: All Newton corrections are obtained as solutions of linear equations with the same matrix and only different right hand sides. In the other cases a nonlinear system of more or less the same complexity as the original discrete problem has to be solved for every iteration (however with increasingly good starting values). In Richardson extrapolation the approximations on an appropriate (coarse) grid are improved by using finer and finer grids. In the correction methods, mentioned above, the corrections may be computed on one (or two, see Allgower-Böhmer-Mc Cormick [1]) appropriate grids. To find this grid usually requires, as in Richardson extrapolation, some tests with very coarse grids, which are unavoidable if one wants to have reliable results.

4. Numerical results

We finally give some numbers to demonstrate the power of the method. Let

$$\Delta z(x,y)+2\sin(x+y) = 0 \quad \text{in } \Omega$$

$$z(x,y)-\sin(x,y) = 0 \quad \text{on } \partial\Omega, \text{ where}$$

$$\Omega := \{(x,y) \mid \left(\frac{x-y}{0.5}\right)^2 + \left(\frac{y-1}{0.78}\right)^2 < 1\}.$$

Then we get for the exact solution $z(x,y)=\sin(x+y)$ and the approximations ζ^h, ζ_1^h, ζ_2^h, ζ_3^h for $h = \frac{1}{25}$ the errors $\|z-\zeta^h\|_\infty,\ldots,\|z-\zeta_3^h\|_\infty$, as given in (2.3), the following table. We want to point out that (3.7) only provides us with good results for ζ_1^h. Nevertheless we include the error ζ_2^h and ζ_3^h to show that for increasing values of k the higher iterates ζ_ℓ^h are excellent approximations, at least for simple cases.

k	$\|z-\zeta^h\|_\infty$	$\|z-\zeta_1^h\|_\infty$	$\|z-\zeta_2^h\|_\infty$	$\|z-\zeta_3^h\|_\infty$
0	3.1, -2	3.3, +0	2.4, +2	1.7, +4
1	1.7, -4	1.3, -4	1.5, -4	1.8, -4
2	2.3, -5	2.4, -6	1.7, -6	8.6, -7
3	2.4, -5	9.2, -8	4.1, -8	3.1, -8
4	2.4, -5	4.2, -9	5.0, -10	2.5, -10
5	2.4, -5	9.9, -9	8.7, -12	3.6, -12
6	2.4, -5	9.9, -9	4.7, -12	5.0, -14

References

1 Allgower,E.,K.Böhmer, S.Mc Cormick: Discrete correction methods for operator equations, in "Numerical solution of nonlinear equations",Eds.:E.Allgower, K.Glashoff, H.-O.Peitgen, 30-97 (1981).

2 Ames, W.F.: Nonlinear partial differential equations in engineering, Academic Press, New York-London (1965).

3 Böhmer,K.: Discrete Newton methods and iterated defect corrections, Numer.Math.37, 167-192(1981).

4 Böhmer,K.: Asymptotic expansions for the discretization error in Poisson's equation on general domains, in "Multivariate Approximation Theory",Eds.W.Schempp, K.Zeller, ISNM 51,30-45 (1979).

5 Böhmer,K.: High order difference methods for quasilinear elliptic boundary value problems on general regions, Math.Research Center, University of Wisconsin-Madison, Technical Summary Report Nr. 2042, 1980.

6 Böhmer,K.: Asymptotic expansion for the discretization error in linear elliptic boundary value problems on general regions, Math.Z. 177, 235-255 (1981).

7 Collatz,L.: Bemerkungen zur Fehlerabschätzung für das Differenzenverfahren bei partiellen Differentialgleichungen, Z. Angew. Math. Mech. 13, 56-57 (1933).

8 Frank, R.: The method of iterated defect-correction and its application to two-point boundary value problems, Part I./II, Numer. Math. 25, 409-419 (1976), 27 407-420 (1977).

9 Frank, R., J. Hertling, C.W. Ueberhuber: A new approach to iterated defect corrections.

10 Gerschgorin, S.: Fehlerabschätzung für das Differenzenverfahren zur Lösung partieller Differentialgleichungen, Z. Angew. Math. Mech. 10, 373-382 (1930).

11 Mikelads, S.: Neue Methoden der Integration von elliptischen und parabolischen Differentialgleichungen, Izv. Akad. Nauk SSSR, Seria Mat. 5, 57-74 (1941) (russian).

12 Pereyra, V.: Iterated deferred corrections for nonlinear operator equations, Numer. Math., 10, 316-323 (1967).

13 Pereyra, V.: Highly accurate numerical solution of casilinear elliptic boundary value problems in n dimensions, Math. Comp., 24, 771-783 (1970).

14 Pereyra, V., W. Proskurowski and O. Widlund: High order fast Laplace solvers for the Dirichlet problem on general regions, Math. Comp., 31, 1-16 (1977).

15 Shortley, G., R. Weller: The numerical solution of Laclace's equation, J. Appl. Phys. 9, 334-348 (1938).

16 Wasow, W.: Discrete approximations to elliptic differential equations, Z. Angew. Math. Phys. 6, 81-97 (1955).

Klaus Böhmer
Fachbereich Mathematik
der Universität Marburg
Lahnberge

D-3550 Marburg

THE USE OF SPARSE MATRIX TECHNIQUES IN ODE - CODES

Per Grove Thomsen

Abstract. Problems involving large systems of ordinary differential equations with sparse Jacobian matrices can be solved efficiently using low-order L-stable one-step methods. Sparse matrix techniques are applied to reduce computational work and to save storage when solving the large systems of linear equations that arise. Variable stepsize strategies have to be used as the systems of ODE's are normally stiff. Iterative refinement is used in connection with incomplete factorisations obtained from the use of drop-tolerances during the factorisation process. This combination leads to reductions in both storage consumption and in the number of evaluations of the Jacobian matrix that has to be performed. Evidence of the efficiency of the strategies involved are given in the form of numerical results from a FORTRAN program package SPARKS that employs a semi implicit Runge-Kutta method.

1 Introduction

In many real-world applications involving the modelling of dynamic systems the final mathematical model consists of a large system of ordinary differential equations that have to be solved numerically. Such large systems may be linear or non-linear but some properties will be shared by almost all systems when they are large in size. Consider the general system of ODE's in autonomous form

$$\underline{y}' = \underline{f}(\underline{y}) \quad , \quad \underline{y} \in R^N \quad , \quad N >> 1 \tag{1}$$

The Jacobian matrix of the system is defined by

$$J = \left\{ \frac{\partial f_i}{\partial y_j} \right\} \quad , \quad i,j = 1,2,\ldots.,N \tag{2}$$

One very important property of the systems we consider is that the Jacobian will have many zero elements, the matrix is sparse. Unless special care is taken by the code to preserve this sparsity it will not give any advantages using normal solution methods.

Another important feature to be noticed is that the eigenvalues of the Jacobian will differ in absolute value by several orders of magnitude thus giving rise to stiff systems of ODE's. Because of this we have to select methods that can solve stiff systems in an efficien way, the selection of method is considered in the first section. The variable stepsize strategy is described and some of the special ways of gaining in efficiency are outlined.

The second section is concerned with some of the features of the sparse techniques that are introduced to take advantage of the sparsity of the Jacobian. The methods involved assume very little about the structure of the matrix and are intended to be used for general sparse matrices. The experience with the methods used has shown that they will be efficient also in cases where one might take advantage of f.ex. band structure.

Evaluation of the Jacobian matrix may be done either from closed form expressions or by means of numerical differentiation. The two cases differ in computational cost and special techniques may be used to reduce the work when numerical differencing has to be used. These techniques are outlined in the third section.

The final section gives some numerical evidence of the efficiency that is the result from implementation of these methods in form of a FORTRAN code. A program package SPARKS has been developed and some test results from running this code is displayed to provide such evidence.

2 Solution method for stiff system of ODE's.

There exists a large number of candidates for the numerical solution of (1). In order to select one that is suitable for the case where N is large and J is sparse we have to consider stability and accuracy requirements and computational work. Most practical applications will only require moderate accuracy and even methods of low order will be suitable. The stability properties will be the best attainable if we require L-stability (for definition see e.g. ref(1)). Only implicit methods of low order can provide the property of L-stability and among the low-order implicit methods we want to select one that lead to a minimum of work in each step. The overhead involved in changes of stepsize will have to be taken into account giving priority to one-step methods.

The Rosenbrock processes of Wolfbrandt (ref.(2)) and the semi-implicit Runge-Kutta methods of Nørsett (ref(3)) are second order accurate methods with all the properties we are looking for and providing reliable estimates of the local error thus enabling us to control the accuracy by means of a variable stepsize strategy. Backward differentiation methods have been considered as candidates but the extra work involved in stepsize-changes has proved too large a disadvantage when very large systems have to be treated.

The semi-implicit Runge Kutta method of Nørsett is defined as follows

$$\underline{k}_i = \underline{f}(\underline{y}_{n,i} + \gamma h \underline{k}_i) \quad , \; i = 1,2,3. \tag{3}$$

$$\underline{y}_{n,i} = \underline{y}_n + h \sum_{j=1}^{i-1} a_{ij} \underline{k}_j \tag{4}$$

$$\underline{y}_{n+1} = \underline{y}_n + h \sum_{i=1}^{3} b_i \underline{k}_i \tag{5}$$

Together with these formulae defining the method the local truncation error is estimated by the formula

$$\underline{LE}_n = h \sum_{i=1}^{3} d_i \underline{k}_i \tag{6}$$

The method has three stages but a special choice of the coefficients lead to a L-stable second order method that has a minimum average amount of work. We specify this method in the notation used by Butcher (ref (4))

$$\begin{array}{|ccc} \gamma & & \\ a_{21} & \gamma & \\ a_{31} & a_{32} & \gamma \\ \hline b_1 & b_2 & b_3 \\ \hline d_1 & d_2 & d_3 \end{array}$$

$\gamma = (1 + \sqrt{2}) / 2$

$a_{21} = 9 - 28 \gamma$

$a_{31} = b_1 = (43 + 10 \gamma) / 62$

$a_{32} = b_2 = (19 - 10 \gamma) / 62$

$b_3 = 0 \quad , \; b_1 + b_2 = 1$

$$d_1 = (28 - 80 \gamma)/93 \; , \quad d_2 = (28 - 18 \gamma)/93 \; , \quad d_3 = 62 \gamma / 93 \; .$$

Because in this formula we have that $a_{31} + a_{32} = 1$ it is realised that

$$\underline{k}_{1,n+1} = \underline{k}_{3,n} \tag{8}$$

when the stepsize is konstant. This means that only two new stages have to be computed in each step unless the stepsize has been changed. Since stepsize changes can be controlled and only happens in a small percentage of the steps if we want that it is seen that on average the method will be two-stage.

The formula for computing $\underline{k}_i$ is a system of non-linear equations in the general case. This implicit system (3) has to be solved by iteration and to ensure good convergence a Newton iteration scheme is used with some modifications. The form of Newton iteration in this case is given by

$$(\mathbf{I} - h \gamma J) \underline{\delta}_i^s = \underline{k}_i^s - \underline{f}(\underline{y}_{n,i} + h \gamma \underline{k}_i^s) \tag{9}$$

$$\underline{k}_i^{s+1} = \underline{k}_i^s + \underline{\delta}_i^s \tag{10}$$

The iteration matrix $(\mathbf{I} - h\gamma J)$ is unchanged from one iteration to the next and also from step to step over large intervals of solution. This is obviously not harmful when h is small while the Jacobian will change very little when h is allowed to grow to large values. This means that it is possible to make one LU-factorisation of the matrix and use this over a long sequence of steps. In this way the very costly factorisation processes will only have to be performed if the convergence-rate of the Newton process becomes unacceptable or if the stepsize is changed.

The stepsize is controlled by the attempt to keep the local truncation error within limits prescribed by the user's accuracy requirements.This is expressed in

$$\varepsilon/\alpha < || \underline{LE} ||_2 < \varepsilon \tag{11}$$

where ε is the prescribed accuracy and α is a design parameter that can be used to tune the method (see ref (5)) . In the next section some strategies are explained with the purpose to reduce even further the number of times that evaluation of J and factorisation of the iteration matrix has to be performed.

3 The application of sparse matrix techniques

When the Jacobian is sparse the iteration matrix will be sparse too and the application of a compact storage scheme may reduce storage requirements and the use of a factorisation method that minimises the total amount of storage needed will be desirable. These objectives are exactly what sparse matrix techniques intend to fulfill at the same time obtaining a minimum operation count for the factorisation process.

If the matrix has a banded structure the choice of a band oriented factorisation method will be appropriate but in the more general case a carefully designed method must be implemented. The following gives a short outline of the features of a sparse matrix solver for linear equations that was developed for this purpose, it is coded in FORTRAN as a subroutine package SSLEST (see ref (6)).

The storage scheme that is used is based on keeping the matrix elements stored in ordered lists , ordered by rows and by coulumns (the elements are stored in one copy only the coulumn ordering is stored symbolically). This data structure has been selected because it makes searches along rows and coulumns easy but as the structure will be compact there will be very little room to store new elements that are generated during the factorisation process. These elements are called fill-in's and their number may be minimised by careful selection of pivots in the elimination and permuting rows and coulumns according to the pivot sequence.

The strategy for selecting pivots in SSLEST is a generalised Markowitz strategy and may be explained looking at the matrix A at the stage in the elimination is taking place in coulumn no. s. The formula for the element a_{ij} is the well known

$$a_{ij}^{s+1} = a_{ij}^{s} - a_{is}^{s} a_{sj}^{s} / a_{ss}^{s} \qquad (12)$$

We now keep track of the following counts:

nr(i,s) = number of non-zero elements in row i with coulumn number larger than s.

nc(i,s) = number of non-zero elements in coulumn j with rownumber larger than s.

The pivot element is chosen among those elements a_{ij} with $i \geq s$ and $j \geq s$ that have a minimum value of the product

$$M_{i,j} = nr(i,s) \cdot nc(i,s) \qquad (13)$$

The strategy may be modified to include only elements of the first few neighbourghing rows to the one with number s. It is also desirable that only pivots that are satisfying a stability criterion can be selected this means that a pivot must satisfy

$$|a_{ij}^{s}| > \max_{s \leq k \leq N} (|a_{ik}^{s}| / u) \qquad (14)$$

where u is a parameter ($u \geq 1$). This test will avoid the problem of small pivots giving rise to growth in the effects of rounding errors.

A special feature that is available and may be used with good effect is the drop-tolerance. If the drop tolerance is greater than zero all elements that are generated from (12) but are smaller in absolute value than this tolerance will be dropped from the list. They do not occupy any storage and are ignored in the rest of the elimination process thus saving storage as well as operations.Of cause the factorisation will be inaccurate but as we are already dealing with an "old" matrix it might not harm our iteration (9-10) too badly. If the convergence becomes too slow the Jacobian is reevaluated and iterative refinement used to improve the solution in (9).

The iterative refinement process is outlined assuming we have an approximate factorisation to the iteration matrix

$$A = (\mathbf{I} - h \gamma J) \quad , \quad \tilde{L} \tilde{U} \simeq A .$$

refinement of the solution is performed according to the iterative scheme

$$\underline{b}^m = \underline{k}_i - \underline{f}(\underline{y}_{n,i} - h\gamma\, \underline{k}_i) \tag{16}$$

$$\tilde{L}\,\tilde{U}\,\underline{\delta}_j^m = \underline{b}^m \tag{17}$$

$$\underline{r}_j = \underline{b} - A\,\underline{\delta}_j^m \tag{18}$$

$$\tilde{L}\,\tilde{U}\,\underline{c}_j = \underline{r}_j \tag{19}$$

$$\underline{\delta}_{j+1}^m = \underline{\delta}_j^m + \underline{c}_j \tag{20}$$

the iteration is continued until: $\|\underline{c}_j\|_2 < \varepsilon_1 \|\underline{\delta}_j\|_2$ (21)

This iteration will converge in a few iterations when the factorisation is sufficiently accurate. If the convergence becomes slow a new factorisation is needed and it will be performed unless this was done in the previous step, in that case the stepsize must be reduced or the drop tolerance ajusted to a smaller value.

The use of iterative refinement makes it necessary to store the iteration matrix in factorised form and also the matrix itself. This may sometimes lead to extra storage consumption but the use of drop tolerance will in many cases save more than is needed for the extra copy and the overall process is using less. A reduction in the number of Jacobian evaluations is also experienced as shown in the example in the last section.

4 Jacobian evaluation techniques

When systems with N very large are considered the main part of the computational work will be involved in the solution of linear systems of equations. Tactics for minimising the work involved has been treated in the section above. The actual evaluation of Jacobians can be very time consuming too especially in the cases where it has to be done by numerical differencing. The case where the elements are available in closed form does only require that the user has to provide a routine that give as results the elements with row and coulumn numbers, in SPARKS they will be accepted in any order as ordering is done internally.

In the case where the elements have to be found from numerical differentiation we will require the positions of the non-zero elements as a minimum of information. First of all this information is often easy to obtain from the model directly. If it was not available the matrix would have to be treated as a full matrix and our sparse techniques would not be applicable. The size of such systems would have to be very limited. When the positions of the nonzero elements are known it enables us to minimise the amount of work needed to estimate the elements by differences. The formula for the element in the i,j'th position is

$$\frac{\delta f_i}{\delta y_j} = \frac{f_i(\underline{y} + \Delta y_j) - f_i(\underline{y})}{\Delta y_j} \tag{22}$$

where Δy_j is a perturbation on the j'th component of the vector $\underline{y}$. Consider now the case where the function $f_r(\underline{y})$ only depends on some of the components in $\underline{y}$ the indices of these form the set I_r , the indices of the components that are in the definition of f_s form the set I_s. If the two sets I_r and I_s have no members in common then it will be possible to evaluate the two elements

$$\frac{\delta f_r}{\delta y_p}, \quad p \in I_r \quad \text{and} \quad \frac{\delta f_s}{\delta y_q}, \quad q \in I_s$$

using the same evaluation of the vector function $\underline{f}$. This will mean great savings in most sparse cases and it is a fairly simple task to examine the matrix structure and identify the disjoint index-sets. Inthe next chapter an example is shown where the numbers of function evaluations with and without use of the structural information shows some of the advantages.

5 Numerical experiments

An example of a small system N=4 is given by Krogh (ref (7)) the details may not be important in this text and we refer to the reference for that.It is intended to supply evidence of the efficiency of the iterative refinement process.

The amount of work is measured by the number of evaluations of the right hand side, of the jacobian and the number of factorisations.

	Without IR	With IR
function evaluations	320	347
Jacobian evaluations	26	13
LU factorisations	50	23

It is seen that due to the fact that more iterations has been needed in the case where iterative refinement was used more function evaluations have been made but the more expensive operations have been halved.

The second example is from a chemical problem taken from Hull (ref (8)) with 33 simultaneous ODE's. The structure of the Jacobian is shown in the figure and the number of function evaluations needed for its evaluation with and without the identification of disjoint index sets are given together with results from applying different drop-tolerances (TOL).

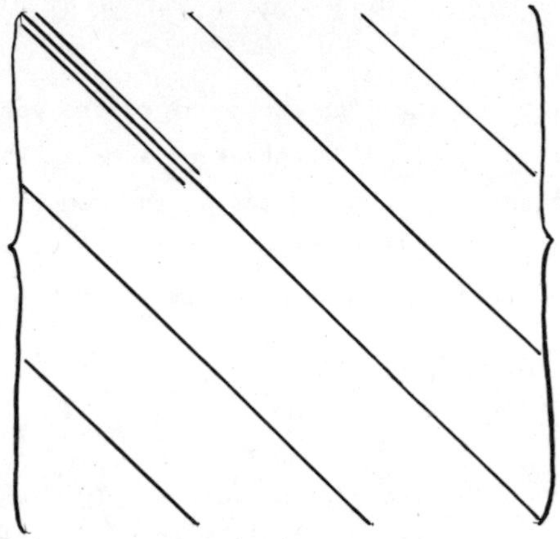

Structure of iteration matrix

Number of f.eval. without index identification	34
Number of f. eval. with index identification	6
Number of fill-in's with TOL=0	275
Number of fill-in's with TOL= 10^{-8}	9
Number of fill-in's with TOL= 10^{-3}	0

6 References

(1) Lambert, J.D."Computational Methods in Ordinary Differential Equations John Wiley & sons , (1973).

(2) Wolfbrandt,A. "Astudy of Rosenbrock processes with respect to order conditions and stiff stability", Computer Sciences 77.01 R, Chalmers University of Technology , Gøteborg ,Sweden.(Thesis).

(3) Nørsett,s.p. "semi explicit Runge-Kutta methods", Mathematics and Computation 6/74 NTH, ISBN 82-7151-009-6.

(4) Butcher,J.C. "Implicit Runge-Kutta Processes", Math.of Comp. vol 18.(1964)

(5) Houbak,N. and Thomsen,P.G. "SPARKS, a FORTRAN subroutine for the solution of large systems of ODE's with sparse Jacobians", rep.NI 79-02, DTH, Inst. for Num Anal.

(6) Zlatev,Z.,Barker,V.A. and Thomsen,P.G. "SSLEST - a subroutine for solving sparse systems of linear equations", NI 78-01, Inst. for Num.Anal. DTH.

(7) Krogh, F. "On testing a subroutine for the numerical integration of ODE's", J. ACM, vol 20, no.4 (1973).

(8) Hull, T.E.,Enright, W.H., Sedgwick,A.E.,"Memorandum on testing of stiff techniques", Dept. of Computer Science , Univ. of Toronto , May 1972.

ON CONJUGATE GRADIENT METHODS FOR LARGE SPARSE SYSTEMS OF LINEAR EQUATIONS

O. Axelsson, University of Nijmegen, The Netherlands

Abstract.

A survey of some recent results on preconditioned conjugate gradient methods is presented.

1. Introduction.

In order to avoid the fill-in (within the envelope) in the factorization of large sparse matrix problems, it is now wellknown that a succesfull method is to use incomplete factorization and accelerate the method by an iterative, gradient type method. Since the rate of convergence of the conjugate gradient method is fastest in energy norm and since it does not need any estimation of parameters (like extreme eigenvalues), this latter method is now widely used. For early references of preconditioned conjugate gradient methods for sparse problems, see Axelsson [1972], [1974], [1976], Tuff and Jennings [1973] and Meijerink, van der Vorst [1977].
In this talk we first shortly survey some preconditioning techniques used. We then describe a generalized preconditioned conjugate gradient method for non-symmetric problems.

2. The preconditioned conjugate gradient method.

Let A be a symmetric positive definite matrix. We want to solve $A\underset{\sim}{u} = \underset{\sim}{a}$. Given an initial approximation $\underset{\sim}{u}^0$ and a nonsingular preconditioning $\tilde{A}$ of A, the preconditioned iterative refinement method has the form

$$\underset{\sim}{u}^{k+1}-\underset{\sim}{u}^k = -\underset{\sim}{r}^k,\ k = 0,1,2,\ldots$$

where

$$\underset{\sim}{r}^k = \tilde{A}^{-1}(A\underset{\sim}{u}^k-\underset{\sim}{a})$$

is the pseudoresidual. In practise it is best to calculate $\underset{\sim}{r}^k$ by the solution of

$$\tilde{A}\underset{\sim}{r}^k = A\underset{\sim}{u}^k-\underset{\sim}{a}.$$

A generalization of this method is the preconditioned conjugate gradient method:

$$\begin{aligned}
&\underset{\sim}{r}^0 := \tilde{A}^{-1}(A\underset{\sim}{u}^0-\underset{\sim}{a});\ \underset{\sim}{d}^0 := -\underset{\sim}{r}^0;\\
&\underset{\sim}{d}^k := -\underset{\sim}{r}^k + \beta_k\underset{\sim}{d}^{k-1};\\
&\underset{\sim}{u}^{k+1}-\underset{\sim}{u}^k := -\lambda_k\underset{\sim}{d}^k; \qquad\qquad k = 0,1,\ldots\\
&\underset{\sim}{r}^{k+1}-\underset{\sim}{r}^k := -\lambda_k\tilde{A}^{-1}A\underset{\sim}{d}^k;
\end{aligned}$$

Here λ_k, β_k are determined by certain innerproducts to be given in Section 3, for instance, in such a way that $\underset{\sim}{d}^j A\underset{\sim}{d}^l = 0,\ j \neq l$

Although, in a finite dimensional space $\mathbb{R}^N$, the method apparently is terminating after at most N steps, it is well known that it should be considered as an iterative method in the presence of round-off errors and because the iteration error is often small enough, even when $k \ll N$. One may stop, for instance when $\| \underset{\sim}{r}^{k+1} \| / \| \underset{\sim}{r}^0 \| \leq \varepsilon$, ε a small enough quantity. The number of iterations needed, depends essentially on the distribution of the eigenvalues of $\tilde{A}^{-1}A$. An upper bound in the symmetric positive definite case is given by $\frac{1}{2}\{H(\tilde{A}^{-\frac{1}{2}}A\tilde{A}^{-\frac{1}{2}})\}^{\frac{1}{2}} \ln \frac{2}{\varepsilon}$ where H is the spectral condition number. The essential number of operations at each iteration comes in general from the matrix vector multiplication by A and from solving linear equations with matrix $\tilde{A}$. A few extra operations per step come from the recursion and the calculation of λ_k, β_k.

From the above we realize that we should try to construct $\tilde{A}$ in such a way that the spectral condition number H is (much) smaller than that of A, but under the constraint that linear systems with matrix $\tilde{A}$ are (much) more easily solved than those with A.

In general, the matrix $\tilde{A}$ is determined as a product of two triangular matrices, $\tilde{L}$, $\tilde{U}$ or, in the symmetric case, as $\tilde{L}\tilde{D}\tilde{L}^T$. A full factorization would lead to an iterative refinement method. (Even in a full factorization, $\tilde{A}^{-1}A$ may in ill conditioned problems be far from the identity.) As remarked in Axelsson [1976], an "infinity" of choices of $\tilde{A}$ exists. Here five, slightly different methods of constructing $\tilde{A}$, which have been used, shall be described.

a. The Generalized SSOR method.

Let $A = D+L+U$, where D is the (block-)diagonal part of A and L and U are the (block-)triangular strictly lower and upper parts of A.
Then we let

$$\tilde{A} = (\tilde{D}+L)\tilde{D}^{-1}(\tilde{D}+U),$$

where $\tilde{D}$ is a suitably chosen diagonal matrix. (See Axelsson [1972] for an application on difference methods and Gustafsson [1979] in a more general context. If $\tilde{D} = \frac{1}{\omega} D$, $0 < \omega < 2$, we have the classical SSOR method.)

The advantage of this choice is that one does not have to construct any new entries, apart from those in $\tilde{D}$. The construction of $\tilde{D}$ is simple, but in many problems this conditioning is not accurate enough.

b. "Chopping off" methods.

When performing a complete LU-decomposition, one often notices that most of the fill-in entries are quite small, in particular in positions away from the non-zero entries of A. Hence an obvious method to preserve sparsity during the solution would be to neglect all entries in the triangular factors for which the relative size of the absolute value is small enough. This naturally demands a full factorization at first, but since this is only done once, it needs only a few data transfers from backing store. One finds that often only 5-10% of the storage of the full factors is needed so each iteration step can be performed much faster that in an iterative refinement method.

Quite a few iterations may be needed however, so the total CPU-time is not smaller. Although most entries in the full factorization will not be used, they have still to be calculated, and this makes the method impossible to use on for instance large 3 dimensional partial differential equation problems. The method has been used, among others by Tuff, Jennings [1973].

c. Incomplete factorization (IC).

A method to avoid a full factorization is described in a particular case by Varga [1960] and in a more general context by Meijerink, van der Vorst [1977], and Beauwens [1976]. In this, fill-in entries are neglected during the factorization. Only fill-in in certain in advance chosen positions is accepted and neglection of fill-in in each pivot-row is done prior to the elimination. Hence the extent of fill-in is controlled.

A problem with incomplete factorization is that the factorized matrix may be instable since the pivot entries may become small or negative, even if A itself is positive definite. For $\mathcal{M}$-matrices, Meijerink and van der Vorst prove, however, that the factorization is always stable. (For a more general class of matrices, see Varga (1980).)

A remedy towards the instability can be to add a positive number to the diagonal, that is to factorize $A + \approx{D}$, where $\approx{D}$ is a non-negative matrix, instead of A. If the entries of $\approx{D}$ become too large, the approximate factorization matrix $\tilde{A}$ is not very close to A, however, which may result in many more iterations.
For discrelized elliptic problems, the asymptotic rate of convergence of this method is slower that that for method a.

d. Modified incomplete factorization (MIC).

In order to get an asymptotic rate of convergence as fast as that for the generalized SSOR method , one can easily modify the IC-methods, by adding neglected fill-in entries to the diagonal in the same matrix row. This method has been examined by Gustafsson [1979]. He finds, that if the same philosophy as in the IC-method, is applied, namely to have certain fixed locations for the allowed fill-in, the total number of operations often starts to increase, if fill-in of more than about say 100% (compared to A) is allowed in each row of $\tilde{L}(\tilde{U})$.

As in the incomplete factorization method, this method may be unstable for general matrices, which are not diagonally dominant.

We now discuss this problem in more detail (see Munksgaard and Axelsson (1980)). It is well known, that for a symmetric positive definite matrix A, a complete factorization $\tilde{L}\tilde{L}^T = A$ is stable. For incomplete factorizations of this kind of matrices, the situation is not as simple however, because the following two operations must be performed for each pivot

(i) the elimination step,

(ii) the modification step.

The elimination step is performed as usual in a Gaussian elimination method, and the symmetry and definiteness of the matrix A^0, given before the elimination step, is unchanged. As well known, this is so, since the matrix $A^{(1)}$ left after the first elimination, defined by

$$L_1 A^{(0)} L_1^T = \begin{bmatrix} 1 & \underline{0}^T \\ \underline{0} & A^{(1)} \end{bmatrix} ,$$

is itself symmetric and positive definite.

The modification step may however destroy positive definiteness. In the modification step we consider the entries generated as fill-ins in the previous elimination step. The small elements can be determined either by a numerical test (see Section 3) or by their geometrical positions in the matrix as proposed by Meijerink and van der Vorst (1977) for the IC-methods. We modify the matrix either by deleting the fill-in - which is small enough - or by moving it to some other location in the same row. If we define the rowsum of a matrix row as the sum of all entries in the row, it is seen that for the MIC-type of modification the rowsum of all rows in $A^{(k)}$, $k = 0,1,\dots,n-1$ will be unchanged by the modification step. This means in particular that if the rowsums of $A^{(0)}$ are non-negative and if $A^{(0)}$ is a positive definite M-matrix, i.e. having all off-diagonal entries non-positive, that the modification step preserves both the positive definiteness and the M-matrix properties. However for positive definite matrices, which are not M-matrices, one may loose positive definiteness due to the modification step.

However it may be expected that it is only the negative entries that are causing the problems because moving a positive entry to the diagonal only increases the dominance of the diagonal elements.

In case the element to move is negative and all other off-diagonal entries in its row are negative as well, diagonal modification as shown in fig. 2.1 seems to be the most reasonable if the rowsum must be preserved. Step 2 in this figure is the symmetry preserving modification.

If however there is a positive element of greater magnitude than the one to be dropped in the row, it may be much more satisfactory to move the element to this location. Fig. 2.2 a and 2.2 b show two variants of this kind of rowsum and symmetry preserving modification. In the full developed version (fig. 2.2 b) the modified off-diagonal elements have been decreased in size, while diagonal entries have only been increased.

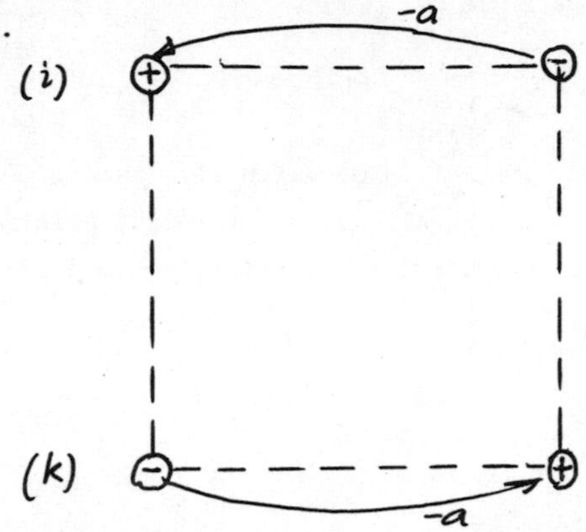

Fig. 2.1 All off-diagonal entries in row (i) are negative. Small elements are moved to the diagonal (a > 0).

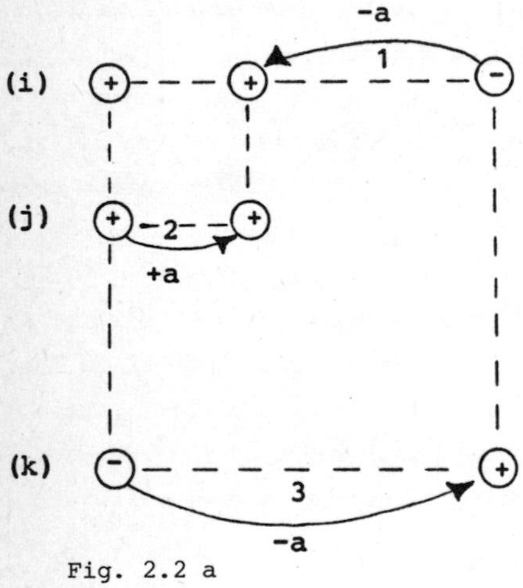

Fig. 2.2 a
Semi-off diagonal modification.

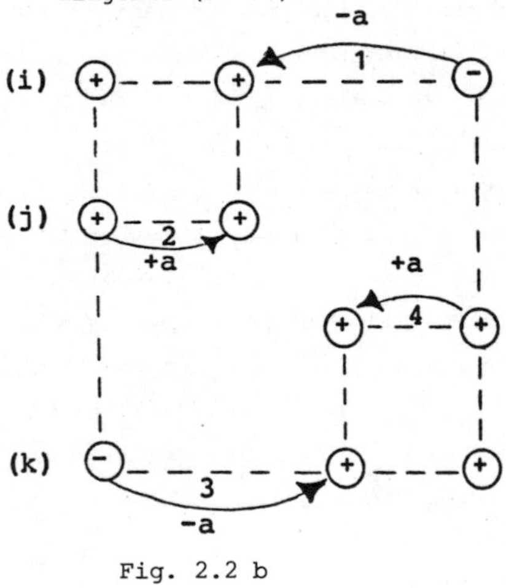

Fig. 2.2 b
Complete-off-diagonal modification.

This way of performing the modification step does still not quarantee positive definiteness of $A^{(k)}$ but may however be quite useful for matrices which contain a mixture of positive and negative off-diagonal elements. Some more profound remedies as proposed by Munksgaard (1979) may be necessary to overcome the definiteness problems.

3. Fixed space factorization.

The technique for dropping small elements may be based on a combination of a geometrical and a numerical criterion. All locations specified as non-zeros in the pattern of A remain in the pattern of the factorized matrix, and only fill-ins which are not members of this pattern, are subject to the numerical criterion.

Reservation of zero-locations in A for fill-ins according to geometrical drop criterions are maintained during the factorization, if they are specified in the "non-zero" pattern of A.

In the numerical criterion we move (or drop) the elements if they are potential fill-ins and if their numerical values relative to the corresponding diagonal elements are less that a relative drop tolerance. In the k'th pivotstep we drop $a_{ij}^{(k+1)}$ if

$$|a_{ij}^{(k+1)}| < c \cdot \sqrt{a_{ii}^{(k)} a_{jj}^{(k)}} \quad . \tag{3.1}$$

The amount of fill-in is determined by the size of c. If c is close to zero we obtain almost a complete factorization while c = 1 corresponds to a factorization which only included the non-zero pattern specified for A.

A major problem in factorizing sparse matrices is to determine the amount of storage needed to hold the factorized form. The incomplete factorization exhibits the same problem if it is performed with a fixed value of c.
We have tried to overcome this problem by (if necessary) increasing c during the factorization so that we never need more space that the user allocates for the arrays holding the matrix. The simplest way to do this is to run with the fixed c value until the space is used and then change c to 1, which means that we complete the factorization without allowing more fill-ins. This factorization is illustrated graphically in Figure 3.1, which shows the number of fill-ins relative to the space allocated for the factorization as function of i_p, where i_p is the relative pivot step number ($0 < i_p \leq 1$). Curve (1) shows the case where c has been chosen so small, e.g. 10^{-5}, that the allocated space has been filled after about 50% of the pivot steps. Consequently we must drop all further fill-ins of which several may be much greater than 10^{-5}, and hence the resulting matrix may be a bad approximation to the coefficient matrix.

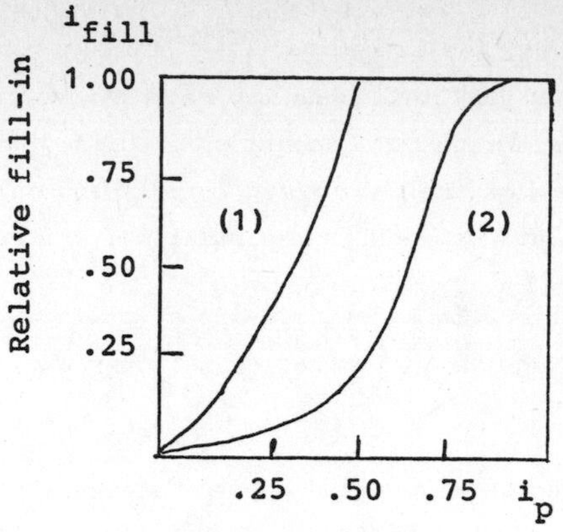

Fig. 3.1 Relative fill-in as a function of relative pivot step.

In Munksgaard, Axelsson (1980) it is described how one may change c adoptivally during the factorization process.

4. Generalized conjugate gradient methods.

For unsymmetric (and for inconsistent) problems, the preconditioned conjugate gradient method can be applied to the "normal" equations

$$A^T A \underset{\sim}{u} = A^T \underset{\sim}{a} ,$$

since here $A^T A$ is positive semi-definite and $A^T \underset{\sim}{a} \in \mathbb{R}(A^T A)$. Furthermore, conjugate gradient algorithms only use the matrix in matrix-vector multiplications, so one obviously does not have to form the matrix $A^T A$ (which could otherwise lead to concellation and loss of sparsity).

However, even so, this approach is not to be recommended in general, since as is well-known, there is usually a serious amplification of the spectral condition number (if is essentially squared). Hence, the number of iterations necessary to solve our problem is $\sim \varkappa(A)$ instead of $\sim \sqrt{\varkappa}(A)$. A similar observation is valid for $\tilde{A}^{-1}A$.

This type of situation is annoying, since if A is "almost" symmetric, i.e. a symmetric positive semi-definite matrix, perturbed by a small skewsymmetric matrix, one would expect about the same number of iterations as for the symmetric case. Hence we look for a method, which in this situation only needs about the same number of iterations as for the symmetric part, but which also applicable in the general case.

In the full version of the method, where, in the unsymmetric case, all previous search directions are used in order to calculate a new approximation, the rate of convergence is determined by the Krylov sequence $B\underset{\sim}{r}^0$, $B^2\underset{\sim}{r}^0,\ldots$ and not by $(B^TB)\underset{\sim}{r}^0$, $(B^TB)^2\underset{\sim}{r}^0,\ldots$, as would have been the case of the normal equations had been used. Here $B = \tilde{A}^{-1}A$ and $\underset{\sim}{r}^0 = B\underset{\sim}{u}^0 - \tilde{A}^{-1}\underset{\sim}{a}$.

Let $a(.,.)$ be a mapping onto $\mathbb{R}$ satisfying:

(i) bilinearity on $V \times V$, V real Hilbert space in particular $V = \mathbb{R}^N$

(ii) coercivity, i.e. $a(\underset{\sim}{u},\underset{\sim}{u}) \geq \rho\|\underset{\sim}{u}\|^2 \;\forall \underset{\sim}{u} \in V,\ \rho > 0$

(iii) boundedness, i.e. $a(\underset{\sim}{u},\underset{\sim}{v}) \leq K\|\underset{\sim}{u}\|\;\|\underset{\sim}{v}\| \;\forall \underset{\sim}{u},\underset{\sim}{v} \in V$

Here

$$\|\underset{\sim}{u}\| = (\underset{\sim}{u},\underset{\sim}{u})^{\frac{1}{2}} = \underset{\sim}{u}^T M\underset{\sim}{u} ,$$

M positive definite, symmetric.

We shall consider the numerical solution of

$$B\underset{\sim}{u} = \underset{\sim}{b} \;,\; \underset{\sim}{u},\underset{\sim}{b} \in V \;,\; B : V \to V$$

and two specific bilinear forms:

$$a_1(\underset{\sim}{u},\underset{\sim}{v}) = (B\underset{\sim}{u},\underset{\sim}{v}) \;,$$
$$a_2(\underset{\sim}{u},\underset{\sim}{v}) = (B\underset{\sim}{u},B\underset{\sim}{v}) .$$

Let $\{\underset{\sim}{d}^j\}_{j=0,1,\ldots}$ be <u>search directions</u>

and $\{\underset{\sim}{u}^j\}_{j=0,1,\ldots}$ <u>successive approximations</u>

of a solution $\hat{\underset{\sim}{u}} \in V$. The following <u>recursion</u> shall be used:

$$\underset{\sim}{d}^k = -\underset{\sim}{r}^k + \beta_{k-1}\underset{\sim}{d}^{k-1}$$
$$\underset{\sim}{u}^{k+1} = \underset{\sim}{u}^k + \sum_{j=0}^{k} \lambda_j^{(k)}\underset{\sim}{d}^j \qquad k = 0,1,2,\ldots$$

with $\beta_{-1} = 0$, $\underset{\sim}{u}^0$ arbitrary and $\underset{\sim}{r}^k = B\underset{\sim}{e}^k = B\underset{\sim}{u}^k - \underset{\sim}{b}$, $\underset{\sim}{e}^k = \underset{\sim}{u}^k - \hat{\underset{\sim}{u}}$.

<u>Determination of the parameters</u> β_k, $\{\lambda_j^{(k)}\}$:

We have

$$\underset{\sim}{e}^{k+1} = \underset{\sim}{e}^k + \sum_0^k \lambda_j^{(k)}\underset{\sim}{d}^j$$
$$\underset{\sim}{r}^{k+1} = \underset{\sim}{r}^k + \sum_0^k \lambda_j^{(k)}B\underset{\sim}{d}^j$$

The parameters β_k, $\{\lambda_j^{(k)}\}$ shall be determined by a Galerkin method.

Let

$$S_k = \mathrm{SPAN}\{\underset{\sim}{r}^0, B\underset{\sim}{r}^0,\ldots,B^k\underset{\sim}{r}^0\}$$

be the so called Krylov sequence. Then

$$\underset{\sim}{d}^k \in S_k \;,\; \underset{\sim}{r}^k \in S_k \;,\; \underset{\sim}{e}^k - \underset{\sim}{e}^0 \in S_{k-1} \;,\text{ and } \underset{\sim}{r}^k - \underset{\sim}{r}^0 \in BS_{k-1} .$$

Galerkin method:

$$a(\underset{\sim}{e}^{k+1},\underset{\sim}{v}) = 0 \;\forall \underset{\sim}{v} \in S_k.$$

Let $\underset{\sim}{v} = \underset{\sim}{d}^l$, $l = k,k-1,\ldots,0$. Then

$$(4.1) \quad \sum_{j=0}^{k} \lambda_j^{(k)} \; a(\underset{\sim}{d}^j,\underset{\sim}{d}^l) = -a(\underset{\sim}{e}^k,\underset{\sim}{d}^l), \; l = k,k-1,\ldots,0.$$

Lemma 4.1: $\Lambda^{(k)} : \Lambda_{lj}^{(k)} = a(\underset{\sim}{d}^j,\underset{\sim}{d}^l)$, $0 < j$, $l \le k$ has a positive definite symmetric part iff $\{\underset{\sim}{d}^j\}_0^k$ are linearly independent.

Proof: $\underset{\sim}{\alpha}^T \Lambda^{(k)} \underset{\sim}{\alpha} = a(\underset{\sim}{u},\underset{\sim}{u}) \ge \rho \| \underset{\sim}{u} \|^2$,

where $\underset{\sim}{u} = \sum_{j=0}^{k} \alpha_j \underset{\sim}{d}^j$; so

$\underset{\sim}{\alpha}^T \Lambda^{(k)} \underset{\sim}{\alpha} > 0$ iff $\| \underset{\sim}{u} \| \neq 0$. □

From (4.1) and Lemma 4.1 follows that

$$\Lambda^k \lambda^{(k)} = \begin{bmatrix} -a(\underset{\sim}{e}^k,\underset{\sim}{d}^l) \\ 0 \\ \vdots \\ 0 \end{bmatrix}$$

has a unique solution iff $\{d^j\}_0^k$ are linearly independent.

By Cramers rule:

$$\lambda_k^{(k)} = \frac{\det(\Lambda^{(k-1)})}{\det(\Lambda^{(k)})} \; a(-\underset{\sim}{e}^k,\underset{\sim}{d}^k).$$

Note that the matrix $\Lambda^{(k)}$ is an extension of $\Lambda^{(k-1)}$, where we have added only one row and a column. The remaining entries are the same.

We shall consider two choices of β_l:

$$(A) \quad \beta_l = \frac{a(\underset{\sim}{d}^l,\underset{\sim}{r}^{l+1})}{a(\underset{\sim}{d}^l,\underset{\sim}{d}^l)} \quad , \; l = 0,1,\ldots,\beta_{-1} = 0$$

$$(b) \quad \beta_l = 0 \quad , \; l = -1,0,1,\ldots .$$

Consider at first the bilinear form a_1:

We have then

$$(\underset{\sim}{r}^{k+1},\underset{\sim}{v}) = 0 \quad \forall \underset{\sim}{v} \in S_k$$

and for $0 \le j \le l-2$:

$$a_1(\underset{\sim}{d}^j,\underset{\sim}{r}^l) = (B\underset{\sim}{d}^j,\underset{\sim}{r}^l) = (\underset{\sim}{r}^l,B\underset{\sim}{d}^j) = 0 \;, \quad \text{since } B\underset{\sim}{d}^j \in S_{j+1} \subset S_{l-1}.$$

Hence

$$a_1(\underset{\sim}{d}^j,\underset{\sim}{d}^l) = a_1(\underset{\sim}{d}^j,-\underset{\sim}{r}^l + \beta_{l-1}\underset{\sim}{d}^{l-1}) = \beta_{l-1}a_1(\underset{\sim}{d}^j,\underset{\sim}{d}^{l-1})$$

and by induction,

$$(4.2) \quad a_1(\underset{\sim}{d}^j,\underset{\sim}{d}^l) = \left(\prod_{m=j+1}^{l-1} \beta_m\right) a_1(\underset{\sim}{d}^j,\underset{\sim}{d}^{j+1}) \;, \; 0 \le j \le l-2.$$

For j = l-1:

$$a_1(\underset{\sim}{d}^{l-1},\underset{\sim}{d}^{l}) = -a_1(\underset{\sim}{d}^{l-1},\underset{\sim}{r}^{l}) + \beta_{l-1}a_1(\underset{\sim}{d}^{l-1},\underset{\sim}{d}^{l-1}).$$

Consider now choice (A):

Then

$$a_1(\underset{\sim}{d}^{l-1},\underset{\sim}{d}^{l}) = 0$$

so

$$a_1(\underset{\sim}{d}^{j},\underset{\sim}{d}^{l}) = 0 \ , \ 0 \le j \le l-1 \ ,$$

i.e. $\Lambda^{(k)}$ is upper triangular, and

$$\lambda_k^{(k)} = -\frac{a_1(\underset{\sim}{e}^k,\underset{\sim}{d}^k)}{a_1(\underset{\sim}{d}^k,\underset{\sim}{d}^k)} = \frac{a_1(\underset{\sim}{e}^k,\underset{\sim}{r}^k)}{a_1(\underset{\sim}{d}^k,\underset{\sim}{d}^k)} = \frac{(\underset{\sim}{r}^k,\underset{\sim}{r}^k)}{a_1(\underset{\sim}{d}^k,\underset{\sim}{d}^k)}$$

Hence $\lambda_k^{(k)} > 0$ (unless $\underset{\sim}{r}^{(k)} = \underset{\sim}{0}$).

Furthermore, we have a unique solution as long as

$$a_1(\underset{\sim}{d}^l,\underset{\sim}{d}^l) \ne 0 \ , \ l = 0,1,\dots,k \ .$$

$\lambda_l^{(l)} > 0$ implies

$$\beta_l = \frac{(B\underset{\sim}{d}^l,\underset{\sim}{r}^{l+1})}{(B\underset{\sim}{d}^l,\underset{\sim}{d}^l)} = \frac{(\underset{\sim}{r}^{l+1}-\underset{\sim}{r}^l,\underset{\sim}{r}^{l+1})}{(\underset{\sim}{r}^{l+1}-\underset{\sim}{r}^l,\underset{\sim}{d}^l)} = \frac{(\underset{\sim}{r}^{l+1},\underset{\sim}{r}^{l+1})}{(\underset{\sim}{r}^l,\underset{\sim}{r}^l)} \quad .$$

The above is a generalization of the classical conjugate gradient method, because if a_1 is a symmetric form (for instance if B is symmetric and M = I), then

$\Lambda^{(k)}$ is diagonal.

Hence since

$$a_1(\underset{\sim}{d}^{j},\underset{\sim}{d}^{l}) = 0 \ , \ 0 \le j \le l-1$$

we have

$$a_1(\underset{\sim}{d}^{j},\underset{\sim}{d}^{l}) = 0 \quad \forall\, j \ne l \ ,$$

which is the conjugacy condition in the classical conjugate gradient method.

Consider now case (B):

Ten $\beta_l = 0$, so

$$a_i(\underset{\sim}{d}^{j},\underset{\sim}{d}^{l}) = 0 \ , \ 0 \le j \le l-2$$

i.e. $\Lambda^{(k)}$ is a upper Hessenberg matrix.

Assume that

$$M = \tfrac{1}{2}(A + A^T)$$

is positive definite and let

$$B = M^{-1}A = M^{-1}(M - N)$$

where

$$N = \tfrac{1}{2}(A^T - A).$$

With the inner product $(\underset{\sim}{u},\underset{\sim}{v}) = \underset{\sim}{u}^T M\underset{\sim}{v}$ we get after some calculations,

$$\underset{\sim}{u}^{k+1} = \underset{\sim}{u}^{k-1} - \lambda_k(\underset{\sim}{r}^k - \underset{\sim}{u}^k + \underset{\sim}{u}^{k-1}) ,$$

$k = 1,2,\ldots$, where $\underset{\sim}{u}^0$ is arbitrary and

$$\lambda_k^{-1} = 1 + \lambda_{k-1}^{-1} \frac{(\underset{\sim}{r}^k,\underset{\sim}{r}^k)}{(\underset{\sim}{r}^{k-1},\underset{\sim}{r}^{k-1})} , \quad k = 1,2,\ldots$$

and $\lambda_0 = 1$. This is the so called "Generalized conjugate gradient method", due to Concus, Golub (1976) and Widlund (1978).

Consider finally the bilinear form a_2:

Here

$$\beta_1 = (B\underset{\sim}{r}^{1+1},B\underset{\sim}{d}^1) / (B\underset{\sim}{d}^1,B\underset{\sim}{d}^1)$$

and leads to the modified minimal resideral method (see Axelsson (1980)). $\Lambda^{(k)}$ is a symmetric matrix and if B is symmetric, then $\Lambda^{(k)}$ is diagonal.

If B is not symmetric but almost symmetric, one finds that in practice few search directions have to be kept along (see Axelsson (1980)).

5. Quasioptimal rate of convexgence

If a(.,.) is a symmetric (and coercive) form we have by Ritz principle,

$$f(\underset{\sim}{u}^k) = \min_{\underset{\sim}{v}\in\underset{\sim}{u}^0\oplus S_k} f(\underset{\sim}{v}) = \min_{\underset{\sim}{v}\in\underset{\sim}{u}^0\oplus S_k}\{\tfrac{1}{2}a(\underset{\sim}{v},\underset{\sim}{v}) - (\underset{\sim}{b},\underset{\sim}{v})\} .$$

Hence

$$a(\underset{\sim}{e}^{k+1},\underset{\sim}{e}^{k+1}) < a(\underset{\sim}{e}^k,\underset{\sim}{e}^k) ,$$

if $\underset{\sim}{r}^k \neq 0$ (monotonicity) and furthermore, we have an optimal rate of convergence with respect to the "energy norm" $\{a(.,.)\}^{\frac{1}{2}}$.

When $a_1(.,.)$ is not symmetric, we have in general only a quasioptimal rate of convergence:

$$a(\hat{\underset{\sim}{u}},\underset{\sim}{v}) = (\underset{\sim}{b},\underset{\sim}{v}) \ \forall\underset{\sim}{v} \in V$$

$$a(\underset{\sim}{u}^k,\underset{\sim}{v}) = (\underset{\sim}{b},\underset{\sim}{v}) \ \forall\underset{\sim}{v} \in S_{k-1} \subset V$$

so

$$a(\underset{\sim}{e}^k,\underset{\sim}{v}) = 0 \quad \forall\underset{\sim}{v} \in S_{k-1}.$$

By coercivity and boundedness,

$$\rho\|\underset{\sim}{e}^k\|^2 \le a(\underset{\sim}{e}^k,\underset{\sim}{e}^k) = a(\underset{\sim}{e}^k,\hat{\underset{\sim}{u}}-\underset{\sim}{v}) \le K\|\underset{\sim}{e}^k\| \ \|\hat{\underset{\sim}{u}}-\underset{\sim}{v}\| \quad \forall\underset{\sim}{v} \in \underset{\sim}{u}^0 \oplus S_{k-1} .$$

Hence

$$\|\underset{\sim}{e}^k\| \le \frac{K}{\rho} \min_{\underset{\sim}{v}\in\underset{\sim}{u}^0\oplus S_{k-1}} \|\hat{\underset{\sim}{u}}-\underset{\sim}{v}\|$$

Hence, the problem of estimating the rate of convergence has been reduced to that of estimating the best approximation of $\underset{\sim}{\hat{u}} \in V$ by $\underset{\sim}{v} \in \underset{\sim}{u}^0 \oplus S_{k-1}$.

We consider now the special case where B is similarly equivalent to a symmetric matrix with nonnegative eigenvalues, i.e.

$$\exists Q \; ; \; Q^{-1} B Q = \tilde{B} \; ,$$

where $\tilde{B}$ is symmetric and positive semidefinite.

Let Π_j be the set of polynomials of degree j or less.

We then have

$$\min_{\underset{\sim}{v} \in \underset{\sim}{u}^0 \oplus S_{k-1}} \| \hat{u} - \underset{\sim}{v} \| = \min_{\underset{\sim}{v} \in S_{k-1}} \| \underset{\sim}{e}^0 - \underset{\sim}{v} \| \le$$

$$\min_{P_{k-1} \in \Pi_{k-1}} \| (1 - B\, P_{k-1}(B)) \underset{\sim}{e}^0 \| \le$$

$$\| Q \| \min_{P_{k-1}} \| (1 - \tilde{B}\, p_{k-1}(\tilde{B})) Q^{-1} \underset{\sim}{e}^0 \|$$

$$\le \; H(Q) \min_{p_{k-1}} \max_{\lambda \in S^+(B)} |1 - \lambda\, p_{k-1}(\lambda)| \; \| \underset{\sim}{e}^0 \| \; ,$$

$$\| \underset{\sim}{e}^k \| \; / \; \| \underset{\sim}{e}^0 \| \le \frac{K}{\rho} H(Q) \min_{p_{k-1}} \max_{\lambda \in S^+(B)} |1 - \lambda\, p_{k-1}(\lambda)| \le \delta^1 \; ,$$

if

$$k \ge \ln\left(\frac{1}{\delta} + \sqrt{\frac{1}{\delta^2}} - 1\right) / \ln \sigma \; ,$$

where

$$\delta = \delta^1 \; \delta \; / \; (K H(Q)) \; , \; \sigma = \frac{1 + H^{\frac{1}{2}}}{1 - H^{\frac{1}{2}}} \; .$$

In general, the dependence on $H(Q)$ is minor.

In the above estimate, $S^+(B)$ is the positive part of the spectrum of B.

References.

1. Axelsson, O, A generalized SSOR method, BIT 12 (1972), 443-467.

2. Axelsson, O, On preconditioning and convergence acceleration in sparse matrix problems, CERN 74-10, Geneve (1974).

3. Axelsson, O, A class of iterative methods for finite element equations, Comp. Meth. Appl. Mech. Eng. 9 (1976), 123-137.

4. Axelsson, O, Conjugate gradient type methods for unsymmetric and inconsistent systems of linear equations, Linear Algebra and its Applications, 24 (1980), 1-16.

5. Beauwens, R and Quenon, L, Existence criteria for partial matrix factorizations in iterative methods, SIAM. J. Numer. Anal. 13 (1976), 615-643.

6. Concus, P and Golub, G.H, A generalized conjugate gradient method for non-symmetric systems of linear equations, Proc. Second Internat. Symp. on Computing Methods in Applied Sciences and Engineering, IRIA (Paris, Dec. 1975), Lecture Notes in Economics and Mathematical Systems, vol. 134, R. Glowinski and J.-L. Lions, eds., Springer-Verlag, Berlin 1976.

7. Gustafsson, I, Stability and rate of convergence of modified incomplete Choleskey factorization methods, Department of Computer Sciences, 79.02 R, Chalmers University of Technology, Göteborg, Sweden.

8. Meijerink, J.A and van der Vorst, H.A, An iterative solution method for linear systems of which the coefficient matrix is a symmetric M-matrix, Math. Comp. 31 (1977), 148-162.

9. Munksgaard, N. (1979): Solving sparse symmetric sets of linear equations by preconditioned conjugate gradients. AERE-Harwell report. CSS 67.

10. Munksgaard, N, and Axelsson, O, Analysis of incomplete factorization with fixed storage allocation, SIAM J. Scientific and Statistical Computing (SISSC), to appear (1980).

11. Tuff, A.D and Jennings, A, An iterative method for large systems of linear structural equations, Int. J. Numer. Math. Engng. 7 (1973), 175-183.

12. Varga, R.S, Factorization and normalized iterative methods, in Boundary problems in differential equations (R.E. Langer, ed.), 121-142, Madison, University of Wisconsin Press, Madison (1960).

13. Varga, R.S, Incomplete factorization of matrices and connections with H-matrices, Proceedings, Conference on elliptic problem solvers, Santa Fc, New Mexico, June 30 - July 2, 1980, SIAM J. Num. 17, 787 (1980).

14. Widlund, O, A Lanczos method for a class of nonsymmetric systems of linear equations, SIAM J. Num. Anal. 15 (1978), 801-812.

A PRECONDITIONED TCHEBYCHEFF ITERATIVE SOLUTION METHOD FOR CERTAIN LARGE SPARSE LINEAR SYSTEMS WITH A NON-SYMMETRIC MATRIX

H.A. van der Vorst
Academisch Computer Centrum
Budapestlaan 6
Utrecht, the Netherlands

Abstract

In this paper methods are described for the solution of certain sparse linear systems with a non-symmetric matrix. The power of these methods is demonstrated by numerical examples. Application of the methods is restricted to problems where the matrix has only eigenvalues with positive real part. An important class of this type of matrices arises from discretisation of second order partial differential equations with first order derivative terms.

The research described in this paper has been supported in part by the European Research Office, London, through Grant DAJA 37 - 80 - C - 0243.

1. Introduction

When an elliptic selfadjoint partial differential equation is discretised over a rectangular region, this results in a linear system Ax=b, where A is a symmetric M-matrix. Efficient algorithms to solve this type of equations iteratively have been proposed by Axelsson [1], Concus, Golub and O'Leary [2], Meijerink and van der Vorst [3,4] and many others.
Another class of problems arises when the partial differential equation includes first order derivatives. Discretisation in this case yields a linear system with a non-symmetric matrix (see section 2). These problems are much more difficult to solve and a lot of research has been done and is currently going on to develop efficient algorithms. Algorithms have been described by Varga [5], Kershaw [6], Manteuffel [7,8], Paige [9], Concus and Golub [10], Widlund [11], a.o..
Manteuffel has compared his algorithm to the bidiagonalisation-method of Paige [9] and to the conjugate gradient algorithm of Kershaw [6]. From this comparison his method seems to be a competitive one.
We consider the use of certain preconditioning techniques for the Manteuffel algorithm. Some specific preconditionings, including fast poisson solvers and incomplete Crout- and Choleski-factorizations are proposed in section 3 and 4. A parameter can be included in the incomplete Crout-factorization in order to avoid the necessity of partial pivoting. From the numerical examples described in section 5 it appears that the use of an incomplete Crout-factorization as a preconditioning (eventually with well-chosen parameter) for the Manteuffel-algorithm leads to highly competitive iterative methods.

2. Description of a class of non-symmetric matrices

In this section we briefly describe a class of non-symmetric matrices that arises from five point finite difference discretisation of certain second order pde's. The preconditioning matrices, presented in section 4, are given in a form that is applicable to these matrices. Most of the ideas can be easily applied to matrices with a different non-zero structure, e.g. matrices arising from finite element approximation.
We consider the partial differential equation

$$(2.1) \qquad -(Du'_x)'_x - (Eu'_y)'_y + Gu'_x + Hu'_y + Cu = F$$

defined on a rectangular region R in the (x,y)-plane, with $D(x,y)>0$, $E(x,y)>0$, $C(x,y)\geq 0$, for $x,y \in R$.
The boundary conditions may be of the form

$$\alpha u + \beta\frac{\partial u}{\partial n} = \gamma \quad , \quad \alpha(x,y)\geq 0 \ , \ \beta(x,y)\geq 0 \ , \quad \alpha + \beta>0 \ ,$$

where $\frac{\partial}{\partial n}$ denotes the outward derivative perpendicular to the boundary δR of R.
The functions $D(x,y)$, $E(x,y)$, $C(x,y)$, $F(x,y)$, $\alpha(x,y)$, $\beta(x,y)$ and $\gamma(x,y)$ should be bounded and continuous over R, δR resp..
A rectangular grid, in each direction equidistant, is imposed on R and (2.1), except for the part $Gu'_x + Hu'_y$, is discretised by an integration method which yields a 5-point discretisation formula for each gridpoint. The matrix resulting from this part of the equation can be shown to be a symmetric positive definite M-matrix M [5], which has the following non-zero structure:

M =
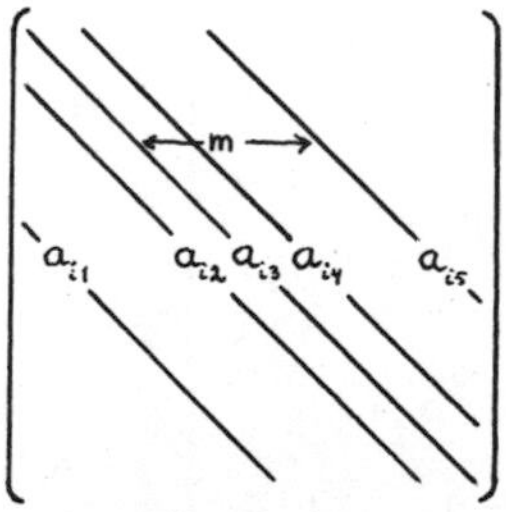

Figure 1

The terms Gu'_x and Hu'_y can be approximated by either central differences or by backward/forward differences. If central differences are used then this results in extra terms in the linear system that build a matrix N, which has a similar structure as M, except for the main diagonal which has zero elements. If $G(x,y)=g(y)$ and $H(x,y)=h(x)$ then N is skew-symmetric.
The choice for backward or forward differences, in the (i,j)-th gridpoint, depends on the sign of the functions G and H in such a way that a positive contribution to the main diagonal results.

3. Manteuffel's algorithm with preconditioning

For the solution of the linear system Ax=b the following two step iterative method can be used:

(3.1) $$x_{k+1} = -\alpha_k A x_k + (1+\beta_k)\, x_k - \beta_k x_{k-1} + \alpha_k b$$

It is shown by Manteuffel [7] that this method converges to the solution of Ax=b if the spectrum of A is enclosed in an ellipse, with focii d-c, d+c, in the right half plane and if α_k and β_k are chosen to be

(3.2) $$\alpha_k = \frac{2}{c}\,\frac{T_k(\frac{d}{c})}{T_{k+1}(\frac{d}{c})} \qquad \beta_k = \frac{T_{k-1}(\frac{d}{c})}{T_{k+1}(\frac{d}{c})}$$

where T_k is the k-th Tchebycheff polynomial $T_k(z) = \cos(k \arccos(z))$, for z complex. The constants d and c should be chosen to define a family of ellipses which contains the ellipse with focii d-c, d+c that encloses all the eigenvalues of A and for which the convergence factor r_c (see (3.4)) is minimal.
For computational purposes the following algorithmic scheme is convenient [7]:

Given x_0, define

$$x_1 = x_0 + p_0 \quad \text{where} \quad p_0 = \frac{1}{d} r_0, \quad r_0 = b - Ax_0$$

$$\alpha_1 = 2d/(2d^2-c^2)$$

$$\beta_1 = d\,\alpha_1 - 1$$

Then

(3.3) $$\left.\begin{aligned} r_i &= b - Ax_i \\ p_i &= \alpha_i r_i + \beta_i p_{i-1} \\ x_{i+1} &= x_i + p_i \\ \alpha_{i+1} &= (d - (\tfrac{c}{2})^2\alpha_i)^{-1} \\ \beta_{i+1} &= d\alpha_{i+1} - 1 \end{aligned}\right\} \quad \text{for } i=1,2,\ldots.$$

The asymptotic convergence factor for this iterative Tchebycheff method is given by

(3.4) $$r_c = \frac{a + \sqrt{a^2-c^2}}{d + \sqrt{d^2-c^2}}$$

where d is the center of the ellipse, d-c and d+c are the focii and a is the length of the axis in the x-direction (see Figure 2).
Manteuffel has provided an algorithm for estimating the parameters d and c adaptively [7,8].

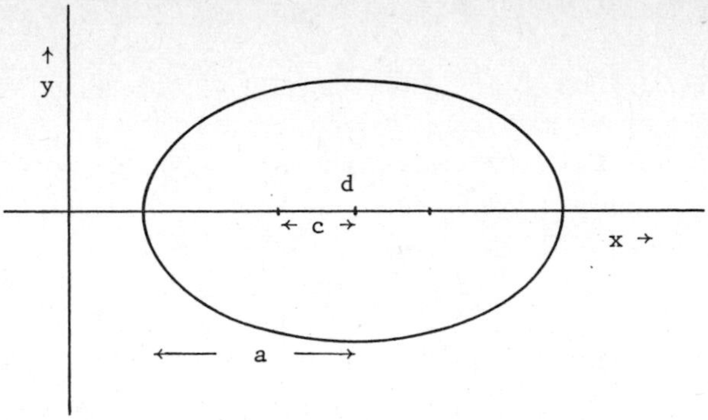

Figure 2

In order to improve the speed of convergence of the algorithm (3.3) one may use a well-chosen preconditioning matrix K and solve the equation $K^{-1}Ax=K^{-1}b$. Well-chosen in this respect means that the ellipse containing the spectrum of $K^{-1}A$ yields a smaller convergence factor r_c than the ellipse that contains all the eigenvalues of A. For successful application of the Manteuffel-algorithm it is necessary for $K^{-1}A$ to have all eigenvalues in the right half-plane, i.e. all eigenvalues should have positive real part. Below a number of possible preconditionings are listed.

3.1 If $A=M+N$ and if M is symmetric positive definite and $N=-N^T$, then $M^{-1}A$ has only eigenvalues with positive real part. In the situation that $M^{-1}x$ can be computed easily, e.g. by a Fast Poisson Solver [13], the matrix M^{-1} can be used as a preconditioning.

3.2 If $A=M+N$ and M is a symmetric M-matrix, then an incomplete Choleski-factorization K of M can be constructed [3]. When $N=-N^T$, then all eigenvalues of $K^{-1}A$ have positive real part which implies that K^{-1} can be used as a preconditioning.

3.3 If A is an M-matrix then an incomplete Crout-factorization K can be constructed [3]. It can be proven that all eigenvalues of $K^{-1}A$ have positive real part, which implies that K^{-1} can be used as a preconditioning.
If, in the terminology of section 2, the first order terms of (2.1) are discretised by central differences, then A is an M-matrix if Δx and Δy are chosen small enough. When the first order terms are discretised by backward or forward differences in such a way that the contribution to the diagonal of A is positive, then A is an M-matrix indepently of the choice of Δx and Δy.

4. Incomplete factorizations of A

The incomplete factorization of the M-matrix A that has a similar sparsity-structure for the upper and lower triangular parts as A has, is denoted by LU(1,1). This notation refers to the fact that in the factorization, except for the main diagonal, only the first and the m-th codiagonal are retained in U and L.
We write this factorization in the form

$$A = L_1 D_1 U_1 + R_1$$

where $\mathrm{diag}(L_1) = \mathrm{diag}(U_1) = D_1^{-1}$. The strict lower triangular part of L_1 and the strict upper triangular part of U_1 are equal to the corresponding parts of A. If the diagonal elements of D_1 are denoted by $\tilde{d}_i$, then in the notation of section 2, figure 1, it follows that

$$(4.1) \qquad \tilde{d}_i^{-1} = a(i,3) - a(i-1,4)a(i,2)\tilde{d}_{i-1} - a(i-m,5)a(i,1)\tilde{d}_{i-m}$$

where non-defined elements should be replaced by a zero.
If A=M+N where M is a symmetric M-matrix and $N=-N^T$ then there are no computational problems with the use of (4.1) for the construction of an incomplete factorization of A, though A itself might not be an M-matrix. However if the elements of N are large then the factors L_1 and U_1 may be very ill-conditioned. This can be prevented by a partial pivoting technique, which has the disadvantage of destroying the sparsity structure, or by replacing (4.1) by formula (4.2), where σ is a well-chosen factor.

$$(4.2) \qquad \tilde{d}_i^{-1} = \sigma\, a(i,3) - a(i-1,4)a(i,2)\tilde{d}_{i-1} - a(i-m,5)a(i,1)\tilde{d}_{i-m}$$

From experiments it follows that σ should be chosen such that $\tilde{d}_i^{-1}$ in magnitude compares with the sum of off-diagonal elements in the i-th row of L_1 or U_1. The factorization defined by (4.2) will be denoted as

$$A = L_\sigma D_\sigma U_\sigma + R_\sigma$$

Another possibility to force the elements $\tilde{d}_i^{-1}$ to be comparable to the sum of off-diagonal elements of L and U is given by the following algorithm for $\tilde{d}_i$.

$$(4.3) \qquad \begin{aligned} \Delta_i^{-1} &= a(i,3) - a(i-1,4)a(i,2)\tilde{d}_{i-1} - a(i-m,5)a(i,1)\tilde{d}_{i-m} \\ \Sigma_{L,i} &= |a(i,1)| + |a(i,2)| \\ \Sigma_{U,i} &= |a(i,4)| + |a(i,5)| \\ \tilde{d}_i^{-1} &= \max\{\Delta_i^{-1}, \Sigma_{L,i}, \Sigma_{U,i}\} \end{aligned}$$

The factorization defined by (4.3) will be denoted as $A = L_{EQ} D_{EQ} U_{EQ} + R_{EQ}$

5. Numerical examples

5.0 General

In this section a number of numerical examples is described. They have all been carried out on the CDC Cyber 73-28 of the Academisch Computer Centrum Utrecht in 48 bits relative working precision. The residuals in all examples, as far as listed, have been computed as $||Ax_i-b||_2$, where x_i is the i-th iterand in the iterative solution process for Ax=b. Central Processor times (CPU-time) have been included to give an impression of the actual behaviour of the different methods. These times are accurate up to about 10%. Other Numerical experiments are described in [15].

5.1 $-u''_{xx} - u''_{yy} + \beta(u'_x + u'_y) + u = 1$

This problem has been discussed extensively by Manteuffel [8] and has also been chosen since a number of properties can easily be checked (e.g. solution, eigenvalues, stability). The equation is discretised over a square region with gridspacing 1.0 in both directions. The boundary conditions are of Dirichlet-type: u=1 along the boundary. The first order derivative terms have been discretised by central differences.
For all the iterative processes the starting vector is chosen to be 0. The initial Manteuffel parameters were d=1, c=0. For a number of unknowns equal to 841 (=29^2) the following iteration results have been obtained.

Method	Number of iterations	Final residual / Initial residual	CPU-time
Manteuffel algorithm without preconditioning	200	$6.2_{10}{-4}$	22.4
id. with $L_\sigma D_\sigma U_\sigma$-preconditioning, σ=2.5	39	$8.1_{10}{-9}$	6.7
id. with $L_{EQ}D_{EQ}U_{EQ}$-preconditioning	39	$8.5_{10}{-9}$	6.7

Table I. Iteration results for 5.1 with β=20.0

For β=20.0 we have checked how the convergence for the $L_\sigma D_\sigma U_\sigma$-preconditioning depends on the choice of σ (formula (4.2)). From straightforward computation it follows that the sum of the off-diagonal elements in a typical row of L_σ in absolute value is equal to the corresponding diagonal element of L_σ when σ=2.48.
The next figure shows how many iterations are required to obtain a final residual $||Au_n -b||_2 < {}_{10}{-6}$ for different values of σ.

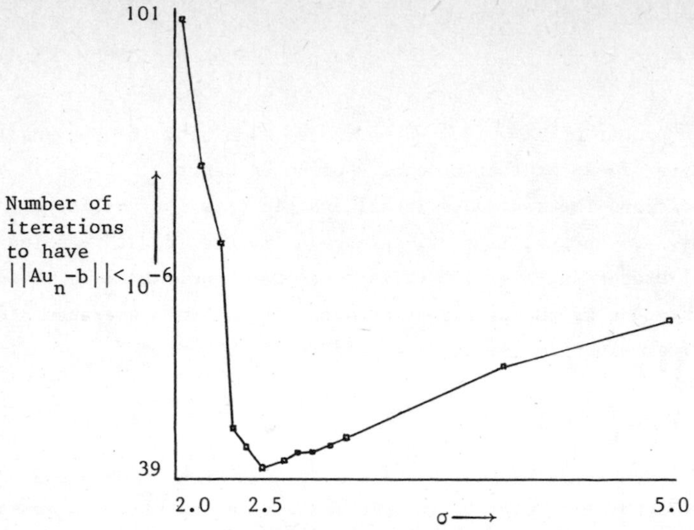

Figure 3

The case $\beta=100.0$ has been selected as an extreme one where the first order derivative terms after discretisation dominate the second order derivative terms. With the choice $\sigma=10.2$ in formula (4.2) the sum of the off-diagonal elements of L_σ compares to the diagonal element of L . This choice is validated by numerical experiments.

Method	Number of iterations	Final residual / Initial residual	CPU-time
Manteuffel algorithm without preconditioning	200	$2.8_{10}{-1}$	22.4
id. with $L_\sigma D_\sigma U_\sigma$-preconditioning, $\sigma=10.2$	153	$1.8_{10}{-9}$	17.1
id. with $L_{EQ} D_{EQ} U_{EQ}$-preconditioning	154	$1.8_{10}{-9}$	17.2

Table II. Iteration results for 5.1 with $\beta=100.0$

5.2 $-u''_{xx} - u''_{yy} + \beta(u'_x + u'_y) = 0$

For this problem the effect of the choice of different preconditionings has been considered. The grid-spacing is still 1.0, the number of unknowns is 961 ($=31^2$). This particular choice was necessary to be able to compare also the Fast Poisson Solver-preconditioning (33 points in each direction, including known boundary points). The following preconditionings are compared.

a) Fast Poisson Solver [12,13,14]

b) Incomplete Choleski on the symmetric part: K_{13} [4,12]

c) $L_\sigma D_\sigma U_\sigma$ with well-chosen σ.

All the eigenvalues of the preconditioned matrix for a) and b) have a positive real part, whereas this property for c) can only be proven for $\beta \leq 2.0$.

Preconditioning	Number of iterations	Final residual / Initial residual	CPU-time
Fast Poisson	57	$5.7_{10}{-8}$	25.2
Inc. Chol. K_{13}	35	$4.8_{10}{-8}$	8.0
$L_\sigma D_\sigma U_\sigma$, σ=0.84	30	$3.7_{10}{-8}$	5.9

Table III. Iteration results for β=0.4

Preconditioning	Number of iterations	Final residual / Initial residual	CPU-time
Fast Poisson	60	$2.7_{10}{-1}$	26.5
Inc. Chol. K_{13}	200	$4.4_{10}{-5}$	45.7
$L_\sigma D_\sigma U_\sigma$, σ=1.216	14	$1.8_{10}{-8}$	2.7

Table IV. Iteration results for β=4.0

5.3 $\quad -u''_{xx} - u''_{yy} + (\frac{\partial}{\partial x}(au) + a\frac{\partial u}{\partial x})/2 = f$

This equation has been taken from Widlund [11]. Since only a first derivative in one direction (the x-direction) is present it can be shown that the incomplete factorization yields well-conditioned factors which implies that (4.1) can be used instead of (4.2).

The equation is discretised over a rectangular grid with equal gridspacing in both directions over the region [0,1]*[0,1], and along the boundaries a Dirichlet boundary condition holds. The function a is chosen as $2e^{3.5(x\uparrow 2+y\uparrow 2)}$ and $20e^{3.5(x\uparrow 2+y\uparrow 2)}$ resp..
The right-hand side is chosen in such a way that $u=\sin \pi x \sin \pi y\, e^{(x/2+y)\uparrow 3}$ satisfies the equation. No efforts have been made to estimate good Manteuffel-parameters d and c, in each case these were set initially to d=1, c=0.

If we compare the results in Table V with those published by Widlund [11, Table 1] then it appears that our method is highly competitive especially for the second choice of a.

Notice that in each iteration of Widlunds algorithm a symmetric linear system of the form $\frac{A+A^T}{2}$ u=v has to be solved.

a	Method	Number of unknowns	Number of iterations	Final residual / Initial residual	CPU-time
$2e^{3.5(x^2+y^2)}$	no precond.	225	200	$4.9_{10}{-2}$	5.9
,,	LU(1,1)	225	29	$2.1_{10}{-8}$	1.2
,,	no precond.	961	200	$5.2_{10}{-8}$	23.0
,,	LU(1,1)	961	47	$3.1_{10}{-8}$	8.1
$20e^{3.5(x^2+y^2)}$	no precond.	225	200	$2.9_{10}{-1}$	5.7
,,	LU(1,1)	225	14	$2.0_{10}{-9}$	.6
,,	no precond.	961	200	$6.5_{10}{-1}$	22.7
,,	LU(1,1)	961	26	$2.2_{10}{-9}$	4.5

Table V. Iteration results for equation 5.3

5.4 Discretisation of 1:st order terms by backward/forward differences

The matrix that arises from standard 5-point discretisation of the second order and linear terms in the pde is a diagonally dominant symmetric M-matrix. If the first order derivative terms are discretised by backward or forward differences depending on the sign of the functions G(x,y) and H(x,y) (see section 2) in such a way that the contribution to the diagonal elements in the matrix is positive, then the resulting matrix A again is a diagonally dominant non-symmetric M-matrix. According to [3] an incomplete stable Crout-decomposition K exists and the eigenvalues of the matrix $K^{-1}A$ have all positive real part. For central differences the equation $-u_{xx}-u_{yy}+\beta(u_x+u_y)=0$ has been considered in section 5.2. The results for backward/forward-differences are listed in Table VI for a matrix of order 1600 ($=40^2$), and for the $L_\sigma D_\sigma U_\sigma$-preconditioning.

β	σ	Number of iterations	Final residual / Initial residual	CPU-time
0.1	0.77	44	$4.1_{10}{-8}$	12.7
4.0	1.0	25	$1.5_{10}{-8}$	7.2
40.0	1.0	11	$5.1_{10}{-10}$	3.1

Table VI. Iteration results for backward/forward diff.

References

1. Axelsson, O., A generalized SSOR method, BIT Vol. 13, 1972, pp. 443-467.

2. Concus, P., Golub, G.H. and O'Leary, D.P.,A generalized conjugate gradient method for the numerical solution of elliptic partial differential equations, Proc. Symp. on Sparse Matrix Computation, ed. Bunch J.R. and Rose D.J., New York, 1975.

3. Meijerink, J.A. and van der Vorst, H.A., An iterative solution method for linear systems of which the coefficient matrix is a symmetric M-matrix, Math. of Comp., Vol. 31, No. 137, 1977, pp. 148-162.

4. Meijerink, J.A. and van der Vorst, H.A., Guide lines for the usage of incomplete decompositions in solving sets of linear equations as occur in practical problems, Technical Report, TR-9, ACCU, Utrecht, 1978.

5. Varga, R.S., Matrix iterative analysis, Prentice Hall, Englewood Cliffs N.J., 1962.

6. Kershaw, D.S., The incomplete Choleski-conjugate gradient method for the iterative solution of systems of linear equations, J. of Comp. Physics, Vol. 26(1), 1978, pp. 43-65.

7. Manteuffel, T.A., The Tchebychev iteration for non-symmetric linear systems, Num. Math., Vol. 28, 1977, pp. 307-327.

8. Manteuffel, T.A., Adaptive Procedure for estimating parameters for the non-symmetric Tchebychev iteration, Sandia Labs. Report SAND 77-8239, Livermore, 1977.

9. Paige, C.C., Bidiagonalisation of matrices and solution of linear equations, Siam J. Num. Anal., Vol. 11, 1974, pp. 197-209.

10. Concus, P. and Golub, G.H., A generalized conjugate gradient method for non-symmetric systems of linear equations, Proc. Second Internat. Symp. on Computing Methods in Applied Sciences and Engineering, IRIA (Paris, Dec. 1975), Lecture Notes in Economics and Mathematical Systems, Vol. 134, R. Glowinski and J.L. Lions eds., Springer Verlag, Berlin, 1976.

11. Widlund, O., A Lanczos method for a class of non-symmetric systems of linear equations, Siam J. Num. Anal. Vol. 15, No. 4, 1978, pp. 801-812.

12. Van Kats, J.M. and van der Vorst, H.A., Software for the discretisation and solution of second order self-adjoint elliptic partial differential equations in two dimensions, Technical Report, TR-10, ACCU, Utrecht, 1979.

13. Buzbee, B.L., Program-description of TBPSDN - Fast Direct Poisson Solver, LASL, 1973.

14. Manteuffel, T.A., private communication.

15. Van der Vorst, H.A. and van Kats, J.M., Manteuffel's algorithm with preconditioning for the iterative solution of certain sparse linear systems with a non-symmetric matrix, Technical Report, TR-11, ACCU, Utrecht, 1979.

On Modified Incomplete Factorization Methods

I. Gustafsson

Katholieke Universiteit
Nijmegen
The Netherlands

Abstract

In this contribution we present the main theoretical results for the modified incomplete factorization (MIC) methods. For a more detailed study including proofs see [1], where also a more complete list of references can be found. A number of numerical results concerning different classes of problems is included.

I. Introduction

We solve the system of linear equations

$$An = f$$

by a preconditioned iterative process, which basic form is

$$C(n^{\ell+1}-n^{\ell}) = -\beta_{\ell}(An^{\ell}-f), \quad \ell = 0,1,\ldots,$$

where C is the preconditioning matrix.

In the case when A and C are symmetric, positive definite we can use the Chebyshev method or a conjugate gradient type of method as an acceleration procedure. As we shall see, for some kind of non-symmetric problems this is still possible.

In order to get an efficient method, the preconditioning matrix C has to have the following properties:

1. It should be easily calculated.
2. It should not need too much storage.
3. Systems with the matrix C should be easily solved. Typically $C = LU$, with sparse lower and upper triangular factors L and U.
4. The spectral condition number $H(C^{-1/2}AC^{-1/2})$ of $C^{-1/2}AC^{-1/2}$ should be much smaller than $H(A)$, the spectral condition number of A.

Our choice of C is based on modified incomplete factorization (defined in the following section) of A and fulfills the desired properties stated above.

In this paper we present the methods for well-structured matrices such as those arising in finite element (FEM) or finite difference (FDM) method. The idea of modification can obviously even be used in a more general context, where the accepted fill-in during the approximate factorization is dynamically controlled, see e.g. [2].

2. Modified Incomplete Factorization

It is well known that a complete Gaussian or Cholesky factorization of a sparse matrix produces fill-in within the band in the upper and lower triangular factors. This leads to a relatively high computational complexity for the factorization and to considerably large storage requirement. By using an incomplete factorization we keep the sparsity and hence need much less computational labor and storage.

The incomplete factorization is used as a preconditioning for the iterative process or, equivalently, one makes an incomplete factorization followed by a number of iterative refinement steps.

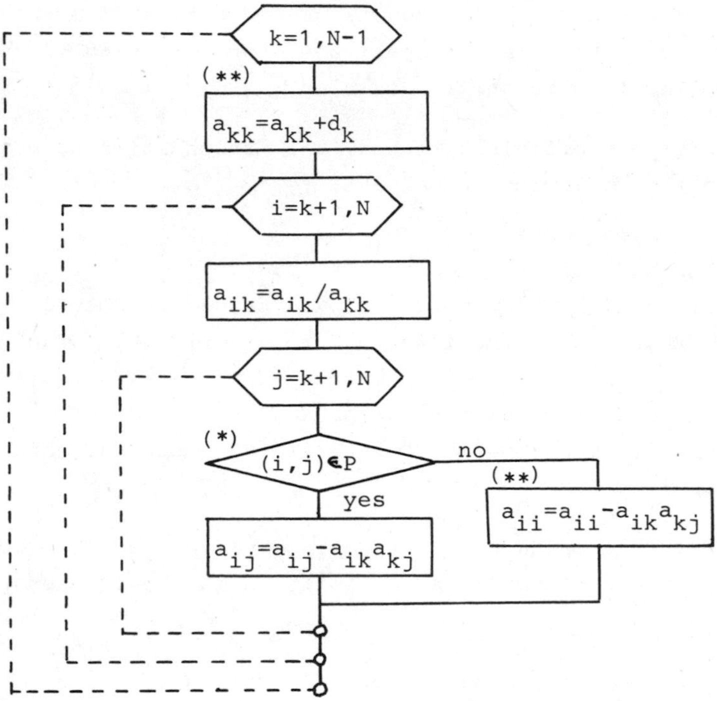

For well structured FDM or FEM matrices the positions, where fill-in is allowed during the eliminating , can be chosen in advance. Let $A = (a_{ij})$ be the $N \times N$ matrix to be factored and let

$$P^* = \{(i,j) \; ; \; a_{ij} \neq 0\}.$$

Further, let P be positions where we allow fill-in in the factors $L = (\ell_{ij})$ and $U = (u_{ij})$ of C, i.e.

$$P = \{(i,j)\ ;\ \ell_{ij} \neq 0 \text{ or } u_{ij} \neq 0\}.$$

In this paper we assume that $P^* \leq P$, that is, we have fill-in in at least positions where A has nonzero entries.

In the following flow-chart the MIC factorization algorithm, with normalization diag(L) = I, is described in a general context. For the definition of the diagonal matrix $D = \mathrm{diag}(d_k)$ see the above analysis (1,1).

From the flow-chart representing a complete factorization (Gaussian-elimination) we obtain an *incomplete* factorization (of our kind) by introducing the test (*) and a *modified incomplete* factorization by further introducing the statements (**). In a MIC factorization we do not drop elements but keep the information by modifying the diagonal.

If P = P* (that is, if we allow no fill-in) we use the notations IC(O) and MIC(O), respectively. We describe these factorizations for the following simple example.

$$\begin{bmatrix} 4 & -1 & 0 & -2 \\ -1 & 4 & -1 & 0 \\ 0 & -1 & 4 & -1 \\ -2 & 0 & -1 & 4 \end{bmatrix} \xrightarrow{\text{elimination}} \begin{bmatrix} 4 & -1 & 0 & -2 \\ 0 & 3\frac{3}{4} & -1 & -\frac{1}{2} \\ 0 & -1 & 4 & -1 \\ 0 & -\frac{1}{2} & -1 & 3 \end{bmatrix}$$

$$\xrightarrow{\text{IC(O)}} \begin{bmatrix} 4 & -1 & 0 & -2 \\ 0 & 3\frac{3}{3} & -1 & 0 \\ 0 & -1 & 4 & -1 \\ 0 & 0 & -1 & 3 \end{bmatrix} \qquad \xrightarrow[\text{D=O}]{\text{MIC(O)}} \begin{bmatrix} 4 & -1 & 0 & -2 \\ 0 & 3\frac{1}{4} & -1 & 0 \\ 0 & -1 & 4 & -1 \\ 0 & 0 & -1 & 2\frac{1}{2} \end{bmatrix}$$

After the three elimination steps we have C = LU = A + R, where

$$R = \begin{bmatrix} 0 & 0 & 0 & 0 \\ 0 & 0 & 0 & \frac{1}{2} \\ 0 & 0 & 0 & 0 \\ 0 & \frac{1}{2} & 0 & 0 \end{bmatrix} \quad \text{for IC(O) and } R = \begin{bmatrix} 0 & 0 & 0 & 0 \\ 0 & -\frac{1}{2} & 0 & \frac{1}{2} \\ 0 & 0 & 0 & 0 \\ 0 & \frac{1}{2} & 0 & -\frac{1}{2} \end{bmatrix}$$

for MIC(O). Notice that for the MIC(O) method, R has rowsum = O and that R is negative semidefinite. This is true in a more general context as will be stated below.

Let A be of order N. After N-1 modified incomplete factorization steps we have

$$C = LU = A + D + R,$$

where R is the defect matrix. Obviously, rowsum (R) = O. Furthermore, if A is an M-matrix ($A^{-1} \geq O$, $a_{ij} \leq O$, $i \neq j$), then R is negative semidefinite. This is so since all off-diagonal entries of R are non-negative.

3. Stability

Definition 3.1. An incomplete LU factorization is said to be *stable* iff diag(U) > O.

Observe that in the case when A is symmetric, positive definite stability means that $C(=LL^T)$ is also symmetric, positive definite.

Theorem 3.1. If A is weakly diagonally dominant ($a_{ii} \geq \sum_{j \neq i} |a_{ij}|$, $i = 1,\ldots,N$) then any MIC factorization of A is stable. The MIC algorithms are in general not stable for M-matrices, which is the case for IC algorithms, see [3].

4. MIC(d) Algorithms for FEM Matrices

Recall that in MIC(O), P = P*, that is, we allow no fill-in. We can obtain more accurate factorizations by admitting fill-in to some extent. We may then use the following strategy:

1. At first let L and U contain non-zero entries in the same positions as A. (This represents the MIC(O) method.)
2. Form C = LU and R = C-A.
3. Re-define L and U in such a way that these matrices are allowed to contain non-zero entries in positions where R has non-zero entries as well.
4. If you are not satisfied, repeat from stage 2.

For well structured matrices such as FDM and FEM matrices we then get MIC(d) algorithms, where d > O indicates that L contains d more non-zero sub-diagonals than the lower part of A. Practical experiments indicate that we get the most efficient method after one or two cycles of the above strategy.

Example 4.1. Consider the 5-point FDM matrix arising from a second order self-adjoint elliptic boundary value problem on the unit square. The matrices A, $C = LL^T$ and R for MIC(0) are defined by the following graphs, where as usual the graph nodes coincide with the FDM nodes. m is the half band-width of A.

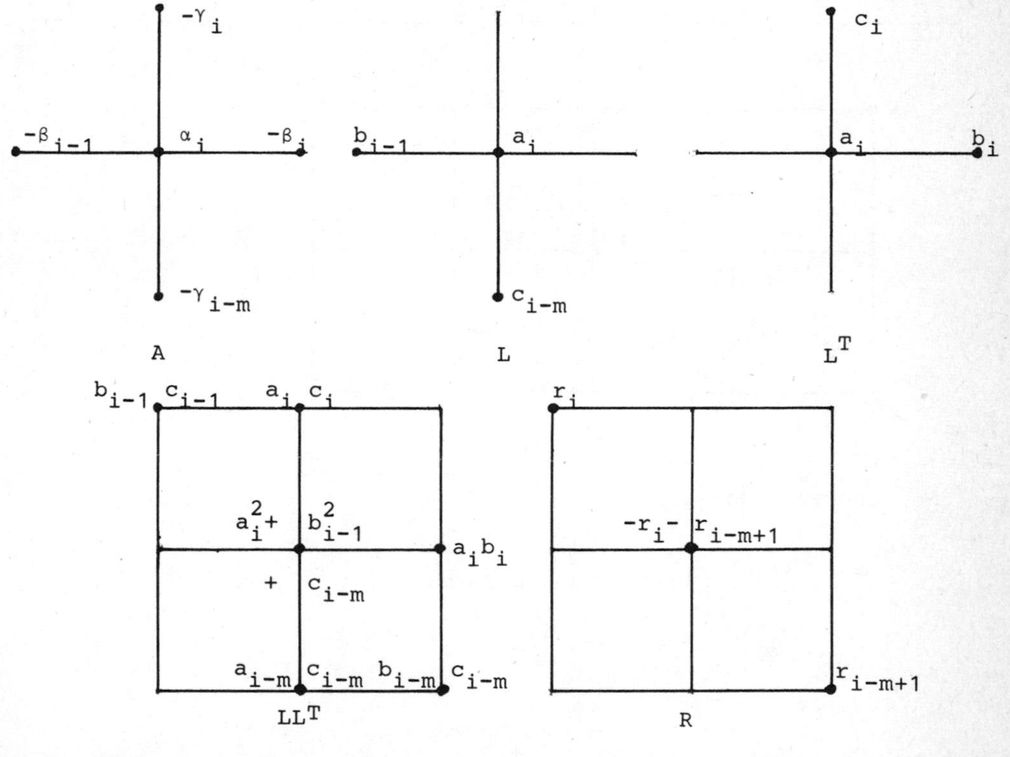

The relation $C = LL^T = A + D + R$ gives the following recursion formulas for the entries of L, defining the MIC(0) algorithms:

$$
\begin{aligned}
a_i^2 &= \alpha_i + \delta_i - b_{i-1}^2 - c_{i-m}^2 - r_i - r_{i-m+1} \\
b_i &= -\beta_i/a_i \\
c_i &= -\gamma_i/a_i \\
r_i &= b_{i-1}\, c_{i-1}.
\end{aligned}
\tag{4.1}
$$

The strategy to obtain a more accurate factorization now leads to the MIC(1) algorithm defined by the graphs

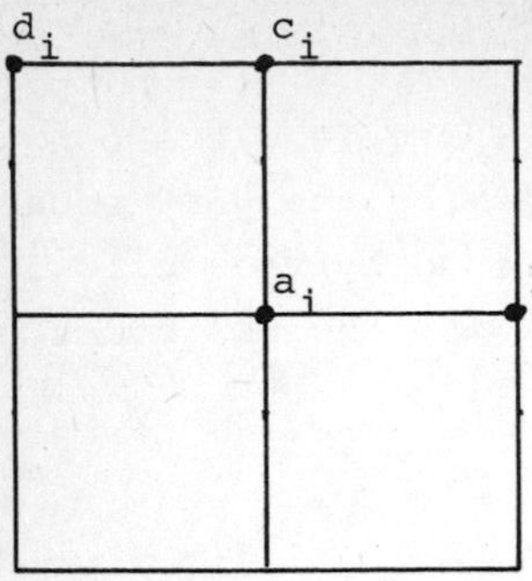

L^T

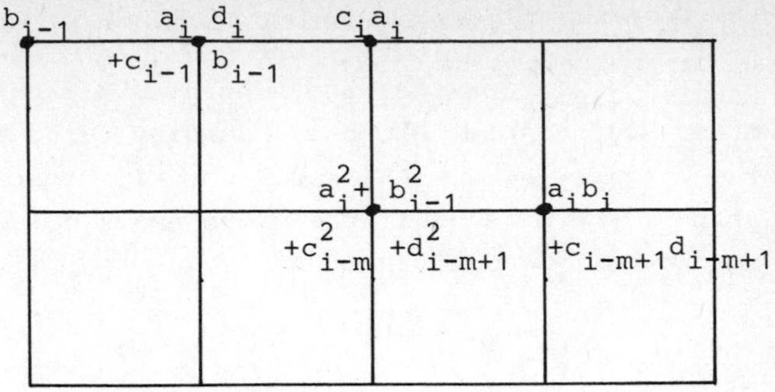

LL^T

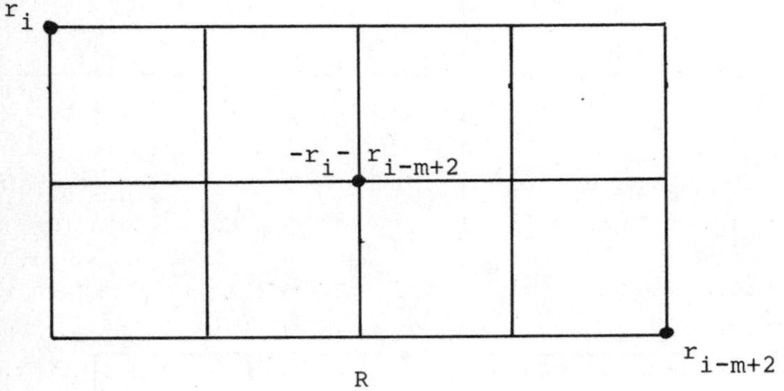

R

and the recursion formulas

$$a_i^2 = \alpha_i + \delta_i - b_{i-1}^2 - c_{i-m}^2 - d_{i-m+1}^2 - r_i - r_{i-m+2}$$

$$b_i = -(\beta_i + c_{i-m+1}\, d_{i-m+1})/a_i$$

$$c_i = -\gamma_i/a_i$$

$$d_i = -b_{i-1}c_{i-1}/a_i$$

$$r_i = b_{i-1}\, d_{i-1}.$$

Continuing in this way we get a sequence of more and more accurate factorizations:

MI(O), MIC(1), MIC(2), MIC(4), MIC(7), MIC(12), etc.

Remark 4.1. In practice we may avoid the square-roots by making a MIC factorization of the type $C = (L+D_1)D_1^{-1}\&L^T+D_1)$, D_1 a diagonal matrix and L strictly lower triangular.

Remark 4.2. In the recursion formulas (4.1) for the MIC(O) algorithm, a_i can be calculated from

$$a_i^2 = \alpha_i + \delta_i - b_{i-1}(b_{i-1}+c_{i-1}) - c_{i-m}(c_{i-m}+b_{i-m})$$

in order to decrease the number of operations. Similarly for MIC(d), d > O.

Remark 4.3. In a Dirichlet problem, we choose $\delta_i = \alpha_i \xi h^2$, i=1,...,N, where $\xi > 0$ is a parameter and h is a mesh parameter. For the choice of $D = \text{diag}(\delta_i)$ in general see the following analysis.

5. The MIC(O)* Algorithm

We also have the following variant of the MIC(O) algorithm, the MIC(O)* algorithm, which can be thought of as generalized SSOR method.

The MIC(O)* algorithm is obtained by disregarding in the MIC(O) algorithm all corrections $-a_{ik}a_{kj}$ to the entries a_{ij}, $i \neq j$, see (1,1). These numbers are instead added to the diagonal of U to get rowsum (R) = O. Apparently, the MIC(O)* algorithm is of the type $C = (\tilde{D}+L)\tilde{D}^{-1}(\tilde{D}+L^T)$, where $D > 0$ is a diagonal matrix and L is the strictly lower triangular part of A, compare with the SSOR method [4].

The advantage with this method is that it needs less storage and less factorization work. On the other hand, the convergence is in general somewhat slower than for the MIC(O) method.

The MIC(O)* algorithm as well as MIC(d) algorithms, d ≥ O, are covered by the following analysis of the rate of convergence of the corresponding MICCG method, that is, the MIC factorization combined with the conjugate gradient (CG) method.

6. Rate of Convergence of MICCG Methods (The symmetric case)

Let h be a mesh parameter and let m_k, k = 1,2,... be independent on h. Further, let n be the space dimension.

Assume that

(i) A is a symmetric M-matrix of order $N = O(h^{-n})$, $n \to 0$,

(ii) Rowsum (A) ≥ O,

(iii) $-2 \sum_{j>i} a_{ij} \leq a_{ii} + m_1 h$, $i \in N_1$,

where $N_1 \subseteq N = \{i;\ 1 \leq i \leq N\}$ and the number of indices in $N_2 = N \setminus N_1$ is $O(h^{-n+1})$, $h \to 0$.

From (i) and (ii) it follows that A is weakly diagonally dominant and hence any MIC factorization is stable.

If we now define $D = \mathrm{diag}(\delta_i)$ in a proper way, namely

(iv) $$\delta_i = \begin{cases} \xi_1 h^2 a_{ii} , & i \in N_1 \\ \xi_2\ h\ a_{ii} , & i \in N_2, \end{cases}$$

where $\xi_i > 0$, $\tau = 1,2$ are parameters, than we have the following result for the spectral condition number κ of $C^{-1/2}\ A\ C^{-1/2}$.

<u>Theorem 6.1.</u> Assume that A satisfies (i), (ii) and (iii). Then, if D is chosen according to (iv), $\kappa = O(h^{-1})$, $h \to 0$.

Since, as is well known, the number of iterations in the MICCG method is of order $O(\sqrt{\kappa})$, we state the following corollary.

<u>Corollary 6.1.</u> Assume that the conditions (i)-(iv) are satisfied. Then the number of iterations in the MICCG method is $O(h^{-1/2})$, $h \to 0$ and the number of arithmetic operations is $O(N^{1+1/2n})$, $N \to \infty$.

If A is a weakly diagonally dominant M-matrix, that is, (i) and (ii) are fulfilled, then (iii) is satisfied for the following types of elliptic PDE problems, if the mesh points are numbered in a natural (rowwise or similar) way:

a) Dirichlet problems with constant coefficients (Laplace equation). In this case $N_1 = N$ and $m_1 = 0$.

b) Dirichlet problems with Lipschitz continuous material coefficients. Here, $N_1 = N$.

c) Neumann problems. Then, N_2 represents points on and/or near the Neumann boundary.

d) Problems with discontinuous material coefficients. Here, N_2 represents points on and/or near an interface over which the coefficients are discontinuous.

<u>Remark:</u> In numberical tests, the number of iterations has turned out to be almost independent of the parameters ξ_1, ξ_2 in a fairly wide range.

In fact, the choice $\xi_1 = \xi_2 = 0$, i.e. $D = 0$, is almost as good as the optimal choice. Hence, in practice, the definition of N_2 offers no problem. For instance, one can use $D = 0$ or $D = \xi h^2 \operatorname{diag}(A)$, $\xi > 0$ (say $\xi = 1$) for all types of problems.

7. MIC for More General FEM Matrices

For matrices that are not diagonally dominant, the MIC as well as the IC factorization may be instable. This can be overcome by using shifted incomplete factorizations (SIC, SMIC), see e.g. [5], [6]:

> Positive numbers are added to the diagonal of U (or A) if non-positive diagonal elements are produced.

In general this approach leads to slower convergence, in practical if this shifting has to be done quite often. Although this author has tested several types of FEM problems with different kinds of approximations, the MIC algorithms have never turned out to be instable. Furthermore, the same rate of convergence as was shown in Section 7 has been measured for more general matrices as well, see the results presented in Section 10. (For the IC algorithms, however, instability has been observed, see also [5].)

For many kinds of problems we can obtain (even theoretically) the same fast rate of convergence as was shown in the previous section, by using the following idea.

Spectral equivalence

Let A_h and B_h of order $N = N(h)$ be two discretizations of a second order elliptic differential operator corresponding to a mesh size parameter h.

Definition 7.1. A_h and B_h are spectrally equivalent if

$$0 < c \le \frac{(A_h x, x)}{(B_h x, x)} \le C , \quad \forall x \in R^N , \quad x \neq 0,$$

where c, C are independent of h.

Assume now that we have an original, coarse FEM mesh consisting of quadrilateral or triangular elements with all angles $\le \Pi/2$ and let the mesh be refined in a uniform way, see e.g. [7]. Further, let $A_h^{(2)}$ and $A_h^{(p)}$ be the matrices corresponding to piecewise polynomial basis functions of degree 1 (linear) and p, respectively. We then have that $A_h^{(1)}$ and $A_h^{(p)}$ are spectrally equivalent, for details see [7].

Example 7.1. Consider the Laplace equation, $-\Delta u = f$ in K_e, $u = 0$ on ∂K_e, K_e the unit square, discretized by FEM based on a uniform rightangled triangular mesh. Let $A_e^{(1)}$ and $A_e^{(2)}$ be elements stiffness matrices corresponding to linear and quadratic basis functions, respectively. $A_e^{(1)}$ is assembled from the four elements corresponding to the finer mesh, see Fig. 7.1.

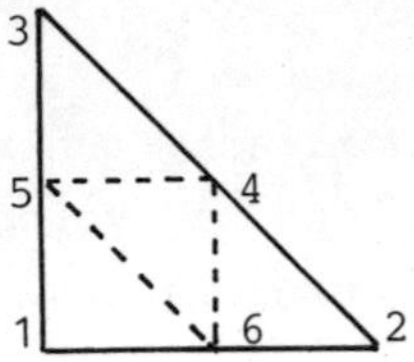

Fig. 7.1: An element

We have
$$A_e^{(1)} = \begin{bmatrix} 2 & 0 & 0 & 0 & -1 & -1 \\ 0 & 1 & 0 & 0 & 0 & -1 \\ 0 & 0 & 1 & 0 & -1 & 0 \\ 0 & 0 & 0 & 4 & -2 & -2 \\ -1 & 0 & -1 & -2 & 4 & 0 \\ -1 & -1 & 0 & -2 & 0 & 4 \end{bmatrix}$$

$$A_e^{(2)} = \begin{bmatrix} 6 & 1 & 1 & 0 & -4 & -4 \\ 1 & 3 & 0 & 0 & 0 & -4 \\ 1 & 0 & 3 & 0 & -4 & 0 \\ 0 & 0 & 0 & 16 & -8 & -8 \\ -4 & 0 & -4 & -8 & 16 & 0 \\ -4 & -4 & 0 & -8 & 0 & 16 \end{bmatrix}$$

and it is easily seen that

$$2(A_e^{(1)}x,x) \le (A_e^{(2)}x,x) \le 4(A_e^{(1)}x,x), \ \forall x \in R^6.$$

Since the global matrices $A_h^{(1)}$ and $A_h^{(2)}$ are "sums" of element matrices, we get

$$2(A_h^{(1)}x,x) \le (A_h^{(2)}x,x) \le 4(A_h^{(1)}x,x), \ \forall x \in R^N.$$

Hence, $A_h^{(1)}$ and $A_h^{(2)}$ are spectrally equivalent and

$\kappa((A_h^{(1)})^{-1/2}A_h^{(2)}(A_h^{(1)})^{-1/2}) = 2.$

Now, $A_h^{(1)}$ is a diagonally dominant matrix so any MIC factorization $C = A_h^{(1)}+D+R$ is stable and $\kappa(C^{-1/2}A_h^{(1)}C^{-1/2}) = O(h^{-1})$, $h \to 0$. Due to the spectrally equivalence we then have

$$\kappa(C^{-1/2}\ A_h^{(2)}\ C^{-1/2}) = O(h^{-1}),\ h \to 0$$

as well. Note that when we use this method we have to assemble both $A_h^{(1)}$ and $A_h^{(p)}$. On the other hand a MIC factorization of $A_h^{(1)}$ is often much simpler to perform than one of $A_h^{(p)}$.

8. A Biharmonic Problem

Let A correspond to the 13-point difference approximation of the biharmonic problem

$$\begin{cases} \Delta^2 u = f & \text{in } K_e \\ u - u_n = 0 & \text{on } \delta K_e \end{cases}$$

and let A_1 correspond to the 5-point difference approximation of the Laplace equation ($\Delta u = g$ in K_e, $n = 0$ on δK_e). Define $\tilde{A} = A_1^2$. Then A and $\tilde{A}$ are spectrally equivalent (Definition 7.1), and

$$\kappa(\tilde{A}^{-1/2}\ A\ \tilde{A}^{-1/2}) = 3, \text{ see [8].}$$

Let $C_1 = A_1 + D + R$ be a MIC factorization of A_1. This is stable since A_1 is diagonally dominant and $\kappa(C_1^{-1/2}A_1C_1^{-1/2}) = O(h^{-1})$, $h \to 0$.

As preconditioning matrix for A we choose $C = \{(C_1^{-1})^2\}^{-1}$. Then $\kappa(C^{-1/2}A\ C^{-1/2}) = \kappa(C_1^{-1}A_1A_1C_1^{-1}) = O(h^{-2})$, $h \to 0$ and the number of iterations in MICCG is $O(h^{-1})$, $h \to 0$. Note that the condition number of A, $\kappa(A) = O(h^{-1})$, $h \to 0$.

In each iteration we solve four triangular systems with the triangular factors of C_1 and the total number of operations is $O(N^{1+1/2})$, $N \to 0$.

<u>Remark</u>: If we use an iterative method on the form

$$(A_1)^2(n^{\ell+1}-n^{\ell}) = -\beta_\ell(A_n{}^{\ell}-f), \quad \ell = 0,1,\ldots.$$

we have only $O(1)$ number of iterations. In each iteration we solve two systems with matrix A_1 by (for instance) a MICCG method (inner iterations). Thus, the number of operations is $O(N^{1+1/4})$, that is, of the same order as for a second order problem. From a practical point of view, however, this seems to be preferable only for fairly small values of h.

9. MIC For A Class of Non-Symmetric Problems

Consider differential equations on the form

$$Lu = -\Delta u + \vec{v}\cdot\nabla u + cu = f, \; u = u(x), \; x \in \Omega$$

with suitable boundary conditions on $\partial\Omega$.

Assume that we use a positive difference scheme (upwind, modified upwind Il'in etc.). Then the associated matrix A is diagonally dominant and the MIC factorization C = A+D+R is stable.

Let A_o be the matrix corresponding to $\vec{v} \equiv 0$ and let $A_s = \frac{1}{2}(A+A^T)$ be the symmetric part of A. We assume that

$$(A_s x,x) \geq (A_o x,x). \quad \forall\, X \in R^N.$$

A sufficient condition for this to be true is that $\operatorname{div}(\vec{v}) \leq 0$, since $\frac{1}{2}(L+L^*)u = -\Delta u - \frac{1}{2}\operatorname{div}(\vec{v})u + cu$.

Further, assume that $\|\vec{v}\|$ is not too large (compare to h^{-1}), say $\|\vec{v}\| \leq c\,h^{-1}$, c independent of h.

Then the eigenvalues of $C^{-1}A$ are situated in an ellipse in the complex plane, centered at (1.0), with (after normalization) semiales; $\alpha = 1 - m_1 h$, $\beta = m_2 h$.

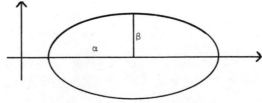

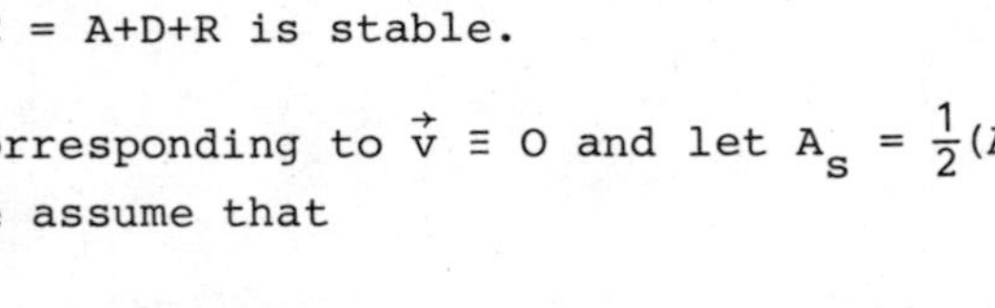

Then the number of iterations, p, in the Chebyshev acceleration method is of order

$$p = O([\ln(1+\sqrt{1-\alpha^2+\beta^2}) - \ln(\alpha+\beta)]^{-1})$$

$$= O(h^{-1/2}),\ h\to 0, \qquad \text{for details see [1].}$$

For the unpreconditioned method the results are; $\alpha = 1-m_3h^2$, $\beta = c_nh$ and $p = O(h^{-1})$, $h\to 0$.

We note that a generalized conjugate direction method, see [9], is appropriate even for non-symmetric problems.

10. Numerical Results

We present some typical results from numerical experiments using MICCG methods. For further tests we refer to [1] and the references therein.

As initial approximation u_o to the conjugate gradient method we have used $u_o = C^{-1}f$ and the iterations were stopped when the residual error (in ℓ_2-norm) was reduced by a factor 2, that is, when $\|r^\ell\| \le \varepsilon \|r^o\|$.

Example 10.1: The model Laplace equation on the unite square,

$$\begin{cases} \Delta u = f & \text{in } K_e \\ u = 0 & \text{on } \partial K_e \end{cases}$$

discretized by bilinear finite elements. The number of iterations for various methods and different values of N, $\varepsilon = 10^{-6}$:

N	MIC(O)	MIC(O)*	SSOR	IC(O)	IC(O)*	MIC(2)
100	8	8	9	8	9	5
400	11	12	12	13	15	7
1600	16	17	18	23	27	10

N=961, the MIC(4) factorization of $A_h^{(2)}$ is only slightly more efficient than the MIC(2) factorization of $A_h^{(1)}$ and that the MIC(0) factorization of $A_h^{(1)}$ is more efficient than the MIC(0) factorization of $A_h^{(2)}$.

<u>Example 10.4:</u> The biharmonic problem described in Section 8. The number of iterations for $\varepsilon = 10^{-3}$ and for MIC factorization of A and A_1, respectively:

h^{-1}	Factorization of A_1	Factorization of A	
	MIC(2)	MIC(0)	MIC(4)
5	4	4	2
10	6	7	3
20	10	16	7
40	17	37	14

The work per unknown for $h^{-1} = 40$ is for the MIC(2) factorization of A_1, the MIC(0) and MIC(4) factorizations of A about 630, 1210 and 620 operations, respectively. One realizes that one needs a quite accurate and hence more complicated factorization of A to reach the same efficiency as for a simpler factorization of A_1.

<u>Example 10.5:</u> The non-symmetric problem

$$-\Delta u + \gamma(u'_x - u'_y) = 1 \quad \text{in } K_e$$

$$u = 0 \quad \text{on } \partial K_e,$$

where $\gamma \geq 0$ is used to vary the degree of non-symmetry. The number of iterations needed in the minimum residual CG method [9] for various preconditionings and different values of h and γ, $\varepsilon = 10^{-3}$:

Example 10.2: A discontinuous problem, $-\frac{\partial}{\partial x}(a\frac{\partial u}{\partial x}) - \frac{\partial}{\partial y}(a\frac{du}{\partial y}) = 1$ on Ω, see the figure, where d is a parameter used to vary the degree of discontinuity.

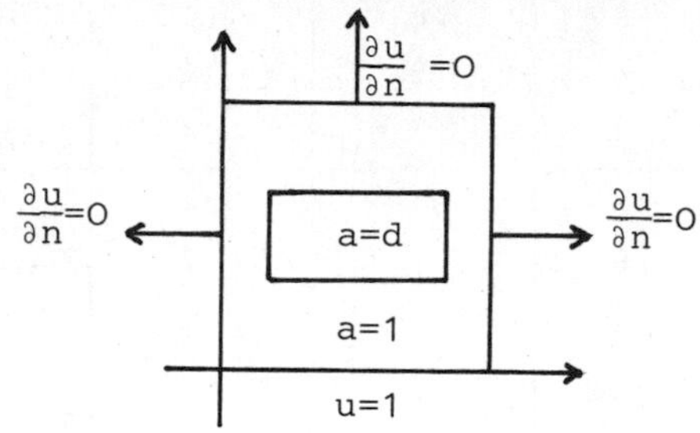

The number of iterations for different methods when the five point FDM approximation was used, $\varepsilon = 10^{-6}$:

d	1			100			10 000		
h^{-1}	SSOR	MIC(0)	MIC(2)	SSOR	MIC(0)	MIC(2)	SSOR	MIC(0)	MIC(2)
6	11	10	6	13	12	7	15	13	7
12	16	14	8	21	17	10	26	18	11
24	26	19	12	34	23	14	42	25	15
27	29	21	13	37	25	15	46	27	16

Notice the remarkable insensivity to the parameter d observed for the MIC methods.

Example 10.3: The problem defined in Example 7.1. Number of iterations for different MIC factorizations of $A_h^{(1)}$ and $A_h^{(2)}$, $\varepsilon = 10^{-4}$:

N	factorizations of $A_h^{(2)}$		factorizations of $A_h^{(1)}$		
	MIC(0)	MIC(4)	MIC(0)	MIC(1)	MIC(2)
49	6	4	7	6	6
225	9	5	9	7	6
961	13	7	13	10	8

If one compares the total work for the methods one finds that for

γ h	C	0	5	10	100	10^5
$\frac{1}{8}$	I	8	18	19	18	18
	MIC(O)	4	5	5	7	7
	MIC(2)	3	3	3	3	3
$\frac{1}{16}$	I	18	34	34	34	35
	MIC(O)	6	6	7	10	11
	MIC(2)	4	4	4	4	4
$\frac{1}{32}$	I	36	72	72	65	62
	MIC(O)	9	9	10	12	13
	MIC(2)	6	6	6	5	5

Observe that, when one uses a more accurate MIC factorization, the number of iterations ist almost independent on X even for fairly large values of γ, that is, for singularly perturbed problems.

11. Conclucions

The results regarding MICCG factorization methods presented in this contribution can be summarized in the following statements:

- The result $\kappa(C^{-1/2}AC^{-1/2}) = O(h^{-1})$, $h \to 0$ has been proved for several classes of FEM matrices corresponding to 2'nd order elliptic differential equations.
- The number of iterations for MICCG is $O(h^{-1/2}\ln 1/\varepsilon)$ to reach a relative residual error ε.
- The number of operation for MICCG is $O(N^{1+1/2n}\ln N)$, $N \to \infty$, if $\varepsilon = O(N^{-\mu})$, $\mu > 0$ and n is the space dimension.
- For a biharmonic problem in two dimensions the number of operations has been shown to be $O(N^{1+1/2}\ln N)$.
- The storage requirement is of optimal order $O(N)$.
- They are easy to program.

References

[1] Gustafsson, I.: Stability and rate of convergence of modified incomplete Cholesky factorization methods, Report 79.02 R, Department of Computer Sciences, Chalwers University of Technology, Göteborg, Sweden (1979)

[2] Munksgaard, N.: Solution of general sparse symmetric sets of linear equations, Report No. NI-78-02, Inst. for Num. Anal., Technical University of Denmark, Lyngby, Denmark (1978)

[3] Meijerink, J.A. and van der Vorst, H.A.: An iterative solution method forlinear systems of which the coefficient matrix is a symmetric M-matrix, Math. Comp. 31 (1977), 148-162

[4] Young, D.: Iterative solution of large linear systems, Academic Press, New York and London (1971)

[5] Kershaw, D.: The incomplete Cholesky conjugate gradient method for the iterative solution of Systems of linear equations. J. Comput. Phys. 26 (1978), 43-65

[6] Manteuffel, T.A.: The shifted incomplete Cholesky factorization, Technical report, Appl. Math. Division 8325, Sandia Laboratories, Livermore, California, USA (1978)

[7] Axelsson, D. and Gustafsson, I.: A preconditionad conjugate gradient method for finite element equations, which is stable for rounding errors, Report 7924, Mathematisch Institnut, Katholieke Universiteit, Nijmegen, The Netherlands (1979)

[8] Gustafsson, I.: On modified incomplete factorizations for a biharmonic problem, in progress

[9] Axelsson, O.: A generalized conjugate direction method and its application on a singular perturbation problem, lecture notes (1979)

Solving Large Sparse Linear Systems Arising in Queuing Problems

L. Kaufman

Bell Laboratories
Murray Hill, New Jersey 07974

ABSTRACT

In this paper we describe iterative methods for finding the null vectors of large sparse unsymmetric singular matrices which arise while modeling queuing networks. We find that these matrices often have some useful algebraic structure, and that classical methods often applied to solving nonsingular problems arising in the study of differential equations can be used for these types of problems.

1. Section 1: Introduction

In this paper we are concerned with finding a vector $\mathbf{p}$ such that

$$A\mathbf{p} = 0$$

$$\sum_{i=1}^{n} p_i = 1 \qquad (1.1)$$

where A is a large, sparse square singular nonsymmetric matrix representing the birth and death equations in a queuing model of a network. Because of its origin all the nonzero off-diagonal elements of A are negative, its column sums are 0 and its rank is $n-1$. The element p_i represents the probability of being in the ith state of the model and the negative of a_{ij} represents the transition rate of going from state i to state j. In essence the matrix A can be generated from the defining parameters of the model.

The present work was stimulated by two distinct problem types neither of which are amenable to standard analytic techniques, e.g. a product form solution.

The first problem type described in [1] concerns a network with overflowing queues. Typically n is about a thousand, but problems with n=30000 have been solved. In the simplest form of the problem, the state space looks like a rectangle, the matrix A has symmetric zero-structure and looks like the standard five point operator that one encounters while discretizing an elliptic partial differential equation. The matrix A also has some algebraic structure which we sometimes can exploit.

The second problem type described in [2] has tandem queues in a network. Typically n is 40,000, the problem is 6 dimensional and is generally conceived as a 2 dimensional problem in which the state space for each dimension is the base of a tetrahedeon. The graph defined by the zero structure of the A matrix has cycles of length 2 and 4 and is hardly reminiscent of problems one usually encounters in modeling in the physical sciences. The zero structure of a typical matrix is given in figure 1. The off diagonal $x's$ should be considered as diagonal matrices. The diagonal $x's$ should be thought of as submatrices having the same zero structure as given in the whole figure.

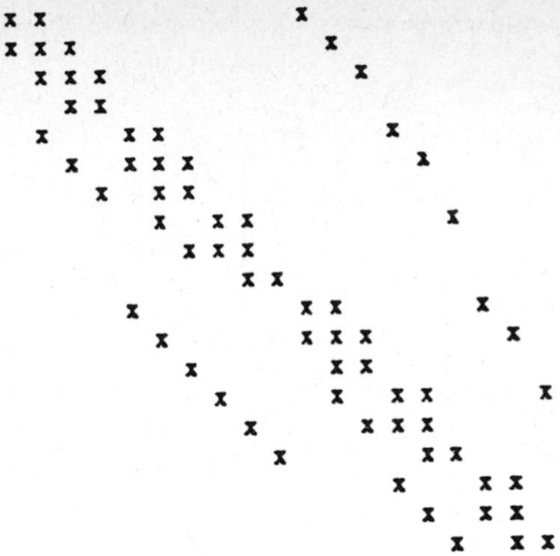

Figure 1: Matrix structure for small problem of type 2.

In this paper we consider 2 classes of methods for solving (1.1). In section 2 we discuss inverse iteration which requires few iterations, but much effort per iteration. Initially we treated problems of the first type using this method and cut down the work by using the algebraic structure of the problem. In section 3 we consider iteration methods based on matrix splittings which normally require little effort per iteration but many iterations. The second problem type was treated exclusively with these techniques.

2. Section 2: Inverse Iteration

Inverse iteration is normally used for finding the null vector of a small dense matrix A. It may be written as

1. Pick $\mathbf{p}^{(0)}$ not orthogonal to the null vector of A

2. For k=1,2,... iterate until convergence as follows

$$\text{Solve } A\,\mathbf{v}^{(k)} = \mathbf{p}^{(k)} \text{ for } \mathbf{v}^{(k)} \tag{2.1}$$

$$\text{Set } \mathbf{p}^{(k+1)} = \frac{\mathbf{v}^{(k)}}{||\mathbf{v}^{(k)}||}$$

Usually only about 2 iterations are necessary. Because of the near singularity of the computer representation of A, the computed v's will not be close to the true solution of (2.1). However after scaling the computed $\mathbf{p}$'s will rapidly approach a null vector of A.

Of course the method assumes one can solve (2.1). In our first problem type the matrix A for the simplest problem has the form

Let $B = \tilde{Q} A \tilde{Q}^{-1}$. The matrix B will have only 3 nonzero diagonals.

3. Determine the permutation matrix P such that

$$C = PBP^T = \begin{pmatrix} T_1 & & & \\ & T_2 & & \\ & & \ddots & \\ & & & T_q \end{pmatrix}$$

The T_i's will have the form

$$T_i = T + \delta_i I$$

where T is tridiagonal with α's on its diagonal, β's from H on its superdiagonal and γ's from G on its subdiagonal. Thus $\tilde{A} = \tilde{Q}^{-1} P^T C P \tilde{Q}$ and since it is easy to solve a linear system with A, P, or C as the coefficient matrix it is easy to solve a linear system with $\tilde{A}$ as the coefficient matrix.

As long as only 2 queues are involved in the queuing model, problems of the first type are amenable to inverse iteration. Although the dimension of the X matrix varies, it usually only represents 1 or 2 lines in a rectangular state space. However when 3 or more queues are considered, A becomes 7-diagonal and the non-separable portion corresponds to planar slices in a 3 dimensional region. For 3 queues when n is 27000, X is a 900×900 dense matrix and the general method considered in this section is no longer viable. For these problems the methods in section 3, which are also easier to implement, are more attractive.

3. Matrix Splitting Techniques

In this section we will consider methods for solving (1.1) based on splitting the matrix A into

$$A = M - N \tag{3.1}$$

where M is nonsingular, has positive diagonal elements and has negative off-diagonal elements, and is an M-matrix (see [4]). For example in the Jacobi method, M is D, the diagonal of A, and in the Gauss Seidel method M is $D+L$ where L is the strictly lower triangular portion of A. Based on (3.1) one can implement the following iterative scheme to determine the probability vector $\mathbf{p}$

Pick $\mathbf{p}^{(1)}$

For $k=1,2,\ldots$ until convergence

$$\text{Solve } M\,\mathbf{p}^{(k+1)} = N\mathbf{p}^{(k)} \quad (3.2)$$

Set $\mathbf{p} = \mathbf{p} / \sum_{i=1}^{n} p_i$

3.1 Convergence Results

Let $T = I - M^{-1}A = M^{-1}N$

Obviously $\mathbf{p}^{(k+1)} = T\,\mathbf{p}^{(k)}$

Let $\rho(X)$ denote the modulus of the largest eigenvalue of the matrix X. If A were nonsingular, as explained in [4], then $\rho(T) < 1$ and an algorithm based on the splitting in (3.1) must converge. However, in our situation, A is singular and the convergence of (3.2) is not automatically assured. The following lemmas govern this case.

$$A = \begin{pmatrix} F_0 & H_0 & & & \\ G_1 & F_1 & & & \\ & \cdot & \cdot & \cdot & \\ & & \cdot & \cdot & H_{m-1} \\ & & & G_m & F_m \end{pmatrix} \tag{2.2}$$

where $G_i = \gamma_i I$; $H_i = \beta_i I$ and the F's are tridiagonal. Moreover for i=0,1...,$m-1$, $F_i = F + \alpha_i I$. Thus except for the last F matrix the A matrix is separable. For small values of n, one can solve (2.1) using either a band solver or a sparse matrix solver, but as n approaches 1000, it pays to consider the algebraic structure of (2.2). We note that A may be partitioned as

$$A = \left(\begin{array}{c|c} \tilde{A} & \begin{matrix} 0 \\ \cdots \\ H_{m-1} \end{matrix} \\ \hline 0 \vdots G_m & F_m \end{array}\right) \tag{2.3}$$

where $\tilde{A}$ is separable. Thus we may write

$$A = \left(\begin{array}{c|c} I & 0 \\ \hline (0 \vdots G_m)\tilde{A}^{-1} & I \end{array}\right) \left(\begin{array}{c|c} \tilde{A} & \begin{matrix} 0 \\ \vdots \\ H_{m-1} \end{matrix} \\ \hline 0 & X \end{array}\right)$$

$$= L\ U \tag{2.4}$$

where $X = F - (0 \vdots G_m)\ \tilde{A}^{-1} \begin{pmatrix} 0 \\ \cdots \\ H_{m-1} \end{pmatrix}$

To solve $A\,\mathbf{v}^{(k)} = \mathbf{p}^{(k)}$, we solve

$$L\mathbf{y} = \mathbf{p}^{(k)}$$

and then

$$U\ \mathbf{v}^{(k)} = \mathbf{y}\ .$$

Because $\tilde{A}$ is separable, solving a linear system with $\tilde{A}$ is easy using a fast direct solver as in [3] and [5]. Moreover when forming X, the sparsity in the algebraic expression involving G_m and H_{m-1} can be used. The matrix $\tilde{A}$ may also be decomposed into simpler factors according to the following algorithm.

1. Determine the eigendecomposition of $F = QDQ^{-1}$. The matrix D will be diagonal and the matrix Q has the form Q_1S where S is diagonal and Q_1 is orthogonal.

2. Let

$$\tilde{Q} = \begin{pmatrix} Q & & & \\ & Q & & \\ & & \cdot & \\ & & & Q \end{pmatrix}$$

Lemma 1: If λ is an eigenvalue of T and $\mathbf{x}$ its corresponding eigenvector and

$$\text{if } \lambda = 1, \text{ then } A\ \mathbf{x}=0 .$$

Proof: The proof is based on the fact that

$$\text{if } T\mathbf{x} = \lambda \mathbf{x}$$

$$\text{then } - M^{-1} A\mathbf{x} = (\lambda - 1)\ \mathbf{x}$$

Lemma 2: $\rho(T) = 1$

Proof:
Let $\mathbf{e}^T = (1,...,1)^T$.
Simple algebra leads to the conclusion that $M^{-T}N^T\mathbf{e} = \mathbf{e}$
Since $M^{-T}N^T$ is a positive matrix,
$\rho(M^{-T}N^T)$ is the maximum column sum, which equals 1.
But $\rho(M^{-T}N^T) = \rho(T)$.
Lemma 3: $\rho(T)$ is simple.

Proof: Uses the Peron-Frobenius theorem and the fact that T is a positive matrix.

From these 3 lemmas, one concludes that asymptotically, the error is decreased each iteration by $\sigma(T)$, the magnitude of the second largest eigenvalue of T. Depending on M and the structure of A, $\sigma(T)$ may be 1 and the process given in (3.2) will not converge without incorporating some averaging process. If A is p-cyclic (i.e. all cycles in the graph defined by the zero structure of A are of length p) and consistently ordered (i.e. the eigenvalues of the matrix

$$K_\eta = \eta D^{-1}L + \eta^{(-p+1)}D^{-1}U \tag{3.3}$$

are independent of η), then according to [4] the Gauss-Seidel process will converge. For example, for the 3X3 matrix

$$A = \begin{pmatrix} 1 & & 1 \\ -1 & 1 & \\ & -1 & 1 \end{pmatrix},$$

condition (3.3) is satisfied, the T matrix for Gauss-Seidel has eigenvalues at (0,0,1), and convergence is assured. However if the states had been reordered so that the A matrix was

$$\begin{pmatrix} 1 & -1 & \\ & 1 & -1 \\ -1 & & 1 \end{pmatrix},$$

then condition (3.3) is not satisfied, the eigenvalues of the T matrix of Gauss Seidel are (0,1, -1), and every other iterate repeats. In general if the underlying graph is p-cyclic, a matrix is consistently ordered if it has the form

$$\begin{bmatrix} x & & & & x \\ x & x & & & \\ & x & x & & \\ & & x & x & \\ & & & x & x \end{bmatrix}$$

The matrices involved in our first problem type are 2-cyclic and the natural ordering is consistent. For the second problem type for certain traffic flows the matrices are 4-cyclic and it is necessary to order the states correctly. Even in situations when convergence is guaranteed, the ordering is important as Table (1) indicates. If under certain parameter settings, the second largest eigenvalue lies on the unit circle, then a slight change in these parameters might yield a convergent algorithm. However, the second largest eigenvalue might be so close to the unit circle that convergence is excruciatingly slow. As a rule of thumb, one should place the largest off diagonal elements below the main diagonal and as close to that diagonal as possible.

Table 1: Effect of ordering on eigenvalues

Ordering Scheme	Matrix Structure	Controlling Eigenvalues
C	4 cyclic λ=0	.21 + .27i
C	4 cycles and 2 cycles λ=.1	.26 ± .24i
R	4 cyclic λ=0	cube roots of unity
R	4 cycles and 2 cycles λ=.1	-.45 ± .838i

C = Consistent ordering

R = Nodes ordered in reverse order from C

The problem of having 2 or more eigenvalues of T on the unit circle can be avoided without reordering by using a step length algorithm. Consider the algorithm

$$\mathbf{p}_{k+1} = \frac{1}{\alpha+1} (\alpha I + T)\mathbf{p}_k$$

for some fixed α and let $\bar{T} = \frac{1}{\alpha+1}(\alpha I+T)$. The matrix $\bar{T}$ has an eigenvalue at 1 whose corresponding eigenvector is a null vector of A; the rest of the eigenvalues of $\bar{T}$ are within the unit circle and the method will always converge. In fact for every eigenvalue of T, there exists an eigenvalue of $\bar{T}$ which lies on the line connecting the eigenvalue of T to 1. This means that if all the eigenvalues corresponding to $\sigma(T)$ have negative real parts, then for small values of α, (3.4) would be faster than (3.2). On the other hand if $\sigma(T)$ corresponded to a positive real eigenvalue, (3.2) would be faster than (3.4) for all values of α.

Because A is singular, another nice property given in [4] for regular splittings of nonsingular M matrices does not hold. If A were nonsingular and one considered two splittings

$$A = M_1 - N_1$$

$$\text{and } A = M_2 - N_2$$

and $N_1 > N_2$ then $\rho(M_1^{-1}N_1) \geq \rho(M_2^{-1}N_2)$. Thus a splitting which supposedly requires more work per iteration would not require more iterations. However for singular matrices the theorem is not true as the following example illustrates.

$$\text{Let } A = \begin{pmatrix} 1 & -1/2 & -1/2 & 0 \\ -1/2 & 1 & 0 & -1/2 \\ -1/2 & 0 & 1 & -1/2 \\ 0 & -1/2 & -1/2 & 1 \end{pmatrix}$$

$$\text{Let } N_1 = \begin{pmatrix} 0 & 1/2 & 1/2 & 0 \\ 0 & 0 & 0 & 1/2 \\ 0 & 0 & 0 & 1/2 \\ 0 & 0 & 0 & 0 \end{pmatrix} \text{ and } N_2 = \begin{pmatrix} 0 & 0 & 1/2 & 0 \\ 0 & 0 & 0 & 1/2 \\ 0 & 0 & 0 & 0 \\ 0 & 0 & 0 & 0 \end{pmatrix}$$

then $\sigma(M_1^{-1}N_1) = 0$ and $\sigma(M_2^{-1}N_2) = 1/9$.

Considering the fact that the convergence theory has all been worked out for nonsingular matrices, it would seem natural to suggest to fix one element in the probability vector, delete one row and column from A and solve a linear system with a nonsingular coefficient matrix for the remaining elements of the probability vector. The key to the attractiveness of this approach is of course the rate of convergence of this scheme compared to the original one based on the singular matrix. When A is symmetric, the interlocking property of eigenvalues guarantees that the Gauss-Seidel method applied to the singular matrix is preferable. This fact was also observed for our problems as Table (2) illustrates. When n is large, the largest eigenvalue for the nonsingular iteration matrix will normally be close to the unit circle and hence that scheme will be excruciatingly slow. In fact in practice the convergence for the nonsingular scheme asymptotically becomes so slow the program may appear to be faulty.

Interestingly enough, the singular scheme does not always perform better. Consider the following matrix

$$A = \begin{pmatrix} \alpha & c & e \\ a & \beta & f \\ b & d & \gamma \end{pmatrix} \text{ where } \begin{matrix} \alpha = -(a+b) \\ \beta = -(c+d) \\ \gamma = -(e+f) \end{matrix} \tag{3.5}$$

If $a=c=e=1$ and $b=d=f=9$ then, the controlling eigenvalue for Gauss-Seidel for the singular scheme is 81/1000 while if one deletes the last row and column, the controlling eigenvalue is 1/100. However, given a matrix of the form (3.5), one can show that there always exists a symmetric permutation of A, so that the singular system is better.

Table 2: Convergence behavior
of Singular vs Nonsingular Scheme

Ordering Scheme	Matrix Structure	controlling eigenvalues for singular	Controlling eigenvalues for nonsingular
C	4 cyclic	.21 ± .27i	.86
			.18 ± .26i
C	4 cycles and 2 cycles	.48 ± .08i	.97
			.47 ± .09i
R	4 cyclic	cube roots of unity	.99
			-.50 ± .86i
		.67 ± .2i	
R	4 cycles and 2 cycles	-.25 ± .66i	.99
			-.25 ± .26i

C = consistent orderings
R = nodes ordered in reverse order from C

3.2 In Practice:

As a first attempt to solve both problem types, point Successive Overrelaxation (SOR) was implemented. In SOR, M in (3.1) is $(L+wD)$ and N is $(-U+(w-1)D)$ where L is the strictly lower part of A, D is the diagonal of A, U is the strict upper-triangular portion, and w is a scalar. Both problem types are sensitive to changes in w and changing w by 10% often doubles the number of iterations. In general, SOR does not fare as poorly as one might expect. In fact for the second problem type, it was not unusual to solve a problem with 40000 variables within 150 iterations.

For the first problem type, the matrix A is 2 cyclic and a formula given in [4] gives ballpark estimates for the optimal w. Usually values of about 1.75 are satisfactory. For the second problem type, the formulae are overestimates. For some traffic flows point SOR diverges with $w=1.2$. For others this value is close to optimal.

For both types, the relative error and the relative residual tend to increase initially when w is increased. Divergence can not be detected just by studying the results of 1 iteration.

Using "line SOR" (see [4]) usually halves the work for the first problem type. In "line" SOR the tridiagonal submatrices on the diagonal of A are incorporated into M. Partitioning the A matrix into separable blocks and combining SOR with the approach outlined in section 2 yields an algorithm which requires far fewer iterations than line SOR with optimal parameters, but the total effort is greater because of the increased work per iteration.

For the second problem type the transition matrix can be arranged so that there are tridiagonal submatrices on the diagonal. These tridiagonal matrices vary in length because of the geometry of the state space and in general are rather small. For certain traffic settings, A is practically 2 cyclic, and using line SOR pays off. For others, the tridiagonal matrices are practically bidiagonal and line SOR is essentially equivalent to point SOR. Although the problem has separable submatrices, but it does not seem advantageous to use that property.

Chebychev acceleration (see [4]), using a Gauss-Seidel preconditioning scheme requires practically the same amount of effort as SOR when computing the residual vector $A\mathbf{p}$ is incorporated into the SOR program. The Chebychev acceleration scheme also involves regulating parameters, but the scheme seems to be much less sensitive to changes in the parameters than SOR. Thus for a production situation it would be preferable. For the first problem type, Chebychev acceleration with an incomplete LU preconditioning also gives favorable results.

In conclusion we find that some of the classical methods used to solve nonsingular problems can be successfully applied to singular systems arising in the modeling of networks. We would also like to emphasize the fact that sometimes one can take advantage of the algebraic structure of a problem.

The author wishes to thank Stanley Eisenstat of Yale University for his observation about the step length algorithm.

Bibliography

1. L. Kaufman, J. B. Seery, and J. A. Morrison, "Overflow Models for Dimension PBX Feature Packages", Bell System Technical Journal, Vol 60, No. 5, May-June 1981.

2. L. Kaufman, B. Gopinath, and E. F. Wunderlich, "Sparse Matrix Algorithms for a Packet Network Control Analysis", Bell Labs Computer Science Technical Report #82, 1980.

3. R. Bank and D. Rose, "Marching Algorithms for Elliptic Boundary Value Problems. I. The Constant Coefficient Case", SIAM J. Numer Anal., Vol 14, No. 5, October 1977, pp. 792-829.

4. R. Varga, "Matrix Iterative Analysis", Prentice Hall, Inc., Englewood Cliffs, New Jersey, 1962.

5. B. L. Buzbee, G. H. Golub and C. W. Nielson, "On Direct Methods for Solving Poisson's Equation," SIAM J. Numer Anal., Vol 7, 1970, pp. 627-656.

LARGE EIGENVALUE PROBLEMS IN QUANTUM CHEMISTRY

Juergen Hinze
Fak. f.Chemie
Univ. Bielefeld
48 Bielefeld, FRG

Introduction:

Eigenvalue problems of the type

$$\underline{H}\,\underline{C}_K = E_K\,\underline{C}_K \tag{1}$$

with an extremely large real and symmetric $\underline{H}$ appear in quantum chemistry commonly in the determination of stationary electronic states of atoms or molecules using the method of configuration interaction (CI), where the desired state function, Ψ_K, is expanded in terms of configuration state functions, (CSF's), Φ_I, as

$$\Psi_K = \Sigma_I^N \,\Phi\, C_{IK}. \tag{2}$$

Here the configuration state functions are constructed in general as specific superpositions of antisymmetrized products, Slater determinants, of single particle functions, φ_i, the orbitals. With this the energy matrix elements, H_{IJ}, can be expressed as

$$H_{IJ} = \langle\Phi_I|\mathcal{H}|\;\Phi_J\rangle$$

$$= \Sigma_{ij}^n(h_{ij}\Gamma_{ij}^{IJ} + 1/2\Sigma_{kl}^n\; g_{ijkl}\;\Gamma_{ijkl}^{IJ}) \tag{3}$$

where the m-electron Hamiltonian operator is given as

$$\mathcal{H} = \Sigma_i^m(-1/2\nabla^2(i) + V(i)) + \Sigma_{i>j}^m \frac{1}{r_{ij}} \tag{4}$$

an operator sum of the kinetic $-1/2\nabla^2(i)$, and potential, $V(i)$, energyoperators for electron i in the field of the fixed nuclei as well as the electron-electron repulsion potential $1/r_{ij}$.

In eq. (3) the Γ_{ij}^{IJ} and Γ_{ijkl}^{IJ} are structure factors, or specifically the first and second order reduced transition density matrix elements between CSF's I and J in the space spanned by the orbitals. These multiply the one and two particle integrals over the orbitals

$$h_{ij} = \langle\varphi_i| - 1/2\nabla^2 + V|\varphi_j\rangle$$

and

$$G_{ijkl} = \langle\varphi_i\langle\varphi_k|1/r_{12}|\varphi_l\rangle\varphi_j\rangle \tag{5}$$

The dimension of the eigenvalue problem, eq. (1), can become extremely large, i.e. N is of the order 10^5. In general only a few, $M < 10$, of the lowest eigenvalues and eigenvectors are desired, thus iterative techniques are used to solve eq. (1), and because of the special structure of H, to be detailed below, special techniques to solve eq. (1) efficiently have been employed by quantum chemists over the last few years. I will give a brief review of these techniques with a few remarks about the experiences gained thus far with these methods. This will be based by and large on reports presented and conclusions reached at a workshop held 1978 by NRCC, the proceedings[1] of which are not generally available.

Before discussing the different procedures used, a few remarks about the special structure of $\underline{H}$. Generally $\underline{H}$ will be diagonally dominant, and the CSF's can be arranged in such an order, that a partitioning of the total N-dimensional space spanned by the CSF's into a part P for I = 1 to L and a part Q for I = L + 1, N, with L of the order of 100, leads to a structure such that H_{QQ} is sparse and less significant; i.e. for the eigenvectors C_K of the roots, E_K, desired we will have

$$\Sigma_{I=1}^{L} C_{IK}^2 > 0.95 \text{ and } \Sigma_{I=L+1}^{N} C_{IK}^2 < 0.05.$$

Thus the structure of $\underline{H}$ will be

$$\left(\begin{array}{c|c} \underline{H}_{PP} & \underline{H}_{PQ} \\ \hline \underline{H}_{QP} & \underline{H}_{QQ} \end{array}\right) \quad \begin{array}{l} \text{with L of order } 100 \\ \text{and N of order } 10^5 \end{array}$$

L N

Furthermore, since the number of orbitals m is generally small, less than loo, it is generally advantageous not to construct $\underline{H}$ explicitely but form the product

$$\underline{b} = \underline{H}\,\underline{C}$$

needed in iterative methods for the solution of eq. (1) directly using eq. (3), i.e. for an element I of $\underline{b}$,

$$b_I = \Sigma_J H_{IJ} C_J$$

$$= \Sigma_J \Sigma_{ij} (h_{ij}\Gamma_{ij}^{IJ} + 1/2\Sigma_{kl} G_{ijkl} \Gamma_{ijkl}^{IJ}) C_J \quad (6)$$

Common to the iterative methods to solve eq. (1) in quantum chemistry is the expansion of the desired solution $\underline{C}_K$ in terms of some basis vectors $\underline{b}_i$ such that in the n'th iteration we have

$$\underline{C}_K \simeq \underline{C}_K^{(n)} = \Sigma_i^k \underline{b}_i^{(n)} a_{iK} \quad (7)$$

with the $\underline{a}_K$'s the eigenvectors of the reduced eigenvalue problem

$$\underline{h}\,\underline{a}_K = \lambda_K \underline{S}\,\underline{a}_K \quad (8)$$

with the reduced matrices defined as

$$h_{ij} = \underline{b}_i^{+} \underline{H}\,\underline{b}_j \quad (9)$$

and

$$S_{ij} = \underline{b}_i^{+} \underline{b}_j \qquad \text{for i and j} = 1,\ldots k$$

The eigenvalue λ_K will be an upper bound $\lambda_K \geqslant E_K$ and approach E_K as the iteration converges.

The different methods in use differ in

a) the number k, of basis vectors used in the expansion, and

b) the method used in constructing the set of basis vectors $\{\underline{b}_1^{(n)}, \underline{b}_2^{(n)} \ldots \underline{b}_k^{(n)}\}$ in the n'th iteration.

Clearly there is also a difference in the number of roots, M, which can be obtained in the different methods. In general we will have

$$M<k<<N$$

Krylov Method: In the classical Krylov or power method[2,3] $\underline{C}_K^{(n)}$ is expanded in terms of $\underline{H}^j\underline{C}_K^{(o)}$ i.e. we have in the n'th iteration

$$\underline{b}_1^{(n)} = \underline{C}^{(n-1)}$$

$$\underline{b}_2^{(n)} = \underline{H}\,\underline{b}_1^{(n)}$$

or

$$\tilde{\underline{b}}_2^{(n)} = \nabla \left. \frac{\underline{b}^+ \underline{H}\,\underline{b}}{\underline{b}^+ \underline{b}} \right|_{\underline{b}=\underline{b}_1^{(n)}} = a\,(\underline{H} - E^{(n-1)})\,\underline{C}^{(n-1)}$$

thus M = 1 and k = 2. Clearly $\{\underline{b}_1, \underline{b}_2\}$ and $\{\underline{b}_1, \tilde{\underline{b}}_2\}$ span the same vector space.

A generalization of this to $M \geqslant 1$ and $k < M$ is the extended gradient method[4] with the n'th iteration given by

$$\underline{b}_{n+1}^{(n)} = \nabla \left. \frac{\underline{b}^+ H \underline{b}}{\underline{b}^+ \underline{b}} \right|_{\underline{b}=\underline{C}^{(n-1)}} = a\,(\underline{H}-E^{(n-1)})\,\underline{C}^{(n-1)}$$

and

$$\underline{b}_i^{(n)} = \underline{b}_i^{(n-1)} \quad \text{for } i \leqslant n$$

Here the vectors $\{\underline{b}_1, \underline{b}_2 \ldots \underline{b}_i\}$ span the space $\{\underline{C}^o, \underline{HC}^o \ldots \underline{H}^{i-1}\underline{C}^o\}$ and several sets of such vectors could be used derived from $\underline{C}_1^o, \underline{C}_2^o \ldots \underline{C}_M^o$ corresponding to the root K = 1 through M desired.

The Lanczos algorithms: In the Lanczos algorithm[5] the expansion is also in terms of $\underline{H}^i\underline{C}^o$, however, the vectors are sequentially orthogonalized to yield a diagonal $\underline{S}$ matrix in eq. (8) and for the case of an implicit orthogonalization a tri-diagonal $\underline{h}$. The algorithm with an implicit orthogonalization is characterized by the sequence of steps

$$\underline{b}_i = \underline{d}_{i-1} - E_{i-1}\,\underline{b}_{i-1} - \beta_{i-2}\,\underline{b}_{i-2}$$

$$s_i = \underline{b}_i^{+}\,\underline{b}_i \qquad\qquad \beta_{i-1} = s_i/s_{i-1}$$

$$\underline{d}_i = \underline{H}\;\underline{b}_i \qquad e_i = \underline{b}_i^{+}\,\underline{d}_i \qquad E_i = e_i/s_i$$

for i = 1 through k starting with $\underline{b}_1 = \underline{C}^{(o)}$ and $\beta_o = 0$.

The elements of the $\underline{S}$ and $\underline{h}$ matrices are then given as

$$S_{ij} = \underline{b}_i^{+}\,\underline{b}_j = s_i\,\delta_{ij}$$

and

$$h_{ij} = \underline{b}_i^{+}\,H\underline{b}_j = \begin{cases} e_i & i = j \\ s_{j+1} & i = j+1 \\ s_{i+1} & j = i+1 \\ 0 & |i-j|>1 \end{cases}$$

Using explicit orthogonalization the sequence of steps becomes

$$\underline{f}_i = \underline{d}_{i-1} - E_{i-1}\,\underline{b}_{i-1} - \beta_{i-2}\,\underline{b}_{i-2}$$

$$\underline{b}_i = \underline{f}_i - \sum_{j=1}^{i-1} \underline{b}_j \cdot \underline{b}_j^{+}\,\underline{f}_j \qquad\qquad s_i = \underline{b}_i^{+}\underline{b}_i$$

$$\underline{d}_i = H\underline{b}_i \qquad\qquad h_{ij} = \underline{b}_j^{+}\underline{d}_i \qquad \text{for } j = 1 \text{ through } i$$

$$E_i = h_{ii}/s_i$$

Note, $\underline{h}$ is not tridiagonal. The power methods do not depend on the diagonal dominance of H, however, the convergence to the first root is often slow, unless $|E_1/E_2|$ is large. Additional roots may be obtained with only little extra work provided $|E_I/E_J|$ is large for I < M and J > k. Care is required since the sequence $\underline{H}^i\underline{C}$ becomes nearly linearly dependent, which leads to a rapid loss of significant figures in the construction of $\underline{h}$. In the Lanczos method with implicit orthogonalization some roots may repeat, and roots may be missed in both forms of the Lanczos algorithm.

Cyclic Methods: The early cyclic methods[6] were restricted to the lowest root, i.e. $M = 1$, and they cycled through a complete set of vector pairs, i.e. $k = 2$ repeatedly using the following sequence of steps:

$$\underline{b}_1^{(n)} = \underline{C}^{(n-1)}$$
$$\underline{b}_2^{(n)} = \underline{\hat{e}}_J \tag{15}$$

starting with $\underline{C}^{(o)} = \underline{\hat{e}}_I$ where I equal to inf. (H_{II}) and cycling J from 1 through N repeatedly to convergence. For each step we have

$$h = \begin{pmatrix} E^{(n-1)}|\underline{C}^{(n-1)}|^2 & [\underline{HC}^{(n-1)}]_J \\ [\underline{H}\,\underline{C}^{(n-1)}]_J & H_{JJ} \end{pmatrix}$$

and (16)

$$s = \begin{pmatrix} \underline{C}^{(n-1)} & C_J^{(n-1)} \\ C_J^{(n-1)} & 1 \end{pmatrix}$$

The approximate ratio of the eigenvector components corresponding to the lowest root of eq. (8) using (16) is obtained as

$$\frac{a_2}{a_1} = \frac{[(\underline{H} - E^{(n-1)}\underline{1})\,\underline{C}^{(n-1)}]_J}{E^{(n-1)} - H_{JJ}} \tag{17}$$

or by solving eq. (8) exactly[7]. With this a new approximation

$$C^{(n)} = a_1\,\underline{b}_1^{(n)} + a_2\,\underline{b}_2^{(n)} \tag{18}$$

and

$$E^{(n)} = \underline{C}^{(n)^+}\,\underline{H}\,\underline{C}^{(n)}/\underline{C}^{(n)^+}\,\underline{C}^{(n)}.$$

is obtained such that

$$E^{(n)} > E^{(n+1)} \ldots > E.$$

To get a higher root M an implicit root shifting procedure[7] can be used, i.e.

The matrix

$$\underset{\sim}{\tilde{H}} = \underline{H} + \sum_{J=1}^{M-1} \mu_J \underline{C}_J \underline{C}_J^{+} \tag{19}$$

is used such that the lowest eigenvalue of $\tilde{\underline{H}}$ is E_M. However, with (19) the advantage of the original sparsenes of $\underline{H}$ is lost, the vectors $\underline{C}_1$ through $\underline{C}_{M-1}$ are required first, and for higher roots there is the danger of error accumulation.

If more than one root is required, i.e. $M > 1$, it is more appropriate to use for example $k = M + 1$ in a cyclic process[8] with

$$\underline{b}_i^{(n)} \doteq \underline{C}_i^{(n-1)} \quad \text{for } i = 1 \text{ through } k$$

and

$$\underline{b}_{k+1} = \hat{\underline{e}}_J$$

cycling through J, each time solving the corresponding k + 1 dimensional eigenproblem, eq. (8).

The cyclic methods are in general rapidly convergent for diagonally dominant matrices, they can be slow otherwise. Near degeneracy of roots can cause difficulties.

<u>Variation - Perturbation Methods</u> are flexible and permit an extensive exploitation of the numerical and physical information available about the problem to be solved. For one root, i.e. $M = 1$ and $k = 2$ the sequence of steps is[9] for the n'th iteration

$$\begin{aligned} \underline{b}_1^{(n)} &= \underline{C}^{(n-1)} \\ \underline{b}_2^{(n)} &= (E^{(n-1)}\underline{1} - \underline{H}^o)^{-1} \, \underline{V} \, \underline{b}_1^{(n)} \end{aligned} \tag{2o}$$

with

$$\underline{V} = \underline{H} - \underline{H}^o \tag{21}$$

or equivalently

$$\tilde{\underline{b}}_2^{(n)} = (E^{(n-1)}\underline{1} - \underline{H}^o)^{-1}(\underline{H} - E^{(n-1)}\underline{1}) \, \underline{b}_1^{(n)} = \underline{b}_2^{(n)} - \underline{b}_1^{(n)} \tag{22}$$

where $(\underline{b}_1^{(n)}, \underline{b}_2^{(n)})$ and $(\underline{b}_1^{(n)}, \tilde{\underline{b}}_2^{(n)})$ span the same space.

Usually the diagonal elements of $\underline{H}$ are chosen to give $\underline{H}^o$, however, other choices are feasable and possibly advantageous. An extension of this procedure[11] to obtain a few of the lowest roots simultaneously, i.e. $M \geqslant 1$ and $k > M$ has been used extensively and successfully in large CI calculations. Using root tracking algorithms[12] (pattern search on the eigenvector) it has even been possible to home in on a specific root E_I, which is not extrem. The basic algorithm is for the n'th iteration

$$\underline{b}_i^{(n+1)} = \underline{b}_i^{(n)} \qquad \text{for } i = 1 \text{ through } k \tag{23}$$

$$\underline{b}_{k+I}^{(n+1)} = (E_I^{(n)}\, \underline{1} - \underline{H}^o)^{-1}(\underline{H} - E_I^{(n)})\, \underline{C}_I^{(n)}$$

for I = 1 through M or the roots which are tracked. Boarder n, increasing k, and solve eq. (8) to obtain a new set of $C_I^{(n-1)}$. If k gets too large before convergence is reached truncate to

$$\underline{b}_i = \underline{C}_i^{(n)} \qquad \text{for } i = 1 \text{ through } M \tag{24}$$

and start over. For numerical stability it is advantageous to orthonormalize the expansion vectors such that the algorithm stated in detail becomes

A) Initialization: for i = 1 to M

$$\underline{f}_i = \underline{C}_i^{(o)} - \sum_{j=1}^{i-1} \underline{b}_j\, \underline{b}_j^+\, \underline{C}_i^{(o)} \qquad \underline{b}_i = \underline{f}_i / (\underline{f}_i^+ \underline{f}_i)^{1/2}$$

$$\underline{d}_i = \underline{H}\, \underline{b}_i \qquad M_{ij} = \underline{b}_i^+ \underline{d}_j \text{ for } j = 1 \text{ to } i \tag{25}$$

B) Iteration n:

solve

$$\underline{h}\ \underline{a} = \underline{a}\ \underline{\lambda} \tag{26}$$

a k x k problem (initially k = M).

Take for the roots i desired

$$E_i^{(n)} = \lambda_i \text{ and } \underline{C}_i^{(n)} = \Sigma_j\, \underline{b}_j\, \underline{a}_{ji}$$

$$\underline{g}_i = (\underline{H} - \underline{1}E_i^{(n)})\, \underline{C}_i^{(n)} = \Sigma_j\, (\underline{d}_j - E_i^{(n)} \underline{b}_j)\, a_{ji}$$

$$\underline{f}_i = (E_i^{(n)}\ \underline{1} - \underline{H}^o)^{-1}\underline{g}_i \tag{27}$$

$$\underline{f}_i = \underline{f}_i - \sum_{j=1}^{k} \underline{b}_j\underline{b}_j^+\ \underline{f}_i \qquad \underline{b}_{k+1} = \underline{f}_i/(\underline{f}_i^+\underline{f}_i)^{1/2}$$

$$\underline{d}_{k+1} = \underline{H}\ \underline{b}_{k+1} \qquad h_{k+1,j} = \underline{b}_{k+1}\underline{d}_j \text{ for } j=1 \text{ to } k+1$$

increase k by 1 and repeat the sequence of steps, eqs.(27), for all roots desired before solving eq. (26) again. For computational efficiency it may be desirable to form the time consuming product $\underline{H} \cdot \underline{b}$ for a set of say M $\underline{b}$ - vectors simultaneously.

Disregard any $\underline{f}$'s with a too small norm, and if k gets too large start over with the initialization step using the $C_i^{(n)}$'s.

This method, which is used extensively and with success in quantum chemistry might possibly be improved by using for $\underline{H}^o$ not only the diagonal elements of $\underline{H}$ but rather the important part of $\underline{H}$, i.e. $\underline{H}_{PP}$ and the diagonal elements of $\underline{H}_{QQ}$. As the strength of this method is surely based on the approximate inverse iteration step, i.e. the calculation of f_i in eq. (27), an improvement of this step by either using a more general $\underline{H}^o$ or an improved calculation of the approximation to the inverse $(E\underline{1}-\underline{H})^{-1}$ using for example partitioning should make this method more general and less dependent on the diagonal dominance of H as it obtains generally in quantum chemistry.

Literature references:

1.) NRCC Workshop Report, "Numerical Algorithms in Chemistry: Algebraic Methods", LBL-8158 UC-32 CONF-78o878 (1978); especially E.R. Davidson, pp 15.
2.) W. Karush, Pacific J. Math. 1, 233 (1951).
3.) M.R. Hestenes, "Simultaneous Linear Equations and the Determination of Eigenvalues", NBS.
4.) J.B. Delos and S.M. Blinder; J. Chem. Phys. 47, 2784 (1967).
5.) C. Lanczos, J. Res. Not. Bur. Stand., 45, 255 (195o).
6.) J.L.B. Cooper, Quart. Appl. Math. 6, 179 (1948); R.K. Nesbet, J. Chem. Phys. 43, 311 (1965); I. Shavitt, J. Comp. Phys. 6, 124 (197o).
7.) D.K. Fadeev and V.N. Faleeva, "Computational Methods of Linear Algebra" Section 61, Freeman (1963); I. Shavitt, C.F. Bender, A. Pipano and R.P. Hosteney, J. Comp. Phys. 11, 9o (1973).
8.) R.C. Raffenetti, J. Comp. Phys. 32, 4o3 (1979).
9.) A. Dalgarno and A.L. Stewart, Proc. Phys. Soc. (London) 77, 467 (1961).
1o.) R. Seeger, R. Krishnan and J.A. Pople, J. Chem. Phys. 68, 2519 (1978).
11.) E.R. Davidson, J. Comp. Phys. 17, 87 (1975).
12.) W. Butscher and W.E. Kammer, J. Comp. Phys. 2o, 313 (1976).

VARIATIONAL PSEUDO-GRADIENT METHOD FOR DETERMINATION OF m FIRST EIGENSTATES OF A LARGE REAL SYMMETRIC MATRIX.

Alojzy Golebiewski

Fakultät für Chemie, Universität Bielefeld, 48 Bielefeld, Germany; Institute of Chemistry, Jagiellonian University, Cracow, Poland*)

1. Introduction

There is a need in many fields of physics and chemistry for efficient ways of solutions of the eigenvalue problem, say

$$HU_m = U_m e_m \tag{1}$$

where H is a large, real, symmetric matrix of dimensions n×n, e_m is a diagonal matrix containing the m lowest eigenvalues, U_m is a rectangular matrix defined by m first eigenvectors. It is assumed throughout this paper that $H_{ii} < H_{jj}$, whenever $i \leq m$ and $j > m$. In principle the algebraic eigenvalue problem has been solved several years ago. For not large n's one could apply, for example, the Householder diagonalization procedure, with the number of operations increasing like $\frac{2}{3} n^3$. For large values of n iterative techniques have been developed, with the number of operations increasing like $N_{iter} \cdot m \cdot n^2$. Description of most standard treatments can be found in books [1].

Our interest is in quantum chemistry. In quantum chemistry H, the energy matrix, has commonly a specific structure. It allows a specific treatment of the eigenvalue problem. For example, in case of the large-scale configuration interaction method, n may be of the order of several thousands. Matrix H is often too large to be stored in the core storage. Sometimes, like in what is called the direct CI method, elements of H are even reconstructed always when required (or a procedure equivalent to that is applied). Thus, advisably, elements of H should not be modified and the number of recalls to H should be small. On the other hand most of the off-diagonal elements of H, in CI, are either equal to zero or are negligible (up to about 95%). Matrix H is sparse. This is another argument to keep H unchanged. Another feature of H is that it is usually diagonally dominant.

A detailed review of the specific methods used in quantum chemistry in this

---

*) present address: 30-060 Kraków, Karasia 3, Poland.

connection has been given in the lecture by J. Hinze [2]. Methods related to the power method require a large number of iterations, very often a significant level shifting, are of restricted utility. A greater interest have here gained the following two methods: the relaxation technique of Shavitt et al. [3] and its simultaneous relaxation version due to Raffenetti [4], the Davidson algorithm, based on the inverse iteration scheme, the variational calculation and the acceleration technique [5]. Attention is also drawn to the method of Miller and Berger [6], with the detailed analysis of the pseudo-convergence problem.

Typical for all known schemes is a direct search for individual eigenvectors, either one after the other, or simultaneously. It seems preferable to us, however, to define iteratively the invariant subspace first, subspace defined by m eigenvectors contained in U_m. Instead of U_m we look then for $\bar{U}_m$, where

$$U_m = \bar{U}_m Q \tag{2}$$

and Q is an orthogonal matrix such that

$$Q^+(\bar{U}_m H \bar{U}_m) Q = e_m \tag{3}$$

Additional degrees of freedom are gained in this way, making the numerical algorithm more flexible.

Few years ago the orthogonal gradient approach has been developed by the author [7], of just the latter property:

$$\bar{D}_m^{(i)} = H\bar{U}_m^{(i)} \tag{4}$$

$$\bar{U}_m^{(i+1)} = \bar{D}_m^{(i)} [\bar{D}_m^{(i)+} \bar{D}_m^{(i)}]^{-\frac{1}{2}} \tag{5}$$

where i enumerates iterations. As follows from equation (4) the approach is closely related to the familiar power method. In the case of typical configuration interaction treatments the convergence of this approach is poor. A completely different approach has been developed which seems to fulfil all requirements in this field of applications.

2. Invariant subspace. Definitions.

Let us consider the sum of m lowest (by assumption) eigenvalues:

$$\begin{aligned} \Delta &= \mathrm{Tr}\ e_m \\ &= \mathrm{tr}\ \bar{U}_m^+ H U_m = \mathrm{tr}\ Q^+(\bar{U}_m^+ H \bar{U}_m)Q \\ &= \mathrm{tr}\ \bar{U}_m^+ H \bar{U}_m \end{aligned} \tag{6}$$

If we extend $\bar{U}_m$ in any way to a full, square, orthogonal matrix $\bar{U}_n$, then

$$H' \overset{def}{=} \bar{U}_n^+ H \bar{U}_n = \left|\begin{array}{c:c} \begin{matrix} xxx \\ xxx \\ xxx \end{matrix} & O^+ \\ \hdashline 0 & \begin{matrix} xx\ldots xxx \\ xx\ldots xxx \\ \ldots\ldots\ldots \\ \ldots\ldots\ldots \\ xx\ldots xxx \end{matrix} \end{array}\right| \begin{matrix} m \\ \\ n \end{matrix} \tag{7}$$

$$\qquad\qquad m \qquad\quad n$$

where x symbolizes an element different in general from zero, 0 is a zero matrix of m columns and (n - m) rows.

Δ is the absolute minimum, by assumption. Suppose that yet $H'_{ij} \neq 0$, where $i \leq m$ and $j > m$. Performing Jacobi-like 2 × 2 rotation, we find that new values of H'_{ii} and H'_{jj} are pushed apart in comparison to the original ones. In result the value of Δ would decrease, in contradiction to the assumption.

Thus the logical way of determining $\bar{U}_m$ is to minimize the trace Δ. At the minimum the appropriate matrix H' achieves the diagonal block form (eq. 7). Δ is then a sum of m eigenvalues. Obviously, however, there is no guarantee that Δ is the sum of the m lowest eigenvalues. If $H_{ii} < H_{jj}$ for all i's not larger than m and all j's larger than m, and if all off-diagonal elements of H are small in absolute sense in comparison to diagonal elements, Δ achieves its absolute minimum. This is just the case typical for configuration interaction calculations.

The problem of (eventually) missing roots is typical for all iterative treatments.

In order to keep $\bar{U}_m$ as a part of an orthogonal matrix $\bar{U}_n$ let us put

$$\bar{U}_m = C_m(C_m^+ C_m)^{-\frac{1}{2}} \tag{8}$$

where C_m is a real matrix of m columns and n rows. Now

$$\Delta = \mathrm{tr}\ [(C_m^+C_m)^{-\frac{1}{2}}\ (C_m^+HC_m)\ (C_m^+C_m)^{-\frac{1}{2}}]$$

$$= \mathrm{tr}\ [(C_m^+HC_m)\ (C_m^+C_m)^{-1}] \tag{9}$$

Our goal is to minimize Δ. Previous to this step let us define a set of matrices, used in this and the forthcoming sections:

$$D_m = HC_m \tag{10}$$

$$X_m = D_m - C_mH_{mm} \tag{11}$$

$$H_{mm} = C_m^+D_m \tag{12}$$

$$X_{mm} = X_m^+X_m \tag{13}$$

$$X_D = X_m^+D_m \tag{14}$$

$$Z_m = HX_m \tag{15}$$

$$W_{mm} = X_m^+Z_m \tag{16}$$

Taking the differential of both sides of eq. (9) we find that

$$\delta\Delta = 2\ \mathrm{tr}\ \{\delta C_m^+[D_m - C_m(C_m^+C_m)^{-1}H_{mm}](C_m^+C_m)^{-1}\} \tag{17}$$

If at the start of any given variational treatment

$$C_m^+C_m = 1_{mm} \tag{18}$$

(a unit matrix) then

$$\delta\Delta = 2\ \mathrm{tr}\ \{\delta C_m^+\ (D_m - C_mH_{mm})\}$$

$$= 2\ \mathrm{tr}\ \delta C_m^+X_m \tag{19}$$

In the standard treatment the Rayleigh quotient is extremized, approaching an eigenvalue at the limit. In the present treatment we minimize the trace Δ. In the standard treatment the residual vector is used to improve the estimated eigenvector. In the present treatment the residual matrix X_m is used to subtract the invariant subspace from the whole space.

3. Variational gradient method

If $\delta C_m = -aX_m$, where $a > 0$, then $\delta\Delta = -2a \operatorname{tr} X_m^+X_m \leq 0$. This would be the simplest version of the gradient method, with a being the variational parameter. More generally we put

$$\delta C_m = - X_m L \tag{20}$$

Passing to finite changes we consider the replacements

$$C_m - X_m L \rightarrow C_m \tag{21}$$

$$D_m - Z_m L \rightarrow D_m \tag{22}$$

where L is a square matrix of variational parameters. The dimensions of L are $m \times m$. Then

$$\Delta(L) = \operatorname{tr} \{(C_m^+ - L^+X_m{}^+)\, H\, (C_m - X_m L)\, [(C_m^+ - L^+X_m^+)\,(C_m - X_m L)]^{-1}\}$$

$$= \operatorname{tr} \{(H_{mm} - L^+X_D - X_D{}^+L + L^+W_{mm}L)\,(1 + L^+X_{mm}L)^{-1}\} \tag{23}$$

In the derivation of eq. (23) use has been made of relation

$$C_m^+X_m = 0 \tag{24}$$

It follows from the definition of X_m (eq.11) and H_{mm} (eq.12), when multiplying both sides of eq. (11) by $C_m{}^+$.

Expression (23) is exact. Under certain restrictions of variational parameters it can be minimized exactly. Restriction of variational parameters in L cannot cause any difficulty; the only consequence is a decrease of Δ not as large as possible at the given step.

With this purpose in mind let us diagonalize the Hermitian matrix X_{mm} (13):

$$U^+X_{mm}U = D_X \tag{25}$$

where U is an orthogonal matrix and D_X a diagonal one. To be consistent let us redefine all the matrices defined in eqs. (10) - (16): $C_mU \rightarrow C_m$, $D_mU \rightarrow D_m$, $U^+H_{mm}U \rightarrow H_{mm}$, $X_mU \rightarrow X_m$, $U^+X_{mm}U \rightarrow D_X$, $U^+X_DU \rightarrow X_D$, $Z_mU \rightarrow Z_m$, $U^+W_{mm}U \rightarrow W_{mm}$. Matrix L is similarly replaced by U^+LU. Now

$$\Delta = \operatorname{tr}\{ (H_{mm} - L^+X_D - X^+{}_DL + L^+W_{mm}L)\,(1 + L^+D_XL)^{-1}\} \tag{26}$$

Some technical problems arise with the inversion of the matrix $(1 + L^+ D_X L)$, for arbitrary L and m > 2. In order to simplify this problem it is tentable to reduce the number of variational parameters in L, taking L as a diagonal matrix:

$$L = D_L \tag{27}$$

Under this assumption minimization of expression (26) simplifies significantly. Now

$$\begin{aligned} y &= \Delta(D_L) - \Delta(0) \\ &= - \sum_{i=1}^{m} \frac{2c_i x_i + b_i x_i^2}{1 + a_i x_i^2} \end{aligned} \tag{28}$$

where *)

$$a_i = (D_X)_{ii} \tag{29}$$
$$b_i = (H_{mm})_{ii}(D_X)_{ii} - (W_{mm})_{ii} \tag{30}$$
$$c_i = (X_D)_{ii} \tag{31}$$
$$x_i = (D_L)_{ii} \tag{32}$$

The variational parameters $(D_L)_{ii}$ can now be optimized independently one after the others. The minimum is obtained for

$$\begin{aligned} x_i &= \frac{b_i + \mathrm{sqrt}(b_i^2 + 4a_i c_i^2)}{2a_i c_i} \\ &= \frac{2c_i}{\mathrm{sqrt}(b_i^2 + 4a_i c_i^2) - b_i} \end{aligned} \tag{33}$$

Having found D_L we redefine C_m and D_m in accordance with eqs. (21)-(22). Prior to proceeding to the text iteration (redefinition of the residual matrix X_m) the columns of C_m have to be reorthonormalized. Reorthonormalization does not influence the value of Δ.

4. Variational pseudo-gradient method.

If H is diagonally dominant a significant improvement of convergence can be obtained when preconditioning the matrix X_m.

---

* In this section $X_m^+ D_m = X_m^+ X_m$, i.e. $X_D = D_X$ and $c_i = a_i$.

Thus instead of transformation (21) let us consider a more general replacement:

$$C_m \; - \; TX_mL \; \rightarrow \; C_m \tag{34}$$

where L is again a small matrix (of dimensions m × m) and T a large one (of dimensions n × n), both containing variational parameters. Restricting to some extent the degrees of freedom L can be treated exactly. It has been done in the previous section. T, on the other hand, can be treated only approximately. For this reason we perform the correction of C_m in two steps: in step 1 we take L = 1 and consider T approximately, transformation $TX_m \rightarrow X_m$ being equivalent to preconditioning of X_m; in step 2 we proceed as in section 3, with X_m (eq.11) replaced by preconditioned X_m. As step 2 is exact Δ can never fall below the sum of m lowest eigenvalues.

Step 1 (preconditioning)

Suppose for the moment that $C_m - X_m' = C_m'$ and

$$X_m' \; = \; TX_m \; - \; C_m(C_m^+TX_m) \; . \tag{35}$$

Then, in analogy to eq. (24),

$$C_m^+X_m' \; = \; 0 \tag{36}$$

and

$$\Delta(T) \; = \; \mathrm{tr}\; \{(C_m - X_m')^+H(C_m - X_m')\;[(C_m - X_m')^+(C_m - X_m')]^{-1}\}$$

$$= \; \mathrm{tr}\; \{(H_{mm} - D^+{}_mX_m' - X_m'^+D_m + X_m'^+HX_m')\;(1 + X_m'^+X_m')^{-1}\} \tag{37}$$

In many applications, as the rule in configuration interaction calculations, the residual matrix X_m (and hence also X_m') consists of elements rather close to zero. In other applications elements of X_m are close to zero at least after a certain iteration. For cases like this it is tentable to assume that

$$(1 + X_m'^+X_m')^{-1} \; \approx \; 1 - X_m'^+X_m' \tag{38}$$

With this replacement, and restricting expansion (37) to linear and bilinear terms in X_m',

$$(T) \; = \; (0) \; + \; \mathrm{tr}\; (-D_m^+X_m' - X_m'^+{}_m + X_m'^+HX_m' - H_{mm}X_m'^+X_m') \tag{39}$$

Let us replace X'_m in accordance with eq. (35), D_m by C_m, X_m by H_{mm} in accordance with eq. (11), let us neglect terms trilinear in X_m. Then, because of hermiticity of H_{mm} and the properties of trace, and restricting T to be Hermitian:

$$\Delta(T) = \Delta(0) + \mathrm{tr}\,(-2p_X T + p_X THT - p_X Th_C T + h_X Tp_C T - h_X T^2) \quad (40)$$

where

$$p_X = X_m X_m^+ \quad (41)$$

$$h_X = X_m H_{mm} X_m^+ \quad (42)$$

$$p_C = C_m C_m^+ \quad (43)$$

$$h_C = C_m H_{mm} C_m^+ \quad (44)$$

The matrices defined in eqs. (41) - (44) are of dimensions n × n. Fortunately mainly diagonal elements will be required in what follows. Restriction of T to a Hermitian matrix is equivalent to reducing the number of independent variational parameters to n(n + 1)/2.

Let us note that for Hermitian matrices A, B, C, D

$$\mathrm{tr}\,(ABCD) = \mathrm{tr}\,(CBAD) \quad (45)$$

Differentiating eq. (40) and taking into account the last property we find that

$$\delta\Delta(T) = 2\ \mathrm{tr}\,\{(-p_X + HTp_X - h_C Tp_X + p_C Th_X - Th_X)\delta T\} \quad (46)$$

The condition for an extremum is

$$(H - h_C)Tp_X + p_X T(H - h_C) + (p_C - 1)Th_X - h_X T(p_C - 1) = 2p_X \quad (47)$$

Unfortunately eq. (47) cannot be solved easily for T. Simplifying the problem we reduce the number of independent variational parameters in T to n, replacing the general matrix T by a diagonal one, D_T. Then

$$\sum_{j=1}^{n} \{[H_{ij} - (h_C)_{ij}](p_X)_{ji} + [(p_C)_{ij} - \delta_{ij}](h_X)_{ji}\}(D_T)_{jj} = (p_X)_{ii} \quad (48)$$

This is still too large a set of linear equations to be solved exactly for large n's. In some fields of application, for example commonly in a configuration interaction treatment in quantum chemistry, diagonal elements of H are dominating at least for $i > m$.

Then approximately, for $i > m$,

$$(D_T)_{ii} = \frac{1}{H_{ii} - E_i} \tag{49}$$

where

$$E_i = (h_C)_{ii} - \frac{(h_X)_{ii}}{(p_X)_{ii}} [(p_C)_{ii} - 1] \tag{50}$$

For $i \leq m$ one possibility is simply to put

$$(D_T)_{ii} = 1 \tag{51}$$

Another simple solution is to solve eq. (48) neglecting coupling terms of $(D_T)_{ii}$ with $(D_T)_{jj}$'s for $j > m$. Then, in this version, variational parameters for $i \leq m$ follow from the set of equations

$$\sum_{j=1}^{m} U_{ij} (D_T)_{jj} = (p_X)_{ii}, \quad i = 1, 2, \ldots, m \tag{52}$$

where

$$U_{ij} = [H_{ij} - (h_C)_{ij}](p_X)_{ji} + (h_X)_{ij}[(p_C)_{ji} - \delta_{ji}] + d_U \delta_{ij} \tag{53}$$

A choosable shift parameter d_U has been introduced, as, in practice, matrix U happens to be almost singular.

Step 2 (minimization)

Having found $T = D_T$ we evaluate X'_m (eq.35). Then, replacing X_m in section 3 by X'_m we proceed in exactly the same way as in the case of the variational gradient method.

5. Acceleration scheme

A straightforward acceleration procedure can be incorporated into above methods, as option. Suppose we know already from r preceding iterations the appropriate matrices:

$$C_m^{(1)},\ C_m^{(2)},\ \ldots,\ C_m^{(r)}$$
$$D_m^{(1)},\ D_m^{(2)},\ \ldots,\ D_m^{(r)} \qquad (54)$$

We can improve $C_m^{(r)}$ and $D_m^{(r)}$ by considering variationally, stepwise for k = 1, 2, ..., r - 1, the contribution from

$$X_m = C_m^{(k)} - C_m^{(r)} (C_m^{(r)+} C_m^{(k)}) \qquad (55)$$

$$Z_m = D_m^{(k)} - D_m^{(r)} (C_m^{(r)+} C_m^{(k)}) \qquad (56)$$

With these definitions the procedure of the variational gradient method (section 3) can be used to accelerate the convergence, without additional recall to H.

6. Program and test calculations

A program has been written for all described versions and tested for n up to 100. Details of the program, its block structure, are given at the end of this work.

Variational gradient method

The method exhibits a good convergence provided the matrix H is sparse, the spread of diagonal elements is not large, the m lowest eigenvalues are well separated from the other ones. In a case like this the program is faster than that of the Householder method from n's slightly smaller than 100. In the configuration interaction method, however, although H is sparse indeed, the spread of diagonal elements is significant. In a case like this the variational pseudo-gradient method is preferrable.

Variational pseudo-gradient method

The method shows a good convergence whenever off-diagonal elements of H are small in absolute sense in comparison to diagonal elements. The spread of diagonal elements can be large and the matrix need not be sparse.

Here is an example for n = 100 and three eigenstates (m = 3): $H_{ii} = i$, $H_{ij} = 1/\mathrm{sqrt}(i \cdot j)$ for $i \neq j$. This is not a very favourable case for iterative diagonalization: all off-diagonal elements are different from zero, most diagonal elements are larger than the three eigenvalues (absolute numbers), there is no distinct gap between the value of H_{mm} and $H_{m+1,m+1}$.

Version	Number of iterations	Accuracy	Number of obtained eigenvectors
No preconditioning	187	0.000 01	2
$(D_T)_{ii} = 1$ for $i \leq m$	13	0.000 007	2
$d_U = 0$ (no shift)	14	0.000 008	3
$d_U = 0.001$	13	0.000 002	2
$d_U = 0.01$	11	0.000 001	2
$d_U = 0.1$	8	0.000 003	3
$d_U = 1.0$	process slowly convergent		2

In the case of sparse matrices the number of iterations is still reduced, in some cases down to 2-3 iterations (without acceleration). Always eigenstates of lowest eigenvalues have been obtained. However, depending on d_U, occasionally a trivial vector (a zero vector) is sometimes obtained, so that the real number of eigenstates obtained is eventually smaller than m. This behaviour is still under investigation.

Acceleration and large-scale CI calculations

The efficiency of the acceleration scheme has also been tested. In the specific example discussed the accuracy of $6 \cdot 10^{-6}$ has been obtained in 6 iterations, the accuracy of $6 \cdot 10^{-8}$ in 8 iterations, for $d_U = 0$. All three eigenstates have been obtained. The utility of the whole scheme in large-scale CI calculations is going to be tested in cooperation with W. Kraemer from the Max Planck Institute (Munich). Results of this investigation will be published elsewhere.

Acknowledgments

The author is indebted to Zentrum für Interdisziplinäre Forschung in Bielefeld for a substancial support of this work. The author also thanks J. Hinze and L. Elsner for several helpful conversations.

A partial support of Institute of Low Temperatures and Structural Research of the Polish Academy of Sciences is also acknowledged.

References

(1) J.H.Wilkinson, "The Algebraic Eigenvalue Problem", Clarendon, Oxford (1972).
(2) J.Hinze, Conference on "Large Linear Systems, Eigenvalue and Linear Equations", April 23 - May 2, 1980, Bielefeld (Germany).
(3) I.Shavitt, C.F.Bender, A.Pipano, R.P.Hosteny, J. Comput. Phys. 11, 90 (1973).
(4) R.Raffenetti, private information.
(5) E.Davidson, J. Comput. Phys. 17, 87 (1975).
(6) H.G.Miller, W.A.Berger, J. Phys. A: Math. Gen. 12, 1693 (1979).
(7) A.Golebiewski, Intern. J. Quant. Chem. 15, 693 (1979).

BLOCK STRUCTURE

```
C  data files: 5-DALI (standard), 13-ACCALI (for acceler.)
   read from 3-INPUT, write on 6-OUTPUT:
      NITER: maximum number of iterations
      IREPL: 1 - final CM to be stored in DALI
             Ø - final CM not stored
      ICØ:   1 - starting CM part of a unit matrix
             Ø - starting CM read from DALI
      ITACC: from which iteration acceleration (if ITACC = NITER, no accele-
             ration)
      ITDOM: from which iteration dominant diagonal approximation
      ITDCI: 1 - H created in program, when used,
             dimension (NH) of H arbitrary
             Ø - H read from 3-INPUT
      N:     length of eigenvectors
      M:     number of desired eigenstates
      EPS:   desired accuracy of eigenvalues
      DU:    parameter used in preconditioning of XM
   if (IDCI ≠ 1) read H from 3 INPUT
   if (ICØ = 1) CM defined as part of unit matrix
   if (ICØ = 0) CM read from DALI
   NCOUNT = 1
   DM = H·CM (if ITDCI = 1, with creation of H_ij's)
   HM_i  =  H_ii for i = 1, 2, ..., N
   HSMALL_ij  =  H_ij for i,j = 1, 2, ..., M
   HMM = CM^+ · DM
   if (ITACC < NITER) write CM, DM on ACCALI
   DELTA = tr HMM
┌---->10 CONTINUE
|  KLOGIC = Ø
|  XM = DM - CM · HMM
|  ┌-----if (NCOUNT < ITDOM) go to 305
|  | C  preconditioning of last N - M rows of XM:
|  |    for i < M:   X = (XM · XM^+)_ii,  Y = (XM · HMM · XM^+)_ii
|  |                 W = (CM · CM^+)_ii,  Z = (CM · HMM · CM^+)_ii
|  |                 if |100 · X| < |Y| then EP = Z
|  |                                    else EP = Z - (W - 1) · Y/X
|  |                 if (EP > HM_M+1) EP = HM_M+1
|  |                 D = 1/(HM_i - EP)
|  |                 XM_ij = D · XM_ij for j = 1, 2, ..., M
|  v
```

```
C   preconditioning of first M rows of XM:
    if (DU ≤ - 1000) go to 110 (no precond.)------------------------>   (down)
    for j,k = 1, 2, ..., M:
       U_jk  =  (XM · XM⁺)_jk · HSMALL_jk  -  (CM · HMM · CM⁺)_jk · (XM · XM⁺)_jk
                 + (XM · HMM · XM⁺)_jk · [(CM · CM⁺)_jk  -  δ_jk] + DU · δ_jk
       DX_k  =  (XM · XM⁺)_kk
    if U singular, go to 110 (no preconditioning)------------------->   (down)
    U  =  U^-1
    for i = 1, 2, ..., M:
       D  =  (U · DX)_i
       XM_ij  =  D · XM_ij for j = 1, 2, ..., M
110 CONTINUE <-------------------------------------------------------
C   extraction of CM from XM:
    XMM  =  CM⁺ · XM
    XM  =  XM  -  CM · XMM                                   (down to (305))
305 CONTINUE
C   optimization of CM, DM, HMM
    XMM  =  XM⁺ · XM
    if (M = 1) go to 57 -------------------------------------   (down)
    call Jacobi: U⁺ · XMM · U  =  DX
    CM  =  CM · U,  DM  =  DM · U,  XM  =  XM · U
 57 CONTINUE <-----------------------------------------------
    HMM  =  CM⁺ · DM,  XD  =  XM⁺ · DM
    if (KLOGIC = 1) go to 601 -------------------------------   (down)
    ZM  =  H · XM (if ITDCI = 1, with creation of H_ij's)
    go to 600 ------------------------------------   (down)
601 CONTINUE <-----------------------------------------------
    ZM  =  ZM · U (provided M > 1)
600 CONTINUE <-----------------------------------
    WMM  =  XM⁺ · ZM
    for i = 1, 2, ..., M:  DH_i  =  HMM_ii · DX_i  -  WMM_ii
    finding DL_i's, which minimize DELTA:
```

$$\mathrm{DELTA} = \mathrm{const} + \sum_{i=1}^{M} (2 \cdot XD_{ii} \cdot DL_i + DH_i \cdot DL_i^2)/(1 + DX_i \cdot DL_i^2)$$

```
    DN  =  (1 + DL · DX · DL)^-1/2
    CM  =  (CM  -  XM · DL) · DN
    DM  =  (DM  -  ZM · DL) · DN                     (up to (305))  (up to (10))
    HMM  =  DN · (HMM + DL · WMM · DL  -  XD⁺ · DL  -  DL · XD) · DN
    if (ITACC = NITER) or (KLOGIC = 1) go to 13 -------------   (down)
    write CM, DM on ACCALI, in sequence
 13 CONTINUE <-----------------------------------------------
```

```
       if (KLOGIC = Ø) NCOUNT  =  NCOUNT  +  1
       if (KLOGIC = 1) go to 306
       if (NCOUNT = NITER) go to 555
       if (NCOUNT < ITACC) go to 44
       rewind ACCALI
       KACC  =  Ø
   306 CONTINUE
       KACC  =  KACC  +  1
       XM  =  CM(KACC) (read from ACCALI)
       ZM  =  DM(KACC) (read from ACCALI)
       if (KACC = NCOUNT) go to 44
       U  =  CM+ · XM
       XM  =  XM  -  CM · U,   ZM  =  ZM  -  DM · U
       KLOGIC  =  1
       go to 305                              (up to (305))
    44 CONTINUE
       DELTA1  =  tr HMM
       CHANGE  =  DELTA  -  DELTA1
       if (KLOGIC = Ø) write: "change" = CHANGE
                else write: "change1"  =  CHANGE
       DELTA  =  DELTA1
       if (CHANGE > EPS) go to 10             (up to (10))
   555 CONTINUE
     C final diagonalization of HMM
       call Jacobi: U+ · HMM · U  =  DX
       XM  =  CM · U
       write on 6-OUTPUT: DX (eigenvalues)
                          XM (eigenvectors)
       if (IREPL = Ø) go to 888
       if (ICØ = 1) go to 333
       rewind DALI
   333 CONTINUE
       write XM (i.e. eigenvectors) on DALI
   888 CONTINUE
       STOP
```

SIMULTANEOUS RAYLEIGH-QUOTIENT ITERATION METHODS FOR LARGE SPARSE GENERALIZED EIGENVALUE PROBLEMS

H.R. Schwarz
Seminar für Angewandte Mathematik
Universität Zürich, Freiestr. 36
CH-8032 Zürich, Switzerland

1. The Problem.

We consider the generalized eigenvalue problem

$$A\underline{x} = \lambda B\underline{x}, \tag{1}$$

where A and B are symmetric matrices, B is assumed to be positive definite, both matrices being sparse and their order n is large, whatever this means. We are interested in computing the p lowest eigenvalues

$$\lambda_1 \leq \lambda_2 \leq \ldots \leq \lambda_p \tag{2}$$

together with the corresponding eigenvectors $\underline{x}_1, \underline{x}_2, \ldots, \underline{x}_p$, such that

$$A\underline{x}_j = \lambda_j B\underline{x}_j \quad (j=1,2,\ldots,p). \tag{3}$$

The number p of required eigenvalues is small compared to the order n of the matrices.

The problem under consideration was studied by the author in connection with finite elements, where the matrices A and B have the same zero-nonzero structure. Moreover, the orders n in practical applications have not been too large, ranging up to only several hundreds. In principle there is no restriction to the order n, however it is the author's opinion that the methods presented below seem to have a natural upper bound for the order n in the region of several thousands just from practical reasons. Hence they may not be well suited for the treatment of eigenvalue problems in quantum chemistry of the extreme large orders of up to 10^5. Therefore the following contribution should be seen as an informative survey of some existing approaches.

2. Short review.

In order to motivate the justification of Rayleigh quotient iteration methods we pass in review known procedures.

a) The usual method of reducing (1) to the eigenvalue problem

$$C\underline{y} = \lambda\underline{y}, \tag{4}$$

where C is symmetric and defined through the Cholesky decomposition $B = LL^T$ by $C = L^{-1} A L^{-T}$ and where $\underline{y} = L^T \underline{x}$, is quite unsuitable, since the sparsity is definitely lost, and C is in general a full matrix. If A and B are bandmatrices with bandwidth m, defined to be the smallest integer such that

$$a_{ik} = 0 \quad \text{for all} \quad |i-k| > m, \tag{5}$$

a highly complex transformation of (1) into (4) is due to Crawford [5] which preserves at least the bandwidth m for C.

b) Inverse vector iteration in its simple or simultaneous version requires the solution of sparse linear equations with A, so A must be assumed to be positive definite, too. If no special techniques are applied such as multigrid or reduction methods or modified incomplete Cholesky conjugate gradient algorithms the necessary Cholesky decomposition of $A = L_A L_A^T$ destroys the sparsity to some extent, too. The procedure can be recommended if the band or even envelope structure of A is exploitable. Whenever enough storage is available for L_A in some sense, the simultaneous vector iteration is indeed an efficient procedure due to the fact that adjacent eigenvalues within the first p do not slow down the convergence, and the convergence rate is determined essentially by the quotient λ_p/λ_{p+1}.

c) Bisection methods for the direct determination of a desired eigenvalue λ_k with given index k and the subsequent computation of the corresponding eigenvector $\underline{x}_k$ by inverse vector iteration are definitely restricted to bandmatrices.

A first realization is based on the fact that the sequence of leading minors of the matrix $(A-\mu B)$ in function of μ forms a Sturm sequence. Hence the number of those eigenvalues λ_j which are smaller than a given value μ is determined by evaluating the sequence of minors with the highly stable algorithm due to Peters and Wilkinson [18]. The procedure has the slight deficiencies to destroy the symmetry and to require a somewhat large working space of about 3mn locations.

A second realization is based on Sylvester's theorem of inertia. In the reduction of the quadratic form for a given μ into a sum of squares

$$\underline{\xi}^T (A-\mu B) \underline{\xi} = \sum_{i=1}^{n} \sigma_i \eta_i^2 \quad (\sigma_i = +1 \text{ or } -1 \text{ or } 0) \tag{6}$$

the number of negative σ_i is equal to the number of those eigenvalues λ_j which are smaller than μ. The reduction process is essentially a Cholesky decomposition combined with auxiliary transformations to increase the numerical stability [27, 28]. The algorithm has the advantages to preserve the symmetry, the required working space is about 2 mn locations, and a unification is achieved, since the bisection and inverse iteration steps use the same basic procedure.

If the reduction of the quadratic form is performed on the base of the algorithm by Bunch et al. using either a single pivot or a submatrix of order two [2, 3], the resulting method is even more efficient, preserves the symmetry, requires a working space of only (m+1)n locations and has the same property of a unified procedure.

d) The Lanczos algorithm due to Golub et al. [9, 10] for the generalized eigenvalue problem requires the Cholesky decomposition of B for an implizit reduction of (1) to (4). The recently proposed spectral transformation Lanczos method by Ruhe [22] requires a routine for solving linear systems of indefinite symmetric equations and a factorization of B.

However, the so far mentioned classes of methods are not fit for solving the problem since they require some manipulations of the given matrices and hence the necessary storage requirements are too high as soon as n gets large. Therefore algorithms are most adequate which exploit the sparsity of A and B to full extent and need no modifications of the given matrices. The Rayleigh quotient minimization methods for instance have these properties.

3. Rayleigh quotient minimization.

The smallest eigenvalue λ_1 of (1) is equal to the minimum of the Rayleigh quotient, which is attained by an eigenvector $\underline{x}_1$.

$$\min_{\underline{x}\neq 0} R[\underline{x}] = \min_{\underline{x}\neq 0} \frac{(\underline{x},A\underline{x})}{(\underline{x},B\underline{x})} = \frac{(\underline{x}_1,A\underline{x}_1)}{(\underline{x}_1,B\underline{x}_1)} = \lambda_1 \tag{7}$$

In principle every algorithm for finding the unconstrained minimum of a function may be applied. From a practical point of view not all of them are quite adequate. In the following only three mainly different classes are outlined.

a) Coordinate relaxation was proposed by several authors, such as Cooper [4], Faddejew/Faddejewa [7], Kahan [13], Nesbet [16], Shavitt [29], Falk [8], Shavitt et al. [30] in chronological order. The basic idea is the following. Starting with an arbitrary initial vector $\underline{x}^{(o)} \neq 0$, a sequence of iterates

$$\underline{x}^{(k)} = \varphi \underline{x}^{(k-1)} + \psi \underline{e}_j \quad \text{with} \quad j \equiv k (\text{mod } n) \tag{8}$$

is generated, where $\underline{e}_j$ is the j-th unit vector, such that in each step the Rayleigh quotient $R[\underline{x}^{(k)}]$ is minimized. The two scalars φ and ψ are usually determined by solving a quadratic equation under the assumption $\varphi=1$, or as the components of an eigenvector of an auxiliary generalized eigenvalue problem of order two [28]. In the normal case we may set $\varphi=1$. A sequence of n single steps for j = 1, 2,..., n forms a cycle, whence the coordinate relaxation is often called simply a cyclic method. Since in each single step just one component of the present approximation $\underline{x}^{(k-1)}$ is changed cyclicly, the process has close analogies with the method of Gauss-Seidel.

b) To improve the convergence coordinate overrelaxation (COR) has been suggested by Nisbet [17], Ruhe [21] and Schwarz [23]. In complete analogy to SOR the correction of the j-th component in (8) is multiplied by a constant relaxation parameter ω. With $\varphi=1$ we have the modified rule

$$\underline{x}^{(k)} = \underline{x}^{(k-1)} + \omega\psi\, \underline{e}_j \quad \text{with } j \equiv k (\text{mod } n). \tag{9}$$

From a practical point of view the following theoretical facts are essential.

α) The optimal choice for ω obeys similar rules as in case of SOR [24]. However, the proper choice of ω must be usually a matter of experience. This may explain to some extent the fact that some people do not consider to apply overrelaxation.

β) In a quite special situation the convergence rate of coordinate relaxation (ω=1) can be described explicitly. If $\underline{x}^{(\nu)}$ (ν=0,1,2,...) denote the iterates after complete cycles, if A has "property A" and having identical diagonal elements a_{ii} = a and if B = I, the convergence quotient is given by [21, 24]

$$q = \left[1-2\frac{\lambda_2^*-\lambda_1}{\lambda_{max}^*-\lambda_1}\right]^2 . \tag{10}$$

In (10) $\lambda_2^* > \lambda_1$ represents the next higher eigenvalue and λ_{max}^* the largest eigenvalue of A. This result might suggest that the convergence behaviour of COR is in general dictated only by the spectrum of (1). Experience contradicts this hypothesis, and an extreme class of matrices could be found, for which the asymptotic convergence rate is completely independent from the spectrum [12]. Reality lies between the two extremes.

γ) The central question of possible wrong convergence could be answered in a satisfactory way. It has been shown [11, 25] that the eigenspaces corresponding to all eigenvalues λ_i of (1) satisfying

$$\lambda_1 < \lambda_i < \min_k (a_{kk}/b_{kk}) =: Q \tag{11}$$

are non-attractive fixed-points of the iteration. The repelling effect is even increased for $\omega > 1$. Whenever the limit of the Rayleigh quotients $R[\underline{x}^{(\nu)}]$ is less than Q, we have

$$\lim_{\nu\to\infty} R[\underline{x}^{(\nu)}] = \lambda_1, \quad \lim_{\nu\to\infty} \underline{x}^{(\nu)} = \underline{x}_1. \tag{12}$$

c) Gradient methods employ the gradient of the Rayleigh quotient for a given approximation $\underline{x}^{(k)}$, that is

$$\underline{g}^{(k)} := \text{grad } R[\underline{x}^{(k)}] = \frac{2}{(\underline{x}^{(k)},B\underline{x}^{(k)})} \{A\underline{x}^{(k)}-R[\underline{x}^{(k)}]B\underline{x}(k)\} \tag{13}$$

as search direction to construct a sequence of iterates

$$\underline{x}^{(k+1)} = \underline{x}^{(k)} + \gamma\underline{g}^{(k)} \quad (k=0,1,2,\dots), \tag{14}$$

where γ is determined such that $R[\underline{x}^{(k+1)}]$ attains its minimal value. Although the gradient seems to be a better choice than the unit vectors in coordinate relaxation, the convergence properties of the gradient Rayleigh quotient minimization method are bad, and the method is much less efficient than COR.

However, the convergence can be improved substantially by damping the corrections $\gamma \underline{g}^{(k)}$ in (14) by a constant factor $\omega < 1$, hence by applying underrelaxation

$$\underline{x}^{(k+1)} = \underline{x}^{(k)} + \omega \cdot \gamma\, \underline{g}^{(k)}, \quad \omega < 1. \tag{15}$$

Experiments indicate that the best relaxation factors lie in the interval (0.8, 0.9) depending on the type of problem.

The essential question, whether the sequence of iterates $\underline{x}^{(k)}$ converges towards an eigenvector $\underline{x}_1$ corresponding to the smallest eigenvalue λ_1 cannot be affirmed in general. If the initial vector $\underline{x}^{(o)}$ is deficient in $\underline{x}_1$, that is if $(\underline{x}^{(o)}, B\underline{x}_1) = 0$ holds, then all iterates fulfill $(\underline{x}^{(k)}, B\underline{x}_1) = 0$, and hence convergence to an eigenvector of a higher eigenvalue occurs.

d) Conjugate gradient methods improve the convergence behaviour. Starting with an initial vector $\underline{x}^{(o)}$ the search direction $\underline{p}^{(o)}$ is chosen to be the negative gradient $\underline{g}^{(o)}$, and hence $\underline{x}^{(1)}$ is defined by

$$\underline{x}^{(1)} = \underline{x}^{(o)} + \gamma \underline{p}^{(o)} \quad \text{with} \quad \underline{p}^{(o)} = -\underline{g}^{(o)},\ R[\underline{x}^{(1)}] = \min! \tag{16}$$

The subsequent search directions are determined as [1, 15, 20]

$$\underline{p}^{(k)} = -\underline{g}^{(k)} + \frac{(\underline{g}^{(k)}, \underline{g}^{(k)})}{(\underline{g}^{(k-1)}, \underline{g}^{(k-1)})}\, \underline{p}^{(k-1)}, \quad (k=1,2,\ldots) \tag{17}$$

and the iterates $\underline{x}^{(k+1)} = \underline{x}^{(k)} + \gamma \underline{p}^{(k)}$ minimize $R[\underline{x}^{(k+1)}]$.

For this (classical) conjugate gradient Rayleigh quotient iteration (CG-RQIT) under- or overrelaxation does not improve the convergence, since this method is very sensitive to the correct line search in each step. A so-called restart of the process is necessary after a number of steps $n_{rest} \geq n$. Since CG-RQIT has indeed an excellent asymptotic convergence [1], the method has to be highly recommended in case of a good initial vector. In case of a general initial vector $\underline{x}^{(o)}$ the previous remark concerning a possible wrong convergence applies to conjugate gradient methods, too.

4. Higher eigenvalues.

All Rayleigh quotient minimization methods yield in the general favourable case λ_1 and a corresponding eigenvector $\underline{x}_1$. In order to get the next higher eigenvalue and eigenvector an appropriate deflation step has to be applied. If the first $\ell-1$ eigenvalues $\lambda_1, \lambda_2, \ldots, \lambda_{\ell-1}$ and the corresponding B-orthonormalized eigenvectors $\underline{x}_1, \underline{x}_2, \ldots, \underline{x}_{\ell-1}$ have been computed, we consider the following eigenvalue problem

$$A_\ell \underline{x} = \lambda B \underline{x} \quad \text{with} \quad A_\ell = A + d \sum_{\mu=1}^{\ell-1} (B\underline{x}_\mu)(B\underline{x}_\mu)^T, \quad d > 0 \tag{18}$$

The eigenvectors of (18) are the same as those of (1), but the eigenvalues are λ_1+d, λ_2+d,..., $\lambda_{\ell-1}$+d; $\lambda_\ell,\ldots,\lambda_n$. The modification of A to A_ℓ in (18) causes a partial shift of the spectrum. For $d > \lambda_\ell - \lambda_1$ it follows that λ_ℓ is indeed the smallest eigenvalue of (18), which can be determined by the previous methods. The deflation step (18) must be realized implicitly in order not to destroy the sparsity of A. For details see e.g. [28].

The Rayleigh quotient minimization methods determining a single eigenvector at a time have the disadvantages that they suffer under a very slow convergence in case of adjacent eigenvalues, and that the higher eigenvalues and eigenvectors are influenced by the previously determined approximate eigenvectors. Hence it corresponds to a more direct approach to apply a simultaneous iteration.

5. Simultaneous Rayleigh quotient iteration.

The basic idea to operate with a set of p vectors $\{\underline{y}_1^{(k)}, \underline{y}_2^{(k)}, \ldots, \underline{y}_p^{(k)}\}$ simultaneously such that

$$\lim_{k\to\infty} R[\underline{y}_i^{(k)}] = \lambda_i, \quad \lim_{k\to\infty} \underline{y}_i^{(k)} = \underline{x}_i, \qquad (i=1,2,\ldots,p) \tag{19}$$

can be performed in a variety of ways.

a) The simultaneous group coordinate overrelaxation (SGCOR) [26, 27, 28] is a generalization of COR. Since p vectors are iterated simultaneously it seems to be more adequate to change a group of g components in each vector. For convenience we consider only the special case of g consecutive components with indices μ, μ+1,..., μ+g-1. Hence a single step of the iteration can be described as follows: With respect to the subspace spanned by the vectors $\underline{y}_1, \underline{y}_2, \ldots, \underline{y}_p, \underline{e}_\mu, \underline{e}_{\mu+1}, \ldots, \underline{e}_{\mu+g-1}$ (the superscript k is deleted for simplicity) we look for the vectors

$$\underline{y}' = \sum_{i=1}^{p} c_i \underline{y}_i + \sum_{j=1}^{g} c_{p+j} \underline{e}_{\mu+j-1} \tag{20}$$

yielding the p smallest stationary values of $R[\underline{y}']$. This task is equivalent to solving the generalized eigenvalue problem

$$\tilde{A} \underline{c} = \Lambda \tilde{B} \underline{c} \tag{21}$$

of order p+g for the p smallest eigenvalues $\Lambda_1 \leq \Lambda_2 \leq \ldots \leq \Lambda_p$ and the corresponding eigenvectors $\underline{c}_1, \underline{c}_2, \ldots, \underline{c}_p$, where the elements of $\tilde{A}$ are defined by

$$\tilde{a}_{ij} = (\underline{y}_i, A\underline{y}_j), \qquad (i,j = 1,2,\ldots,p) \tag{22}$$

$$\tilde{a}_{i,p+j} = (\underline{y}_i, A\underline{e}_{\mu+j-1}) = (A\underline{y}_i)_{\mu+j-1} \qquad \begin{cases} i=1,2,\ldots,p \\ j=1,2,\ldots,g \end{cases} \tag{23}$$

$$\tilde{a}_{p+i,p+j} = (\underline{e}_{\mu+i-1}, A\underline{e}_{\mu+j-1}) = a_{\mu+i-1,\mu+j-1} \qquad i,j=1,2,\ldots,g \tag{24}$$

Similar formulas hold for the elements of $\tilde{B}$. The eigenvalue problem (21) is solved by the sequence of steps of reducing it to a special eigenvalue problem, Householder transformation, bisection, inverse iteration for the tridiagonal matrix and back-substitution. The eigenvectors $\underline{c}_i = (c_1^{(i)}, c_2^{(i)}, \ldots, c_{p+g}^{(i)})^T$ satisfy the relations

$$(c_i, \tilde{B}\underline{c}_j) = \delta_{ij}, \quad (\underline{c}_i, \tilde{A}\underline{c}_j) = \Lambda_i \, \delta_{ij}. \tag{25}$$

The set of iterated vectors $\underline{y}'_\ell$ of a single step are computed by

$$\underline{y}'_\ell = \sum_{i=1}^{p} c_i^{(\ell)} \underline{y}_i + \omega \sum_{j=1}^{g} c_{p+j}^{(\ell)} \underline{e}_{\mu+j-1}, \quad (\ell=1,2,\ldots,p), \tag{26}$$

where we have already introduced the relaxation factor ω for improving the convergence. For $\omega \neq 1$ we have of course $R[\underline{y}'_\ell] \neq \Lambda_\ell$.

If it happens by chance that the p+g vectors $\underline{y}_1, \underline{y}_2, \ldots, \underline{y}_p, \underline{e}_\mu, \underline{e}_{\mu+1}, \ldots, \underline{e}_{\mu+g-1}$ are linearly dependent, the matrix $\tilde{B}$ is singular, and the reduction step for (21) is impossible. In this case the step is just skipped for the group of coordinates.

For preparing the next step the submatrices of order p in the left upper corner in $\tilde{A}'$ and $\tilde{B}'$ may be computed recursively for the sake of efficiency according to

$$\tilde{a}'_{ij} = (\underline{y}'_i, A\underline{y}'_j) = (\tilde{\underline{c}}_i, \tilde{A}\tilde{\underline{c}}_j), \quad \tilde{b}_{ij} = (\tilde{\underline{c}}_i, \tilde{B}\tilde{\underline{c}}_j), \tag{27}$$

where $\tilde{\underline{c}}_i = (c_1^{(i)}, \ldots, c_p^{(i)}, \omega c_{p+1}^{(i)}, \ldots, \omega c_{p+g}^{(i)})^T$. The slight numerical instability of the recursive computation makes it necessary to recompute the elements $\tilde{a}_{ij}$ and $\tilde{b}_{ij}$ through (22) from time to time. For $\omega = 1$ the submatrices in $\tilde{A}'$ and $\tilde{B}'$ are diagonal due to (25) with $\tilde{a}'_{ii} = \Lambda_i$, $\tilde{b}'_{ii} = 1$.

The described single step forms an element of a complete cycle which is built up in a similar fashion to block SOR. The SGCOR algorithm is summarized as follows:

<u>Start</u>: Choose the number ν of groups of disjoint sets of g_ℓ consecutive unit vectors with $\sum_{\ell=1}^{\nu} g_\ell = n$, the relaxation factor ω and p independent initial vectors $\underline{y}_1, \underline{y}_2, \ldots, \underline{y}_p$.

$$\left.\begin{array}{l} \tilde{a}_{ij} = (\underline{y}_i, A\underline{y}_j), \quad \tilde{b}_{ij} = (\underline{y}_i, B\underline{y}_j) \\ R_i = \tilde{a}_{ii}/\tilde{b}_{ii} \end{array}\right\} \quad (i=1,2,\ldots,p; \; j=1,2,\ldots,i)$$

<u>Cycle</u>: For $\ell = 1,2,\dots,\nu$:

a) Remaining elements of $\tilde{A}$, $\tilde{B}$.

b) $\tilde{A}\,\underline{c} = \Lambda\,\tilde{B}\,\underline{c} \implies \begin{cases} \Lambda_1 \le \Lambda_2 \le \dots \le \Lambda_p \\ \underline{c}_1,\ \underline{c}_2,\ \dots,\ \underline{c}_p \end{cases}$

c) $c_j^{(i)} := \omega \cdot c_j^{(i)}$ $\quad (i=1,2,\dots,p;\ j=p+1,\dots,p+g_\ell)$

d) $\underline{y}_i := \sum_{j=1}^{p} c_j^{(i)} \underline{y}_j + \sum_{j=1}^{g_\ell} c_{p+j}^{(i)} \underline{e}_{\mu+j-1}$ $\quad (i=1,2,\dots,p)$

e) $\tilde{a}_{ij} := (\underline{c}_i, \tilde{A}\underline{c}_j)$, $\quad \tilde{b}_{ij} := (\underline{c}_i, \tilde{B}\underline{c}_j)$ $\quad (i=1,2,\dots,p;\ j=1,2,\dots,i)$

f) $R_i = \tilde{a}_{ii} / \tilde{b}_{ii}$

Each cycle requires the solution of ν eigenvalue problems of orders $p+g_\ell$. The number of arithmetic operations per cycle can be estimated with the simplifying assumption $n \doteq \nu g$ and disregarding the problem dependent effort for the needed p multiplications of A and B with vectors $\underline{y}_i$ to get the remaining elements of $\tilde{A}$ and $\tilde{B}$ to be [28]

$$Z \approx \frac{n}{g}\left[\frac{1}{6}(p+g)^2(29p+8g) + p^2 n + p(p+g)(p+51)\right]. \tag{28}$$

The following table illustrates the dependency of Z from g for some given values of n and p.

n=	p=	g=1	g=4	g=8	g=12	g=16	g=20	g=24	factor.
300	1	12.9	<u>6.11</u>	7.59	11.2	—	—	—	10^4
300	4	19.3	6.10	4.41	<u>4.34</u>	4.78	5.49	6.40	10^5
300	8	8.01	2.34	1.48	1.26	<u>1.22</u>	1.25	1.33	10^6
1000	1	11.3	3.79	<u>3.40</u>	4.33	5.90	7.99	10.6	10^5
1000	4	17.6	4.83	2.87	2.38	2.29	2.39	2.60	10^6
2000	1	42.6	12.6	9.31	10.3	13.1	17.0	21.9	10^5
2000	4	67.2	17.7	9.74	7.43	6.58	6.38	6.53	10^6

In finite element applications the convergence of SGCOR does not seem to be sensitive to the size g of the groups. Hence it makes sense to choose the medium group size such that Z is minimized. The table indicates that it is more efficient to take g>1 even in case a single eigenvalue is wanted. However, it is likely that a proper problem oriented choice of groups may improve the convergence properties.

The SGCOR method has the property that adjacent eigenvalues within the p smallest do not slow down the convergence. Moreover, due to the theoretical fact about COR there is some evidence that no eigenvalue is missed, although no proof could be found. The convergence rate seems to depend very much from the type of the problem, but not only from the spectrum of (1). The optimal choice of ω is in general unsolved and can be overcome only by experience. The situation is again not satisfactory.

Due to the high complexity of each cycle one might think about possible improvements or simplifications of SGCOR. As soon as the iterates $\underline{y}_i^{(k)}$ are good approximations for $\underline{x}_i$ in an advanced stage of the process, the p eigenvectors $\underline{c}_i$ of (21) are approximately equal to the i-th unit vectors $\underline{e}_i$. Perturbation techniques to solve (21) reduce indeed the required amount of arithmetic operations to about 50% [12].

b) The simultaneous coordinate overrelaxation (SICOR) represents another simplification of SGCOR to avoid the generalized eigenvalue problem (21) in each step of a cycle. The idea is to decouple completely the corrections for the p iterates during each cycle, to reduce the groups again to a single unit vector and to return to the common COR applied independently to the p iterates. To avoid convergence of the iterates $\underline{y}_i^{(k)}$ towards the same eigenvector $\underline{x}_1$ for all i=1,2,...,p a Ritz-Step after each cycle has to be applied. The SICOR algorithm may be described as follows:

<u>Start</u>: Choose the relaxation factor ω and p independent initial vectors

$\underline{y}_1, \underline{y}_2, \ldots, \underline{y}_p$ which form the (n×p) matrix Y.

$\hat{A} = Y^T AY, \quad \hat{B} = Y^T BY, \quad R_i = \hat{a}_{ii} / \hat{b}_{ii}$

<u>Ritz-step</u>: $\hat{A}\underline{c} = \Lambda \hat{B} \underline{c} \implies C = (\underline{c}_1, \underline{c}_2, \ldots, \underline{c}_p)$ with $\Lambda_1 \leq \Lambda_2 \leq \cdots \leq \Lambda_p$

$Y \cdot C =: Y$

$\hat{A} := Y^T AY = \mathrm{diag}(\Lambda_1, \Lambda_2, \ldots, \Lambda_p)$

$\hat{B} := Y^T BY = I, \qquad R_i := \Lambda_i$

<u>Cycle</u>: For $\ell = 1,2,\ldots,n$:

Determine the p corrections of the ℓ-th components of the $\underline{y}_i$ according to COR (9). To do this, the ℓ-th rows of A and B are required.

Update the (p×p)-matrices $\hat{A}$, $\hat{B}$.

For $\omega = 1$ this is precisely Raffenetti's algorithm [19]. He reports an improved convergence in comparison with COR, a reduction of input/output demands, but the requirement of the algorithm that the matrices A and B must be accessible ordered by rows may be a disadvantage in certain applications. The same remark holds of course also for SGCOR.

Recent numerical experiments indicate two things. First of all SICOR has the weak property that the vectors corresponding to the higher values of the Rayleigh quotient may be changed during a cycle so much that the Rayleigh quotients get even smaller than λ_2. The Ritz-step throws the vectors back again, but this cycling is repeated, and no convergence takes place. To prevent the higher vectors from such a cycling the ℓ-th component of $\underline{y}_i$ (i>1) should be changed only if the new i-th Rayleigh quotient is still larger than $R[\underline{y}_{i-1}]$. For an easy test the corrections should be computed using the above mentioned generalized eigenvalue problem of order two (see e.g. [28]). Hence the Rayleigh quotients of the iterates are ordered in increasing size in each stage of the process. With this simple modification the SICOR method turns out to be much superior to SGCOR and to yield the desired eigenvalues and eigenvectors much more efficiently.

c) The simultaneous gradient iteration (SIRQGR) may be seen as a direct generalization of SGCOR, when the group of unit vectors is replaced by the p gradient vectors for the present iterates. On a first sight it might be promising to use the gradient vectors $\underline{g}_i^{(k)}$ or simpler the p residual vectors

$$\underline{r}_i^{(k)} = A\underline{y}_i^{(k)} - R[\underline{y}_i^{(k)}]B\underline{y}_i^{(k)}, \quad (i=1,2,\ldots,p), \tag{29}$$

which appear to be more appropriate for minimizing the Rayleigh quotient. However, the complexity of the resulting algorithm is much higher due to the computation of the residual vectors (29), to the more complicated structure of the matrices $\tilde{A}$ and $\tilde{B}$ of order 2p in (21) and to the more general linear combination (20) with 2p full vectors. Even with the essential improvement of convergence by damping all the components of the gradient directions with a constant factor $\omega \in (0.8, 0.9)$ the SIRQGR method is much less efficient than SGCOR. The most serious deficiency of SIRQGR is the fact that the 2p vectors $\underline{y}_1^{(k)},\ldots,\underline{y}_p^{(k)}, \underline{r}_1^{(k)},\ldots, \underline{r}_p^{(k)}$ tend to get linearly dependent as is pointed out in [15], too. The matrix $\tilde{B}$ becomes singular, and the usual reduction is impossible. Therefore SIRQGR has no practical meaning.

d) The simultaneous gradient iteration with continuous B-orthogonalization (SIRQGRO) is an efficient simplification of SIRQGR to reduce the complexity [15]. First of all the auxiliary eigenvalue problem (21) of order 2p should be eliminated. This may be achieved similar to Raffenetti's algorithm by uncoupling the corrections of the iterates $\underline{y}_i^{(k-1)}$ and $\underline{y}_j^{(k-1)}$ for $i \neq j$ and by using only one, but individual search direction for each iterate. Further we postulate that the p iterated vectors $\underline{y}_1^{(k)},\ldots,\underline{y}_p^{(k)}$ are B-orthonormal at each stage, thus satisfying the condition of the eigenvectors. This postulate is accomplished by proper B-orthonormalizations of the current iterates and the corresponding residual vectors. The last goal gives the method some similarities with the simultaneous vector iteration [28], which is sometimes also called subspace iteration. Hence additional Ritz steps are appropriate to improve convergence. They should be applied more often in the beginning and more seldom in later stages of the process, since the effort does not pay out. The Ritz steps require the solution of a common eigenvalue problem of order p only due to the B-orthogonality of the iterates. Underrelaxation of the (modified) residual vectors improves the convergence again essentially.

The SIRQGRO algorithm reads now as follows:

Start: Choose $\omega \in (0.8, 0.9)$ and p B-orthonormal initial vectors

$$(\underline{y}_1^{(o)}, \underline{y}_2^{(o)}, \ldots, \underline{y}_p^{(o)}) =: Y^{(o)}, \quad Y^{(o)T} B\, Y^{(o)} = I.$$

Iteration: (for k=1,2,...):

a) $\tilde{A} = Y^{(k-1)T} A\, Y^{(k-1)}, \quad R[\underline{y}_i^{(k-1)}] = \tilde{a}_{ii}$

b) Ritz step, if k = 1, 2, 4, 8, 16, 32, 48, 64,...

$$\tilde{A}\underline{c} = \Lambda \underline{c} \Longrightarrow \Lambda_1 \leq \Lambda_2 \leq \ldots \leq \Lambda_p$$

$$(\underline{c}_1, \underline{c}_2, \ldots, \underline{c}_p) =: C \quad \text{(orthogonal)}$$

$$Y^{(k-1)} := Y^{(k-1)} C, \quad R[\underline{y}_i^{(k-1)}] := \Lambda_i$$

c) $\underline{r}_i^{(k-1)} = A\underline{y}_i^{(k-1)} - R[\underline{y}_i^{(k-1)}]\, B\underline{y}_i^{(k-1)} \quad (i=1,2,\ldots,p)$

(Test for convergence)

d) for i = 1,2,...,p:

$$\gamma_i \;:\; R[\underline{y}_i^{(k-1)} - \gamma_i \underline{r}_i^{(k-1)}] = \min!$$

$$\underline{y}_i^{(k)} := \underline{y}_i^{(k-1)} - \omega\, \gamma_i\, \underline{r}_i^{(k-1)}$$

$$\underline{y}_i^{(k)} := \underline{y}_i^{(k)} / (\underline{y}_i^{(k)}, B\underline{y}_i^{(k)})^{\frac{1}{2}} \quad \text{(B-normal)}$$

for j = i+1, i+2,...,p:

$$\underline{y}_j^{(k-1)} := \underline{y}_j^{(k-1)} - (\underline{y}_j^{(k-1)}, B\underline{y}_i^{(k)})\underline{y}_i^{(k)}$$

$$\underline{r}_j^{(k-1)} := \underline{r}_j^{(k-1)} - (\underline{r}_j^{(k-1)}, B\underline{y}_i^{(k)})\underline{y}_i^{(k)}$$

The mentioned deficiency of SIRQGR is completely eliminated. The method is indeed fairly efficient, in general it is slightly less efficient than SGCOR if an almost optimal relaxation factor ω is chosen. On the other hand the convergence of SIRQGRO is not very sensitive to the choice of $\omega \in (0.8, 0.9)$. This fact may be an advantage in comparison with SGCOR from the point of view of the uncertainty of the proper choice of ω. In all matrix-vector operations the elements of the two matrices A and B may be accessible randomly, a possibly advantageous fact in some applications. The storage requirements are increased in comparison with SGCOR, since two sets of p vectors each of length n are needed.

e) The simultaneous conjugate gradient iteration (SIRQCG) follows the same philosophy as SIRQGRO to operate with a set of B-orthonormal iterates, but the above used p residual vectors are replaced by conjugate gradient directions defined according to

(17). Again the process has to be restarted after a certain number n_{rest} of iteration steps. A rough version of SIRQCG can be defined as follows:

<u>Start</u>: Choose n_{rest} and p B-orthonormal initial vectors

$$(\underline{y}_1^{(o)}, \underline{y}_2^{(o)}, \dots, \underline{y}_p^{(o)}) =: Y^{(o)}, \quad Y^{(o)T} B Y^{(o)} = I.$$

<u>Iteration:</u> (for k = 1,2,...):

a) if $k \equiv 1 \pmod{n_{rest}}$: (Restart with Ritz step)

$$\tilde{A} = Y^{(k-1)T} A Y^{(k-1)}, \quad \tilde{A}\underline{c} = \Lambda \underline{c} \Rightarrow C := (\underline{c}_1, \dots, \underline{c}_p)$$

$$Y^{(k-1)} := Y^{(k-1)} C$$

b) $R[\underline{y}_i^{(k-1)}]; \underline{g}_i^{(k-1)} := 2\{A\underline{y}_i^{(k-1)} - R[\underline{y}_i^{(k-1)}]B\underline{y}_i^{(k-1)}\}$

$$\underline{p}_i^{(k)} = \begin{cases} -\underline{g}_i^{(k-1)} & \text{(if restart)} \\ -\underline{g}_i^{(k-1)} + \dfrac{(\underline{g}_i^{(k-1)}, \underline{g}_i^{(k-1)})}{(\underline{g}_i^{(k-2)}, \underline{g}_i^{(k-2)})} \underline{p}_i^{(k-1)} & \end{cases} \quad (i=1,2,\dots,p)$$

c) for i = 1,2,...,p:

$$\underline{y}_i^{(k)} = \underline{y}_i^{(k-1)} - \gamma_i \underline{p}_i^{(k)} \quad \text{with } R[\underline{y}_i^{(k)}] = \min!$$

$$\underline{y}_i^{(k)} := \underline{y}_i^{(k)} / (\underline{y}_i^{(k)}, B\underline{y}_i^{(k)})^{\frac{1}{2}} \quad \text{(B-normal)}$$

for j = i+1, i+2,....,p:

$$\underline{y}_j^{(k-1)} := \underline{y}_j^{(k-1)} - (\underline{y}_j^{(k-1)}, B\underline{y}_i^{(k)})\underline{y}_i^{(k)}$$

$$\underline{p}_j^{(k)} := \underline{p}_j^{(k)} - (\underline{p}_j^{(k)}, B\underline{y}_i^{(k)})\underline{y}_i^{(k)}$$

The SIRQCG algorithm is more efficient than SIRQGRO. It is comparable or superior to SGCOR in finite element applications at least. The proper choice of the value n_{rest} is again left to experience. However, the convergence is less sensitive on the value n_{rest} in comparison with the sensitivity on ω in SGCOR. Experience indicates that the best value for n_{rest} decreases with increasing p. It seems that the process itself with the B-orthogonalization of the search directions $\underline{p}_j^{(k)}$ weakens or even destroys the essential property of conjugacy. According to a comment of Ph. Toint the convergence property of the above given algorithm may be improved essentially by applying finer methods for the construction of such conjugate directions which are less sensitive to the optimal line search, thus probably avoiding the expensive matrix-vector multiplications, to include an appropriate test for restart and to do the restart differently (see.e.g.[14]).

6. Example.

As a typical problem of finite element applications the frequencies and modes of an accoustic field in the interior of a car are to be computed. Under some simplifications we have to solve the Neumann eigenvalue problem [28]

$$\Delta u + \lambda u = 0 \quad \text{in } G, \tag{30}$$

$$\frac{\partial u}{\partial n} = 0 \quad \text{on } \partial G. \tag{31}$$

G denotes the plane section of the car in figure 1, where the triangularization into 139 triangles is shown. Using cubic triangles with nine degrees of freedom (see [28]), the resulting generalized eigenvalue problem (1) is of the order n = 285. About 5% of the matrixelements in A and B are different from zero. In the following table some representative informations are given for SGCOR, SICOR and SIRQCG. The number of simultaneously iterated vectors is p, the number of desired eigenvalues is $p' < p$, the iteration is stopped as soon as the maximal absolute value of the residuals is smaller than $\varepsilon = 10^{-6}$. The initial vectors are chosen randomly.

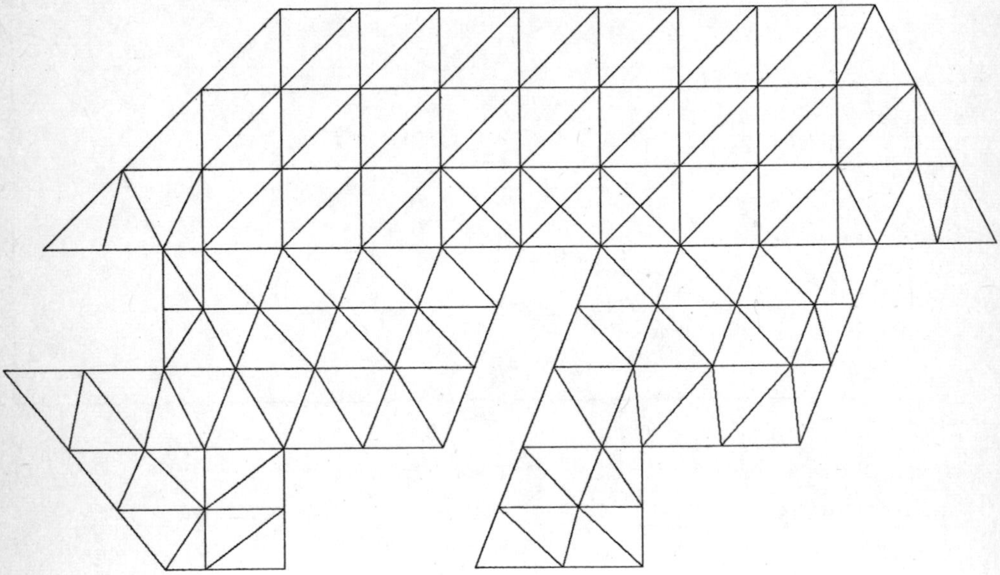

Figure 1. The model problem

For this type of problem the SICOR method solves the problem in the most efficient way concerning computing time and storage requirements.

Table 1. Performance of the methods

SGCOR:	$p = 10$, $p' = 8$, $\nu = 19$, $g = 15$						
ω =	1.20	1.25	1.30				
# iter =	31	28	25				
CPU(sec)	181	163	145				
SICOR:	$p = 10$, $p' = 8$						
ω =	1.0	1.2	1.3	1.4	1.5	1.6	
# iter =	55	36	38	36	39	45	
CPU(sec)	35	23	24	23	25	28	
SIRQCG:	$p = 7$, $p' = 5$						
n_{rest} =	10	20	30	40	50	100	150
# iter =	95	84	92	95	101	129	157
CPU(sec)	67	60	65	67	71	91	111
SIRQCG:	$p = 10$, $p' = 8$						
n_{rest} =	5	10	15	20	25	30	50
# iter	140	92	83	80	82	110	116
CPU(sec)	151	99	87	85	87	115	122

References

[1] Bradbury, W.W. and Fletcher, R.: New iterative methods for solution of the eigenproblem. Numer. Math. 9 (1966) 259-267.

[2] Bunch, J.R., Kaufman, L. and Parlett, N.P.: Decomposition of a symmetric matrix. Numer. Math. 27 (1976) 95-109.

[3] Bunch, J.R. and Kaufman, L.: Some stable methods for calculating inertia and solving symmetric linear systems. Math. Comp. 31 (1977) 163-179.

[4] Cooper, J.L.B.: The solution of natural frequency equations by relaxation methods. Quart. Appl. Math. 6 (1948) 179-183.

[5] Crawford, C.R.: Reduction of a band-symmetric generalized eigenvalue problem. Comm. ACM 16 (1973) 41-44.

[6] Ericsson, Th. and Ruhe, A.: The spectral transformation Lanczos method for the numerical solution of large sparse generalized symmetric eigenvalue problems. Report UMINF-76.79, University of Umeå, 1979.

[7] Faddejew, D.K. and Faddejewa, W.N.: Computational methods of linear algebra. Freeman & Co., 1963.

[8] Falk, S.: Berechnung von Eigenwerten und Eigenvektoren normaler Matrizenpaare durch Ritz-Iteration. ZaMM 53 (1973) 73-91.

[9] Golub, G.H., Underwood, R. and Wilkinson, J.H.: The Lanczos algorithm for the symmetric $Ax=\lambda Bx$ problem. STAN-CS-72-270, 1972.

[10] Golub, G.H.: Some uses of the Lanczos algorithm in numerical linear algebra. In Muller, J. (ed): Topics in numerical analysis. Academic Press, 1974.

[11] Gose, G.: Relaxationsverfahren zur Minimierung von Funktionalen und Anwendungen auf das Eigenwertproblem für symmetrische Matrizenpaare. Diss. Techn. Univ. Braunschweig, 1974.

[12] Hanzal, A.: Varianten der Koordinatenüberrelaxation. Dissertation, Universität Zürich, 1978.

[13] Kahan, W.: Relaxation methods for an eigenproblem. Technical Report CS-44, Computer Science Department Stanford University, 1966.

[14] Lenard, M.L.: Accelerated conjugate direction methods for unconstrained optimization. J. Optim. Theory Appl. 25 (1978) 11-31.

[15] Longsine, D.E. and McCormick, S.F.: Simultaneous Rayleigh quotient minimization methods for Ax=λBx. **Lin. Alg. Appl. 34, 195 (1980).**

[16] Nesbet, R.K.: Algorithm for diagonalization of large matrices. J. Chem. Physics 43 (1965) 311-312.

[17] Nisbet, R.M.: Acceleration of the convergence in Nesbet's algorithm for eigenvalues and eigenvectors of large matrices. J. Comp. Phys. 10 (1972) 614-619.

[18] Peters, G. and Wilkinson, J.H.: Eigenvalues of Ax=λBx with band symmetric A and B. Comp. J. 12 (1969) 398-404.

[19] Raffenetti, R.C.: History of relaxation algorithms and theoretical chemistry. In Moler, C. and Shavitt, I. (Ed.): Report on the Workshop "Numerical algorithms in chemistry: Algebraic methods", August 9-11, 1978, at Lawrence Berkeley Laboratory.

[20] Ruhe, A.: Iterative eigenvalue algorithms for large symmetric matrices. Report UMINF-31.72, Umeå University, 1972.

[21] Ruhe, A.: SOR-methods for the eigenvalue problem. Report UMINF-37.73, Umeå University, 1973. Math. Comput. 28 (1974) 695-710.

[22] Ruhe, A.: Computation of eigenvalues and eigenvectors. In Barker, V.A. (ed.), Sparse matrix techniques, Lecture Notes Math. 572, Springer, Berlin-Heidelberg-New York, 1977, 130-184.

[23] Schwarz, H.R.: The eigenvalue problem (A-λB)x=0 for symmetric matrices of high order. Comp. Meth. Appl. Mech. Eng. 3 (1974) 11-28.

[24] Schwarz, H.R.: The method of coordinate overrelaxation for (A-λB)x=0. Numer. Math. 23 (1974) 135-151.

[25] Schwarz, H.R.: La méthode de surrelaxation en coordonnées pour (A-λB)x=0. Séminaire d'analyse numérique, Université de Grenoble, report no. 223, 1975.

[26] Schwarz, H.R.: Praktische Erfahrungen mit Varianten der Koordinatenüberrelaxation zur Lösung von Eigenwertaufgaben. ISNM 31 (Birkhäuser, Basel und Stuttgart, 1976), 199-222.

[27] Schwarz, H.R.: Two algorithms for treating Ax=λBx. Comp. Meth. Appl. Mech. Eng. 12 (1977) 181-199.

[28] Schwarz, H.R.: Methode der finiten Elemente, Teubner, Stuttgart, 1980.

[29] Shavitt, I.: Modification of Nesbet's algorithm for the iterative evaluation of eigenvalues and eigenvectors of large matrices. J. Comput. Phys. 6 (1970),124-130.

[30] Shavitt, I., Bender, C.F., Pipano, A. and Hosteny, R.P.: The iterative calculation of several of the lowest or highest eigenvalues and corresponding eigenvectors of very large symmetric matrices. J. Comput. Phys. 11 (1973) 90-108.

LARGE SPARSE UNSYMMETRIC EIGENVALUE PROBLEMS

Françoise CHATELIN, Professeur
Mathématiques Appliquées, IMAG
Tour des Mathématiques, B.P. 53 X
38041 GRENOBLE cédex (France)

INTRODUCTION

A rich variety of problems in Natural, Engineering and Social Sciences yield a large eigenvalue problem to be solved, either $A\phi = \lambda\phi$, or more often $A\phi = \lambda B\phi$, where A and B are sparse. Structured, often symmetric matrices commonly arise from finite difference and finite element discretizations of continuous problems. And unstructured, unsymmetric matrices arise in Social Sciences. We cite, for example, a Markov chain model of a queueing system and the dynamic stability of macroeconomic models. We give a review of current iterative numerical methods to compute some of the eigenvalues of $A\phi = \lambda\phi$, when A is large, unstructured, and in general unsymmetric. As a rule, these methods only require us to perform matrix by vector multiplications of the type Ax. All these methods amount to a projection on subspaces. Depending on the choice of the subspaces we get the simultaneous iteration method (section 2), and the Arnoldi method (section 5). When A is hermitian, the Arnoldi method reduces to the Lanczos method (section 3). We give an analysis of the rate of convergence of these methods, which is based on estimates of the acute angle between ϕ, the eigenvector to be computed, and the considered subspace. This presentation of the convergence analysis of the methods is borrowed from Saad (1980, a,b).

1. PRINCIPLE OF THE METHODS

The idea is to approximate the eigenelements of A, of order N, by those of a matrix of much smaller order ν, obtained by *orthogonal* projection on a ν-dimensional subspace X_n of $X = \mathbb{C}^N$. Let π_n be the matrix of the orthogonal projection of X on-to X_n. $A\phi = \lambda\phi$ is approximated in X_n by :

$$\pi_n A \phi_n = \lambda_n \phi_n \quad , \quad 0 \neq \phi_n \in X_n .$$

λ_n, ϕ_n are therefore eigenelements of $A_n = \pi_n A \pi_n$, the *Galerkin* approximation of A, associated with the subspace X_n. We set $\hat{A}_n := A_n : X_n \to X_n$.

The computational scheme consists in generating an orthonormal basis for X_n, along with the $\nu \times \nu$ matrix B_n wich represents $\hat{A}_n$ in this basis of X_n. We are then led to an eigenvalue problem in $\mathbb{C}^\nu$: $B_n \xi_n = \lambda_n \xi_n$, $0 \neq \xi_n \in \mathbb{C}^\nu$, the vector ξ_n represents the components of ϕ_n in the basis of X_n.

$\|\cdot\|_2$ denotes the euclidean norm on $\mathbb{C}^N$.

Given an eigenvalue λ, with associated eigenvector ϕ, $\|\phi\|_2 = 1$, we wish to know if there exists a sequence of eigenelements λ_n, ϕ_n of A_n, which would converge rapidly to λ, ϕ when n increases. When λ is simple, the quantities $|\lambda - \lambda_n|$ and

$\|\phi-\phi_n\|_2$ may be bounded in terms of the residual for A_n : $u_n = A_n\phi-\lambda\phi = (A_n-A)\phi$. $(A_n-A)\phi = \pi_n A(\pi_n-I)\phi + (\pi_n-I)A\phi$, therefore $\|u_n\|_2 \leq (\|A\|+|\lambda|)\|(I-\pi_n)\phi\|_2$, since $\|\pi_n\|_2 = 1$. The bounds of $|\lambda-\lambda_n|$ and $\|\phi-\phi_n\|_2$ in terms of $\|(I-\pi_n)\phi\|_2$ will be explicited for each particular method. When λ is of algebraic multiplicity m , the analysis is more delicate and the exponent $1/\ell$ appears , where ℓ is the ascent of λ : $1 \leq \ell \leq m$ and $\text{Ker}\,(A-\lambda I)^\ell \equiv \text{Ker}\,(A-\lambda I)^m$.

$\alpha_n := \|(I-\pi_n)\phi\|_2$ represents the distance between ϕ and the subspace X_n , $\alpha_n = \sin\theta(\phi,X_n)$, where $\theta(\phi,X_n)$ is the acute angle between ϕ and X_n. With appropriate choices of X_n , there exists one or several eigenvectors ϕ of A such that α_n is small And the rates of convergence for the eigenelements will be derived from *estimates of* α_n .

Remark : If A is hermitian, so is A_n (and B_n) because π_n is an orthogonal projection. λ_n is the Rayleigh quotient of A , based on ϕ_n , $\|\phi_n\|_2 = 1$. Indeed : $\phi_n^H A \phi_n = \phi_n^H A_n \phi_n = \lambda_n$, since $\pi_n \phi_n = \phi_n$. $|\lambda-\lambda_n|$ is then of the order of $(\|A\phi_n - \lambda_n\phi_n\|_2)^2$, by the Kato-Temple inequality , λ being multiple or simple.

In practice the subspaces X_n are generated from Krylov vectors obtained from a given starting vector x , or from a set of r vectors $\{x_i\}_1^r$. Let U be the span of $x_1,x_2,\ldots,x_r$. We present an analysis of the rate of convergence of the two main classes of methods derived from the following choices of X_n :

a) $X_n = A^n U$, for n=1,2,... , $\dim X_n = r$ is constant throughout the iterations. This choice yields the simultaneous iteration method for $r > 1$ and the power method for r=1.

b) $X_n = \{U,AU,\ldots,A^{n-1}U\}$, for n=1,2,...,N , $\dim X_n = nr$ increases over the iterations. This choice corresponds to the block-Arnoldi method. If A is hermitian, the method reduces to the block-Lanczos method. In practice, the method is *incomplete*, which means that n is kept much smaller than N .

2. THE SIMULTANEOUS ITERATION METHOD

We suppose that the N repeated eigenvalues of A (counted with their multiplicities) are such that : $|\mu_1| \geq \cdots \geq |\mu_r| > |\mu_{r+1}| \geq \cdots \geq |\mu_N|$: (*) ; $\{\phi_i\}_1^N$ is the associated Jordan basis.

The first r eigenvalues are the r *dominant* eigenvalues. The associated invariant subspace, $M = \{\phi_1,\ldots,\phi_r\}$ is the *dominant* invariant subspace. dim M = r , and M is unique by assumption(*). M is the direct sum of the invariant subspaces associated with the dominant distinct eigenvalues. Let P be the associated spectral projection. P is a projection on M , along S : $\mathbb{C}^N = M \oplus S$.

2.1. The algorithm

Let U be the span of r *orthonormal* vectors $\{x_k\}_1^r$ and let Q_o be the $N \times r$ matrix $Q_o := (x_1,\ldots,x_r)$. The simultaneous iteration algorithm produces a sequence of matrices Q_n , whose columns are an orthonormal basis of $X_n = A^n U$ in the following way:

for $n \geq 1$:

1. Compute the $N \times r$ matrix $Y_n := A\,Q_{n-1}$
2. Orthonormalize the column vectors of Y_n :

   $Y_n = Q_n R_n$, that is $Q_n := Y_n \cdot R_n^{-1}$, where R_n is an $r \times r$ upper triangular matrix.

The $r \times r$ matrix $B_n := Q_n^H A\,Q_n$ represents $\hat{A}_n$ in the orthonormal basis of X_n given by the columns of Q_n . If $r = 1$, this is the power method with the starting vector $x_1 = x$.

2.2. Convergence

Under the assumption (*), it may be proved that $X_n = A^n U$ tends to M as $n \to \infty$, if and only if the r vectors $\{Px_k\}_1^r$ are independant (cf. Parlett-Poole (1973)). This means that $Q_n \to Q$, an orthonormal basis of M , and $B_n \to B := Q^H A\,Q$, which represents the matrix $A_{\restriction M}$ in the basis Q of M . The eigenelements of B_n converge to the eigenelements of $A_{\restriction M}$.

If $r = 1$, the condition $\{Px_i\}_1^r$ independant reduces to $\Psi_1^H \cdot x_1 \neq 0$ where Ψ_1 is the left eigenvector of A associated with μ_1, $|\mu_1| > |\mu_2|$.

2.3. Estimate of $\|(I-\pi_n)\phi_i\|_2$

The vectors $\phi_1,\dots,\phi_r$ associated with $\mu_1,\dots,\mu_r$ are either eigenvectors or generalized eigenvectors. We suppose that ϕ_i is an *eigenvector* of A .

Lemma 1. *If the* r *vectors* $\{Px_k\}_1^r$ *are independant, then for any* ϕ_i , $1 \leq i \leq r$, *which is an eigenvector of* A , *there exists a unique* $u_i \in U$ *such that* $Pu_i = \phi_i$ *and*

$$\|(I-\pi_n)\phi_i\|_2 \leq \|\phi_i - u_i\|_2 \cdot \left(\left|\frac{\mu_{r+1}}{\mu_i}\right| + \varepsilon_n\right)^n \quad \textit{where } \varepsilon_n \to 0 \textit{ as } n \to \infty .$$

Proof : Any $u \in U$ may be written $u = \sum_{k=1}^{r} t_k x_k$. Then $Pu = \sum_{k=1}^{r} t_k Px_k$. Given $\phi_i \in M$ there exists a unique $u_i \in U$ such that $Pu_i = \phi_i$, because of the independance of the $\{Px_k\}_1^r$. Then u_i may be written $u_i = \phi_i + v_i$, where $v_i = (I-P)u_i \in S$.

By definition $\|(I-\pi_n)\phi_i\|_2 = \min_{x \in X_n} \|\phi_i - x\|_2 \leq \|\phi_i - \hat{x}_i\|_2$, where $\hat{x}_i = \frac{1}{\mu_i^n} A^n u_i = \phi_i + \frac{1}{\mu_i^n} A^n v_i$, because ϕ_i is the eigenvector associated with μ_i .

If A is not diagonalisable $\left\|\frac{1}{\mu_i^n} A^n v_i\right\|_2 \leq |\mu_{r+1}/\mu_i|^n \cdot \|v_i\|_2$.

If A is not diagonalizable, we set $B := \frac{1}{\mu_i} A_{\restriction S}$; $r_\sigma(B) = |\mu_{r+1} / \mu_i|$. And

$$\|B^n v_i\|_2 \leq \|B^n\|_2 \cdot \|v_i\|_2 \leq (\|B^n\|_2^{1/n})^n \cdot \|v_i\|_2 \leq (r_\sigma(B)+\varepsilon_n)^n \cdot \|v_i\|_2 , \text{ where } \varepsilon_n \to 0$$

as $n \to \infty$. $\|v_i\|_2 = \|\phi_i - u_i\|_2$. □

The constant $\|\phi_i - u_i\|_2$ expresses how good the initial subspace U is with respect to ϕ_i , cf. the case r = 1 on fig. 1. If A is hermitian, $\|v_i\|_2 = \tan \theta_i$ where $\theta_i = \theta(\phi_i, u_i)$ is the acute angle between the eigenvector ϕ_i and the corresponding vector u_i in U .

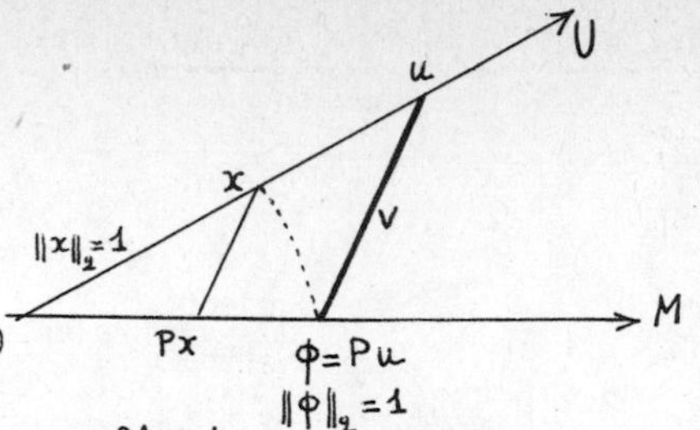

fig. 1

2.4. Rate of convergence.

Let λ (resp. λ_n) be a *simple* eigenvalue of A (resp. A_n), with associated eigenvector ϕ , (resp. ϕ_n) $\|\phi\|_2 = \|\phi_n\|_2 = 1$. λ is chosen among the r dominant eigenvalues of A . k is a generic constant.

Lemma 2. *Given* λ , ϕ , *then for* n *large enough, there exist* λ_n , ϕ_n *such that* $|\lambda-\lambda_n| \leq k\,\alpha_n$, $\|\phi-\phi_n\|_2 \leq k\,\alpha_n$. *If* A *is hermitian, then* $|\lambda-\lambda_n| \leq k\,\alpha_n^2$.

Proof. Let U be such that $\tilde{Q} := (\phi, U)$ is unitary. We consider $\tilde{Q}^H A_n \tilde{Q}$ and set $\rho_n := \phi^H A_n \phi = \lambda - \phi^H u_n$ with $u_n = (A-A_n)\phi, \|u_n\|_2 \leq k\,\alpha_n$. Let $C_n := U^H A_n U$; when $n \to \infty$, $\|(C_n-\rho_n I)^{-1}\|_2 = \|(B_n-\rho_n I)^{-1}\|_2 \to \|(B-\lambda I)^{-1}\|_2 < \infty$ because λ is a simple eigenvalue of A . $\|A_n^H \phi\| \leq \|A\|_2$; $|\lambda-\rho_n| \leq \|u_n\|_2$ and $\|A_n\phi-\rho_n\phi\|_2 \leq 2\|u_n\|_2$. For α_n small enough, there exists λ_n, ϕ_n such that $|\lambda_n-\rho_n| \leq 2\|(C_n-\rho_n I)^{-1}\|_2 \cdot \|A_n\phi-\rho_n\phi\|_2 \cdot \|A\|_2$ and $\|\phi-\phi_n\|_2 \leq 2\|(C_n-\rho_n I)^{-1}\|_2 \|A_n\phi-\rho_n\phi\|_2$, (Cf. Chatelin (1981)). It follows that, for n large enough, $|\lambda-\lambda_n| \leq k\,\alpha_n$ and $\|\phi-\phi_n\|_2 \leq k\,\alpha_n$.

If A is hermitian, $|\lambda-\lambda_n| \leq k\|(A-A_n)\phi_n\|_2^2$, and from $(A-A_n)\phi_n =$

$= \lambda(I-\pi_n)\phi + (I-\pi_n)A(\phi_n-\phi)$, follows that $|\lambda-\lambda_n| \leq k\,\alpha_n^2$, λ being possibly multiple □

Theorem 3. *If* $|\mu_r| > |\mu_{r+1}|$ *and if* $\{Px_k\}_1^r$ *are independant, the simultaneous iteration method on* r *vectors is convergent. If moreover, the* i^{th} *dominant eigenvalue is simple, the rate of convergence of the* i^{th} *eigenpair is of the order of* $\left|\frac{\mu_{r+1}}{\mu_i}\right|$, i=1,...,r . *If* A *is hermitian, the rate of convergence for the* i^{th} *eigenvalue is squarred to* $\left|\frac{\mu_{r+1}}{\mu_i}\right|^2$.

Proof. It is a straight-forward application of lemma 1 and 2. Note that when A is hermitian, the hypothesis that the ith eigenvalue is simple may be dropped. □

3. THE LANCZOS METHOD FOR A HERMITIAN MATRIX

3.1. The algorithm

We suppose that A is hermitian. Given $x \neq 0$, the sequence $\{X_n\}$ is now defined by $X_n = \{x, Ax, \ldots, A^{n-1}x\}$ for $n=1,2,\ldots,\ \nu < N$. The Lanczos method provides a simple way of realizing the Ritz projection of A on X_n. If exact computation were performed, the Lanczos algorithm generates *iteratively* an orthonormal basis $\{v_i\}_1^n$ of X_n, in which $\hat{A}_n$ is represented by a *tridiagonal* matrix T_n :

- $v_1 := x / \|x\|_2$, $a_1 := v_1^H \cdot Av_1$, $b_1 := 0$
- for $j=1,2,\ldots,n-1$ do :

$$x_{j+1} := A\,v_j - a_j\,v_j - b_j\,v_{j-1}\ ,\ b_{j+1} := \|x_{j+1}\|_2\ ,$$

$$v_{j+1} := x_{j+1} / \|x_{j+1}\|_2\ ,\ a_{j+1} := v_{j+1}^H \cdot A\,v_{j+1}\ .$$

T_n is an $n \times n$ tridiagonal hermitian matrix with diagonal elements a_i , $i=1,\ldots,n$ and sub-diagonal elements b_i , $i=2,\ldots,n$.

This algorithm was proposed by Lanczos as a means for tridiagonalizing a hermitian matrix, by performing N steps ($X_N = X$). But the numerical loss of orthogonality caused it to be abandoned for the Householder method. However for large matrices, the use of the above incomplete tridiagonalization (n steps) has given a revival to this old method.

This algorithm is feasible if $x_j \neq 0$ for $j=2,\ldots,n$. This is achieved if the starting vector x is such that its annihilating polynomial is of degree $\geq n$. T_n is then an unreduced tridiagonal matrix, and, as a consequence, all its eigenvalues are *simple*. The eigenelements of A_n are often referred to as the *Ritz values* and *vectors* of A.

Let $\{\lambda_i\}_1^K$ be the K *distinct* (possibly multiple) eigenvalues of A. For $1 \leq k \leq n < N$, we are interested in the k largest positive or smallest negative eigenvalues of A.

The positive eigenvalues of A (resp. T_n) are ordered by decreasing magnitude : $\lambda_1 > \lambda_2 > \ldots > \lambda_K$ (resp. $\lambda_1^n > \lambda_2^n > \ldots > \lambda_n^n$) . Similarly the negative eigenvalues are ordered by increasing magnitude.

For $i=1,\ldots,K$, P_i is the orthogonal eigenprojection associated with λ_i , ϕ_i^n is the eigenvector of A_n associated with λ_i^n , P_i^n is the corresponding eigenprojection. The estimate of $\|(I-\pi_n)\phi_i\|_2 = \sin\theta(\phi_i, X_n)$ will be given below through bounds on $\tan\theta(\phi_i, X_n)$.

3.2. Estimate of $\tan\theta(\phi_i, X_n)$

$\mathbb{P}_{n-1}$ is the set of polynomials of degree $\leq n-1$.

We suppose that $P_i x \neq 0$, and set $\phi_i := \dfrac{P_i x}{\|P_i x\|_2}$; $\hat{x}_i := \dfrac{(I-P_i)x}{\|(I-P_i)x\|_2}$ if $(I-P_i)x \neq 0$, $\hat{x}_i := 0$ otherwise.

Lemma 4 : $\tan\theta(\phi_i,X_n) = [\min_{\substack{p\in\mathbb{P}_{n-1}\\ p(\lambda_i)=1}} \|p(A)\hat{x}_i\|_2].\tan\theta(\phi_i,x)$

Proof : Because $X_n = \{x,Ax,\ldots,A^{n-1}x\}$, any $u \in X_n$ may be written :

$u = q(A)x$, $q \in \mathbb{P}_{n-1}$. Now from $\sum_{j=1}^{K} P_i = I$ follows that $x = P_i x + \sum_{j\neq i} P_j x$,

$u = q(\lambda_i) P_i x + \sum_{j\neq i} q(\lambda_j) P_j x$.

$$\tan^2\theta(P_i x,u) = \frac{\sum_{j\neq i} q^2(\lambda_j)\|P_j x\|_2^2}{q^2(\lambda_i)\|P_i x\|_2^2}$$

If $(I-P_i)x \neq 0$, $\sum_{j\neq i} q^2(\lambda_j)\|P_j x\|_2^2 = \|q(A)\hat{x}_i\|_2^2 \cdot \|(I-P_i)x\|_2^2, \|\hat{x}_i\|_2 = 1$.

If $x = P_i x$, we set $\hat{x}_i = 0$: $\theta(P_i x,x) = 0$.

We define $p(t) = \frac{q(t)}{q(\lambda_i)}$: $p \in \mathbb{P}_{n-1}$ and $p(\lambda_i) = 1$.

$$\tan\theta(P_i x,X_n) = \min_{u\in X_n}\tan\theta(P_i x,u) = \underbrace{\min_{\substack{p\in\mathbb{P}_{n-1}\\ p(\lambda_i)=1}} \|p(A)\hat{x}_i\|_2}_{t_{in}} . \frac{\|(I-P_i)x\|_2}{\|P_i x\|_2}$$

and $\frac{\|(I-P_i)x\|_2}{\|P_i x\|_2} = \tan\theta(P_i x,x)$. □

We now define for $i \leq k$, $K_1 := 1$, $K_i := \prod_{j=1}^{i-1} \frac{\lambda_j-\lambda_K}{\lambda_j-\lambda_i}$, $i > 1$, and

$$\gamma_i := 1 + 2\,\frac{\lambda_i-\lambda_{i+1}}{\lambda_{i+1}-\lambda_K}$$

$C_m(t) = \frac{1}{2}[(t+\sqrt{t^2-1})^m + (t-\sqrt{t^2-1})^m]$ is the mth degree Chebyshev polynomial of the first kind in t .

Theorem 5 : *If* $P_i x \neq 0$, *then, with* $\phi_i = \frac{P_i x}{\|P_i x\|_2}$,

$$\tan\theta(\phi_i,X_n) \leq \frac{K_i}{C_{n-i}(\gamma_i)} \tan\theta(\phi_i,x) \text{ , for } i \leq k \text{ .}$$

Proof : We wish to bound t_{in} . We define $\beta_j = \|P_j \hat{x}_i\|_2$, $\beta_i = 0$, hence $\sum_{j\neq i}\beta_j^2 = 1$.

a) case $i = 1$. $t_{1n} = \min_{\substack{p\in\mathbb{P}_{n-1}\\ p(\lambda_1)=1}} [\sum_{j\neq 1}\beta_j^2 p^2(\lambda_j)]^{1/2}$

$$(\sum_{j\neq 1}\beta_j^2 p^2(\lambda_j))^{1/2} \leq (\sum_{j\neq 1} p^2(\lambda_j))^{1/2} \leq \max_{t\in[\lambda_K,\lambda_2]} |p(t)|$$

Now $\min\limits_{\substack{p\in \mathbb{P}_{n-1}\\ p(\lambda_1)=1}} \max\limits_{t\in[\lambda_K,\lambda_2]} |p(t)| = \dfrac{1}{C_{n-1}(\gamma_1)}$ with $\gamma_1 = 1 + 2\,\dfrac{\lambda_1-\lambda_2}{\lambda_2-\lambda_K}$ (cf. Cheney (1966)).

b) case $i > 1$. $\sum\limits_{j\neq i} \beta_j^2\, p^2(\lambda_j) \le \max\limits_{j\neq i} |p(\lambda_j)|^2$

$$\min_{\substack{p\in\mathbb{P}_{n-1}\\ p(\lambda_i)=1}} \max_{j\neq i} |p(\lambda_j)| \le \min_{\substack{p\in\mathbb{P}_{n-1}\\ p(\lambda_1)=\ldots=p(\lambda_{i-1})=0\\ p(\lambda_i)=1}} \max_{j\neq i} |p(\lambda_j)|$$

Now such a p may be decomposed : $p(t) = \left(\prod\limits_{\ell=1}^{i-1} \dfrac{t-\lambda_\ell}{\lambda_i-\lambda_\ell}\right) \dfrac{q(t)}{q(\lambda_i)}$ where $q \in \mathbb{P}_{n-i}$. Then

$$\max_{j>i} |p(\lambda_j)| = \max_{j>i} \left|\left(\sum_{\ell=1}^{i-1} \frac{\lambda_j-\lambda_\ell}{\lambda_i-\lambda_\ell}\right) \frac{q(\lambda_j)}{q(\lambda_i)}\right| \le \left(\sum_{\ell=1}^{i-1} \frac{\lambda_\ell-\lambda_K}{\lambda_\ell-\lambda_i}\right) \max_{j>i} \frac{|q(\lambda_j)|}{|q(\lambda_i)|}$$

Therefore $t_{in} \le K_i \cdot \min\limits_{\substack{q\in\mathbb{P}_{n-i}\\ q(\lambda_i)=1}} \max\limits_{t\in[\lambda_K,\lambda_{i+1}]} |q(t)| = \dfrac{K_i}{C_{n-i}(\gamma_i)}$. □

We define $\bar{t}_{in} := \dfrac{K_i}{C_{n-i}(\gamma_i)}$, $i \le k$.

Theorem 5 shows that $\theta(\phi_i, X_n)$ decreases at least as rapidly as $\bar{t}_{in}$. γ_i depends on the gap $\lambda_i-\lambda_{i+1}$, $\gamma_i > 1$, we define $\tau_i := \gamma_i + \sqrt{\gamma_i^2 - 1} > 1$. For n large enough $C_{n-i}(\gamma_i) \approx \frac{1}{2}\,\tau_i^{n-i}$: the rate of decrease of $\theta(\phi_i, X_n)$ is $\dfrac{1}{\tau_i}$.

Theorem 5 also indicates that for any eigenvalue λ_i , $i < k$, there exists at least one vector in X_n which is close to the eigenvector $\phi_i = \dfrac{P_i x}{\|P_i x\|_2}$. We show now that there is only one. This means that a multiple eigenvalue λ_i can be approximated by at most *one* simple eigenvalue λ_i^n .

Let E be the invariant subspace spanned by the K vectors $\{P_i x\}_1^K$ that we suppose non zero. dim E = K . Let A' be the matrix representing $A_{\restriction E}$ in an orthonormal basis of E .

Proposition 6 . *The Lanczos process amounts to approximating the eigenelements of* A' *whose eigenvalues are simple.*

Proof : Let $x = \sum\limits_{i=1}^{K} P_i\, x$. A is hermitian, then $A = \sum\limits_{i=1}^{K} \lambda_i\, P_i$ and we get

$A^k x = \sum\limits_{i=1}^{K} \lambda_i^k\, P_i x$, for $k = 1,\ldots,n-1$. Therefore $X_n \subseteq E$ for all n . Thus the Lanczos method applied to A or A' yields the same matrices $\hat{A}_n$ and T_n .

$A'P_i x = AP_i x = \lambda_i P_i x$: λ_i is an eigenvalue of A' corresponding to the eigenvector $P_i x \neq 0$. A' which is of order K , has then K distinct eigenvalues which have to be simple. □

3.3. Rate of "convergence"

Because n takes a finite number of values, we cannot, rigorously speaking, talk about the convergence of the method. But $\lambda_i - \lambda_i^n$ and $\|\phi_i - \phi_i^n\|_2$ will be bounded in the proposition to follow by means of $\beta_{in} := \tan(\phi_i, x) \cdot \bar{t}_{in}$, the bound of $\tan(\phi_i, X_n)$. This gives the accuracy of the Lanczos method as function of n and x.

We set $K_1^n := 1$, $K_i^n := \prod_{j=1}^{i-1} \frac{\lambda_j^n - \lambda_K}{\lambda_j^n - \lambda_i}$ for $1 < i \leq k$ (defined if $\lambda_{i-1}^n > \lambda_i$) ,

$d_{in} := \min_{j \neq i} |\lambda_i - \lambda_j^n|$.

Lemma 7 : *If* $P_i x \neq 0$, *then* : $0 \leq \lambda_i - \lambda_i^n \leq (\lambda_i - \lambda_K)\left(\frac{K_i^n}{K_i}\right)^2 \beta_{in}^2$

$$\|(I-\pi_n)\phi_i\|_2 \leq \|(I-P_i^n)\phi_i\|_2 \leq \sqrt{1 + \frac{\|A\|_2^2}{d_{in}^2}} \cdot \beta_{in}$$

Proof : It is based on the minimax characterization of the eigenvalues, cf. Saad (1980 a). □

Proposition 8 : *If* $P_i x \neq 0$ *then* $0 \leq \lambda_i - \lambda_i^n \leq k\, \beta_{in}^2$ *and* $\|(I-P_i^n)\phi_i\|_2 \leq k\ \beta_{in}$, *where* k *is a generic constant.*

Proof : The constants in lemma 7 can be bounded independently of n if β_i^n is small enough. For the eigenvalues, this may be done by induction on i : for i=1 , $K_1^n = K_1 = 1$, then $\lambda_1^n \to \lambda$. Now assuming that $\lambda_j^n \to \lambda_j$ for j=1,...,i-1 implies that $\lambda_{i-1}^n > \lambda_i$ and $K_i^n \to K_i$, $\lambda_i^n \to \lambda_i$. As for the eigenvectors, if $\lambda_j^n \to \lambda_j$ for j=1,...,i+1 , $d_{in} \to d_i := \min_{j \neq i} |\lambda_i - \lambda_j|$. □

The above bounds may be weak in the case where λ_i is close to λ_{i+1} . The bounds may be sharpened by taking advantage of the structure of the spectrum, that is by choosing more appropriately the polynomials used in the proof of theorem 5 (cf. Saad, 1980 a).

The above theorem shows that only a few extreme eigenvalues are computed to a reasonable accuracy by this basic method. Several variants of the method have been designed to :

- avoid full reorthogonalization (Kahan-Parlett (1976), Parlett-Scott (1979))
- compute interior eigenvalues, by performing n > N iterations (Cullum-Willoughby (1978), Parlett-Reid (1980)).

The computation of the eigenvectors normally requires to store all the vectors $\{v_i\}_1^n$. A variant is given in Saad (1978) which requires only the storage of 5 vectors, instead of n .

Another way to cope with a cluster of dominant eigenvalues or a multiple eigenvalue is to use the block-Lanczos method.

4. THE BLOCK-LANCZOS METHOD

4.1. The algorithm

Given a set of r orthonormal vectors $\{x_i\}_1^r$, let U be the span $U = \{x_1,\dots,x_r\}$ and $X_n = \{U, AU, \dots, A^{n-1}U\}$. The block-Lanczos algorithm realizes a projection on X_n in the following way. Let Q_o be the $N \times r$ matrix $Q_o := (x_1,\dots,x_r)$, the algorithm produces a sequence of orthonormal $N \times r$ matrices Q_j , $j=1,\dots,n-1$ such that the columns of $Q_o, Q_1,\dots,Q_{n-1}$ are an orthonormal basis of X_n , in which $\hat{A}_n$ is represented by a *block-triangular* matrix, the blocks being $r \times r$:

- $\tilde{A}_1 := Q_o^H A Q_o$; $\tilde{B}_1 := 0$
- for $j=1,2,\dots,n-1$ do : $D_j := A Q_{j-1} - Q_{j-1} \tilde{A}_j - Q_{j-2} \tilde{B}_j^H$

perform the orthonormalization of D_j : $D_j := Q_j R_j$, where R_j is an $r \times r$ regular triangular matrix, and set $\tilde{B}_{j+1} := R_j$, $\tilde{A}_{j+1} := Q_j^H A Q_j$.

If $\dim X_n = n \times r$, the algorithm is feasible (R_j regular for $j = 1,\dots,n-1$). This condition is satisfied if the starting vectors $\{x_k\}_1^r$ are such that $\sum_{k=1}^{r} p_k(A)\, x_k \neq 0$

for all $p_k \in \mathbb{P}_{n-1}$, $k=1,\dots,r$. A result similar to proposition 6 can be proved : the block-triangular matrix has eigenvalues of multiplicities not larger than r , and we can assume, without loss of generality, that the eigenvalues of A are of multiplicities *not larger than* r . Let $\{\mu_i\}_1^N$ be the *repeated* eigenvalues of A, the associated eigenvectors (resp. eigenprojections) are $\{\phi_i\}_1^N$ (resp. $\{P_i\}_1^N$). Notice the difference with the notations of § 3.1. We suppose that

$$\mu_1 \geq \dots \geq \mu_{i-1} > \mu_i \geq \mu_{i+1} \geq \dots \geq \mu_{i+r-1} > \mu_{i+r} \geq \dots \geq \mu_N .$$

Let I be the set of indices $\{i, i+1, \dots, i+r-1\}$, $P := \sum_{j \in I} P_j$.

The rates of convergence can be studied again by means of an estimate of $\tan \theta(\phi_\ell, X_n)$, $\ell \in I$. We define for $\ell \in I$,

$\hat{\gamma}_\ell := 1 + 2 \dfrac{\mu_\ell - \mu_{i+r}}{\mu_{i+r} - \mu_N}$, K_i as introduced in theorem 5 is well defined since $\mu_{i-1} > \mu_i$.

4.2. Estimate of $\tan \theta(\phi_\ell, X_n)$

Lemma 9 : *If the* r *vectors* $\{P x_k\}_1^r$ *are independant, then given* ϕ_ℓ , $\ell \in I$, *there exists a unique* $u_\ell \in U$ *such that* $P u_\ell = \phi_\ell$.

Proof : $u \in U$ may be written $u = \sum_{k=1}^{r} t_k x_k$, $P u = \sum_{k=1}^{r} t_k P x_k$. The existence and uniqueness of u_ℓ follows from the independance of $\{P x_k\}_1^r$. We set $v_\ell := (I-P)u_\ell$: $u_\ell = \phi_\ell + v_\ell$ and $\|\phi_\ell - u_\ell\|_2 = \tan \theta(\phi_\ell, u_\ell)$. □

Theorem 10 : *We suppose that* $\{P\ x_k\}_1^r$ *are independant, and* $\mu_{i-1} > \mu_\ell > \mu_{i+r}$ for $\ell \in I = \{i,i+1,\dots,i+r-1\}$.

$$\tan\theta(\phi_\ell,X_n) \le \frac{K_i}{C_{n-i}(\hat{\gamma}_\ell)} \tan\theta(\phi_\ell,u_\ell)\ , \quad \textit{for}\ \ \ell \in I\ .$$

Proof : Given ϕ_ℓ, we write $u_\ell = \phi_\ell + \sum_{j\notin I} P_j\, u_\ell$. We consider $u \in X_n$ of the form $u = q(A)u_\ell$, with $q \in \mathbb{P}_{n-1}$. Then :

$u = q(\mu_\ell)\,\phi_\ell + \sum_{j\notin I} q(\mu_j)\, P_j\, u_\ell$.

a) case i=1 : $\dfrac{\|(I-P_1)u\|_2^2}{\|P_1\ u\|_2^2} = \sum_{j\ge 1+r} \dfrac{q^2(\mu_j)\|P_j\ u_1\|_2^2}{q^2(\mu_1)}$. The minimum of the right-hand side over $q \in \mathbb{P}_{n-1}$ is achieved for $p \in \mathbb{P}_{n-1}$.

We set $\bar{u} = p(A)u_1 \in X_n$, $\alpha_1 := \dfrac{2}{\mu_{1+r}-\mu_N}$, $\beta_1 := \dfrac{\mu_{1+r}+\mu_N}{\mu_{1+r}-\mu_N}$.

Then for $j \ge 1+r$, $\alpha_1\,\mu_j - \beta_1 = 1 - 2\,\dfrac{\mu_{1+r}-\mu_j}{\mu_{1+r}-\mu_N} = \theta_j$, $|\theta_j| \le 1$ and $|C_{n-1}(\theta_j)| \le 1$ because $|C_{n-1}(t)| \le 1$ on $[-1,1]$.

$$\tan^2\theta(\phi_1,X_n) \le \frac{\|(I-P_1)\bar{u}\|_2^2}{\|P_1\ \bar{u}\|_2^2} = \sum_{j\ge 1+r} \frac{p^2(\mu_j)\|P_j\ u_1\|_2^2}{p^2(\mu_1)} \le \sum_{j\ge 1+r} \frac{C_{n-1}^2(\alpha_1\mu_j-\beta_1)\|P_j u_1\|_2^2}{C_{n-1}^2(\alpha_1\mu_1-\beta_1)} \le$$

$$\le \sum_{j\ge 1+r} \frac{\|P_j u_1\|_2^2}{C_{n-1}^2(\hat{\gamma}_1)}\ , \text{ and } \sum_{j\ge 1+r} \|P_j u_1\|_2^2 = \|(I-P)u_1\|_2^2 = \|\phi_1-u_1\|_2^2\ .$$

b) case i > 1 : we now set $\alpha_i := \dfrac{2}{\mu_{i+r}-\mu_N}$, $\beta_i := \dfrac{\mu_{i+r}+\mu_N}{\mu_{i+r}-\mu_N}$, then $\alpha_i\mu_\ell - \beta_i = \hat{\gamma}_\ell$.

We define $p_i(t) := [\prod_{j=1}^{i-1}(t-\mu_j)]\cdot C_{n-i}(\alpha_i t-\beta_i)$, $\bar{u} = p_i(A)u_\ell \in X_n$, with $p_i(\mu_j) = 0$ for $j=1,\dots,i-1$.

$$\tan^2\theta(\phi_\ell,X_n) \le \frac{\|(I-P_\ell)\bar{u}\|_2^2}{\|P_\ell \bar{u}\|_2^2} \le \sum_{j\ge i+r} \frac{p_i^2(\mu_j)\|P_j\ u_\ell\|_2^2}{p_i^2(\mu_\ell)} \le$$

$$\le \left(\prod_{j=1}^{i-1} \frac{\mu_j-\mu_N}{\mu_j-\mu_i}\right) \frac{\sum_{j\ge i+r}\|P_j\ u_\ell\|_2^2}{C_{n-i}^2(\hat{\gamma}_\ell)} \le \frac{K_i^2}{C_{n-i}^2(\hat{\gamma}_\ell)}\ \|\phi_\ell-u_\ell\|_2^2\ . \ \square$$

Remark : The above bound reduces to that of theorem 5 if $r = 1$. $\theta(\phi_\ell, X_n)$ decreases at least as rapidly as $1/C_{n-i}(\hat{\gamma}_\ell)$. This quantity depends on the gap $\mu_\ell - \mu_{i+r}$. The extension from the Lanczos to the block-Lanczos method is, in many respects, similar to the extension from the power method to the simultaneous iteration method.

From the above estimate of $\tan\theta(\phi_\ell, X_n)$, one may derive the rates of convergence of $|\mu_\ell - \mu_\ell^n|$ and $\|\phi_\ell - \phi_\ell^n\|_2$ in a manner similar to that of proposition 8 (cf. Saad 1980a). Different approaches may be found in Cullum-Donath (1974) and Underwood (1975).

5. The Arnoldi method for a non-hermitian matrix

We now describe a method which generalizes the Lanczos method to non-hermitian matrices. The biorthogonalization Lanczos algorithm is numerically unstable and does not possess, as we shall see, the convergence property which makes the Lanczos method on a hermitian matrix, behave as a rapidly converging iterative method. The simplest generalization of the Lanczos method is by means of the Arnoldi algorithm, cf. Wilkinson (1965).

5.1. The algorithm

Consider again $X_n = \{x, Ax, \dots, A^{n-1}x\}$, the Arnoldi algorithm computes *iteratively* an orthonormal basis $\{v_i^n\}$ of X_n in which $\hat{A}_n$ is represented by an *upper-Hessenberg* matrix $H_n = (h_{ij})$, $h_{ij} = 0$ for $i > j+1$.

$$v_1 := x/\|x\|_2 \ , \ h_{11} := v_1^H \cdot A\, v_1 \ , \ h_{21} := \|x\|_2 \ ,$$

for $j=1,\dots,n-1$ do :

$$x_{j+1} := A\, v_j - \sum_{i=1}^{j} h_{ij}\, v_i \quad , \quad h_{j+1,j} := \|x_{j+1}\|_2 \ ,$$

$$v_{j+1} := x_{j+1} / \|x_{j+1}\|_2 \ , \ h_{ij+1} := v_i^H \cdot A\, v_{j+1} \ \text{for } i \le j+1 \ .$$

It is feasible if $x_j \neq 0$ for $j=1,\dots,n-1$, that is $p(A)x \neq 0$ for all $p \in \mathbb{P}_{n-1}$. If this condition is satisfied, H_n is an unreduced Hessenberg matrix : $h_{j+1\ j} \neq 0$ for $j=1,\dots,n-1$. Therefore if A is diagonalizable, so are A_n and H_n, and H_n has n simple eigenvalues.

5.2. "Convergence"

Let $\{\lambda_i\}_1^K$ be the K *distinct* eigenvalues of A , $\{P_i\}_1^K$ are the associated spectral projections. If A is *diagonalizable*, $A = \sum_{i=1}^{K} \lambda_i P_i$, and propostion 6 holds for the Arnoldi process. We may then suppose , without loss of generality, that the N eigenvalues of A are *simple*.

Given the eigenpair (λ_n, ϕ_n) for A_n , $\|\phi_n\|_2 = 1$, we consider the residual $v_n = (A-\lambda_n)\phi_n = (I-\pi_n)A\phi_n$. ϕ, $\|\phi\|_2 = 1$ is an eigenvector of A, $\alpha_n = \|(I-\pi_n)\phi\|_2$.

Lemma 11 : *Given* λ_n, ϕ_n and ϕ , $\|v_n\|_2 \le k(n)\alpha_n$, $\|\phi-\phi_n\|_2 \le k(\|v_n\|_2 + \alpha_n)$, *and* $|\lambda-\lambda_n| \le k\|\phi-\phi_n\|_2$, *where* k *is a generic constant,* $k(n)$ *depends on* n .

Proof : If $\phi_n \in X_{n-1}$, $A\phi_n \in X_n$ and $(I-\pi_n)A\phi_n = 0$, that is $A\phi_n = \lambda_n\phi_n$, (λ_n,ϕ_n) is an eigenpair for A. We suppose that $\phi_n \notin X_{n-1}$. By hypothesis $X_{n-1} \subset X_n$ strictly and $\pi_n\phi \notin X_{n-1}$, unless $\phi \in X_{n-1}$, then $\pi_n\phi = \xi_n\phi_n + v_n$, where $v_n \in X_{n-1}$, and $\xi_n \neq 0$. $(I-\pi_n)A\phi_n = \frac{1}{\xi_n}(I-\pi_n)A\pi_n\phi$, because $(I-\pi_n)Av_n = 0$, and $(I-\pi_n)A\phi_n = \frac{1}{\xi_n}(I-\pi_n)(\lambda I-A)(I-\pi_n)\phi$. Then $\|v_n\|_2 \le \frac{1}{|\xi_n|}\|A-\lambda I\|_2\cdot\alpha_n = k(n)\alpha_n$.

$(I-\pi_n)A\phi_n = \lambda(I-\pi_n)\phi + (I-\pi_n)A(\phi_n-\phi)$, and

$\|\phi_n-\phi\|_2 \le k(\|v_n\|_2+\alpha_n)$. $\lambda-\lambda_n = \phi^HA\phi - \phi_n^HA\phi_n = (\phi-\phi_n)^HA\phi + \phi_n^HA(\phi-\phi_n)$,

then $|\lambda-\lambda_n| \le k\|\phi-\phi_n\|_2$. □

For n fixed such that $\xi_n \neq 0$ then $|\lambda-\lambda_n|$ and $\|\phi-\phi_n\|_2$ are of the order of α_n .
For α_n small enough, $\|\phi_n-\pi_n\phi\|_2 \to 0$ and $\xi_n \to 1$.

5.3. Estimate of $\|(I-\pi_n)\phi_i\|_2$

$x = \sum_1^K P_i x$. If $P_i x \neq 0$, we define $C_i := \dfrac{\sum_{j\neq i}\|P_jx\|_2}{\|P_i x\|_2} \ge \dfrac{\|(I-P_i)x\|_2}{\|P_i x\|_2}$

$\varepsilon_i^{(n)} := \min\limits_{\substack{p\in\mathbb{P}_{n-1} \\ p(\lambda_i)=1}} \max\limits_{j\neq i} |p(\lambda_j)|$ is the degree of approximation of the null function on the set $\{\lambda_j\}_1^K$, $j\neq i$, by polynomials p of degree $\le$ n-1 satisfying $p(\lambda_i) = 1$ (cf. Lorentz (1966)).

Theorem 12 : *If* A *is diagonalizable and* $P_ix \neq 0$ *then* $\|(I-\pi_n)\phi_i\|_2 \le C_i\,\varepsilon_i^{(n)}$, *with* $\phi_i = \dfrac{P_ix}{\|P_ix\|_2}$.

Proof : Any $u \in X_n$ may be written $u = q(A)x$, $q \in \mathbb{P}_{n-1}$.

$x = P_ix + \sum_{j\neq i} P_jx$, $u = q(\lambda_i)P_ix + \sum_{j\neq i} q(\lambda_j)P_jx$.

For $p \in \mathbb{P}_{n-1}$ such that $p(\lambda_i) = 1$, $\dfrac{1}{\|P_ix\|_2}u - \phi_i = \sum_{j\neq i} p(\lambda_j)\dfrac{P_jx}{\|P_ix\|_2}$.

$\min\limits_{u\in X_n}\|u-\phi_i\| \le \|\sum_{j\neq i}\dfrac{P_jx}{\|P_ix\|_2}p(\lambda_j)\|_2 \le \max\limits_{j\neq i}|p(\lambda_j)|\,(\sum_{j\neq i}\|P_jx\|_2)\,/\,\|P_ix\|_2$. □

C_i expresses how good x is, with respect to ϕ_i .

The convergence analysis reduces to the estimation of $\varepsilon_i^{(n)}$.
This is a difficult problem of approximation theory in the complex variable. Except for some particular shapes of spectra, such as purely real or almost real spectra, it is not easy to establish bounds on $\varepsilon_i^{(n)}$ which are both sharp and simple. For example, with the notations of theorem 5 we have the :

<u>Proposition 13</u> : *We suppose that* A *is diagonalizable with real eigenvalues* : $\lambda_1 > \lambda_2 > \ldots > \lambda_K$, *then* $\varepsilon_i^{(n)} \leq \dfrac{K_i}{C_{n-i}(\gamma_i)}$.

The proof is analogous to that of theorem 5. A similar result holds if the spectrum lies on a straight line of the complex plane.

But if we only know that the dominant eigenvalue λ_1 is real and simple : $|\lambda_1| > |\lambda_i|$, $i \geq 2$, it may be shown that $\varepsilon_1^{(n)} \leq \left|\dfrac{\lambda_2}{\lambda_1}\right|^{n-1}$, by considering the disk $(0,|\lambda_2|)$. This is not better than the power method ! If the spectrum is almost real, elliptic domains may be used (cf. Saad (1979)). Nevertheless, these theoretical bounds seem to be much too pessimistic. And numerical experiments indicate that the Arnoldi generalization of the Lanczos method retains much of the features of the Lanczos method (cf. Saad 1980 b).

5.4. <u>The unsymmetric Lanczos method.</u>

We consider the incomplete Lanczos biorthogonalization. For $y \neq 0$, $x^H y = 1$, we define $X'_n = \{y, A^H y, \ldots, (A^H)^{n-1} y\}$. Let π'_n be the orthogonal projection on X'_n . The method consists in computing the eigenelements of the tridiagonal matrix which represents $\pi'_n A \pi_n : X_n \to X'_n$ in the computed bases of X_n and X'_n . For the eigenvector ϕ to be well approximated it is necessary that : $\pi'_n A \pi_n \phi = \lambda\phi + \varepsilon_n$, then $0 = \lambda(I-\pi'_n)\phi + (I-\pi'_n)\varepsilon_n$ and $\|(I-\pi'_n)\phi\|_2 \leq \dfrac{1}{|\lambda|} \|\varepsilon_n\|_2$: ϕ should be well approximated by a vector in $X'_n = \{y, A^H y, \ldots, (A^H)^{n-1} y\}$.
This is not possible in general if A is non-normal.

5.5. <u>A practical algorithm</u>

The incomplete Arnoldi algorithm has been suggested by Saad (1980 b) to deal with large non hermitian matrices. The method is stable when one uses reorthogonalization but it requires a lot of storage. In the same paper, Saad has devised several variants with incomplete orthogonalization which partially overcome this drawback. They are based on the observation that the upper-right corner of H_n tends to vanish when n increases. In case of close or multiple eigenvalues, a block-Arnoldi method $(r > 1)$ can be considered as well.

REFERENCES

Arnoldi, W. (1951) : "The principle of minimized iterations in the solution of the matrix eigenvalue problem" Quart. Appl. Math. 9, 17-29.

Chatelin, F. (1981) "Linear spectral approximation in Banach spaces". Academic Press, to appear.

Cheney, W. (1966) : "Introduction to approximation theory", Mac Graw Hill.

Clint, M. ; Jennings, A. (1971) : "A simultaneous iteration method for the unsymmetric eigenvalue problem".
J. Inst. Math. Appl. 8, 111-121.

Cullum, J. ; Donath. W.(1974) : "A block-Lanczos algorithm for computing the q algebraically largest eigenvalues and a corresponding eigenspace of large sparse, real symmetric matrices". IEEE Conference on Decision and Control.

Cullum, J. ; Willoughby, R. (1978) : "Lanczos and the computation of the total spectrum". Fall SIAM Meeting.

Kahan, W. ; Parlett, B. (1976) : "How far should you go with the Lanczos process" in Bunch & Rose (ed.) Sparse matrix computations, 131-144, Academic Press, N.Y.

Lorentz, G. (1966) : "Approximation of functions". Holt, Rinehart & Winston, N.Y.

Parlett, B. ; Poole, W. (1973) : "A geometric theory for the QR, LU and power iterations". SIAM J. Num. Anal. 10, 389-412.

Parlett, B. ; Scott, D. (1979) : "The Lanczos algorithm with selective orthogonalization". Math. Comp. 33, 217-238.

Parlett, B. ; Reid, J. (1980) : "Tracking the progress of the Lanczos algorithm for large symmetric eigenproblems".
Techn. Rep. CSS 83, Harwell.

Saad, Y. (1978) : "Calcul de vecteurs propres d'une grande matrice creuse par la méthode de Lanczos".
Int. Congress on Num. Meth. for Eng. Paris, december.

Saad, Y. (1979) : "Etude de la convergence du procédé d'Arnoldi pour le calcul des éléments propres de grandes matrices creuses non symétriques".
Sém. An. Num. N° 321, Labo. IMAG, Université de Grenoble.

Saad, Y. (1980 a) : "On the rate of convergence of the Lanczos and block-Lanczos methods". SIAM J. Num. Anal. 17, 687 (1980).

Saad, Y. (1980 b) : "The generalized Lanczos method for computing the eigenelements of large sparse unsymmetric matrices".
Lin. Alg. Appl. (special issue on sparse matrices) to appear.

Stewart, G. (1979) : "SRRIT - a Fortran subroutine to calculate the dominant invariant subspaces of a real matrix". ACM TOMS.

Underwood, R. (1975) : "An iterative block Lanczos method for the solution of large sparse symmetric eigenproblems".
Ph. D. Thesis, Stanford Report STAN-CS-496.

Wilkinson, J. (1965) : "The algebraic eigenvalue problem". Clarendon Press, Oxford.

Vol. 817: L. Gerritzen, M. van der Put, Schottky Groups and Mumford Curves. VIII, 317 pages. 1980.

Vol. 818: S. Montgomery, Fixed Rings of Finite Automorphism Groups of Associative Rings. VII, 126 pages. 1980.

Vol. 819: Global Theory of Dynamical Systems. Proceedings, 1979. Edited by Z. Nitecki and C. Robinson. IX, 499 pages. 1980.

Vol. 820: W. Abikoff, The Real Analytic Theory of Teichmüller Space. VII, 144 pages. 1980.

Vol. 821: Statistique non Paramétrique Asymptotique. Proceedings, 1979. Edited by J.-P. Raoult. VII, 175 pages. 1980.

Vol. 822: Séminaire Pierre Lelong–Henri Skoda, (Analyse) Années 1978/79. Proceedings. Edited by P. Lelong et H. Skoda. VIII, 356 pages, 1980.

Vol. 823: J. Král, Integral Operators in Potential Theory. III, 171 pages. 1980.

Vol. 824: D. Frank Hsu, Cyclic Neofields and Combinatorial Designs. VI, 230 pages. 1980.

Vol. 825: Ring Theory, Antwerp 1980. Proceedings. Edited by F. van Oystaeyen. VII, 209 pages. 1980.

Vol. 826: Ph. G. Ciarlet et P. Rabier, Les Equations de von Kármán. VI, 181 pages. 1980.

Vol. 827: Ordinary and Partial Differential Equations. Proceedings, 1978. Edited by W. N. Everitt. XVI, 271 pages. 1980.

Vol. 828: Probability Theory on Vector Spaces II. Proceedings, 1979. Edited by A. Weron. XIII, 324 pages. 1980.

Vol. 829: Combinatorial Mathematics VII. Proceedings, 1979. Edited by R. W. Robinson et al.. X, 256 pages. 1980.

Vol. 830: J. A. Green, Polynomial Representations of GL_n. VI, 118 pages. 1980.

Vol. 831: Representation Theory I. Proceedings, 1979. Edited by V. Dlab and P. Gabriel. XIV, 373 pages. 1980.

Vol. 832: Representation Theory II. Proceedings, 1979. Edited by V. Dlab and P. Gabriel. XIV, 673 pages. 1980.

Vol. 833: Th. Jeulin, Semi-Martingales et Grossissement d'une Filtration. IX, 142 Seiten. 1980.

Vol. 834: Model Theory of Algebra and Arithmetic. Proceedings, 1979. Edited by L. Pacholski, J. Wierzejewski, and A. J. Wilkie. VI, 410 pages. 1980.

Vol. 835: H. Zieschang, E. Vogt and H.-D. Coldewey, Surfaces and Planar Discontinuous Groups. X, 334 pages. 1980.

Vol. 836: Differential Geometrical Methods in Mathematical Physics. Proceedings, 1979. Edited by P. L. García, A. Pérez-Rendón, and J. M. Souriau. XII, 538 pages. 1980.

Vol. 837: J. Meixner, F. W. Schäfke and G. Wolf, Mathieu Functions and Spheroidal Functions and their Mathematical Foundations Further Studies. VII, 126 pages. 1980.

Vol. 838: Global Differential Geometry and Global Analysis. Proceedings 1979. Edited by D. Ferus et al. XI, 299 pages. 1981.

Vol. 839: Cabal Seminar 77 – 79. Proceedings. Edited by A. S. Kechris, D. A. Martin and Y. N. Moschovakis. V, 274 pages. 1981.

Vol. 840: D. Henry, Geometric Theory of Semilinear Parabolic Equations. IV, 348 pages. 1981.

Vol. 841: A. Haraux, Nonlinear Evolution Equations- Global Behaviour of Solutions. XII, 313 pages. 1981.

Vol. 842: Séminaire Bourbaki vol. 1979/80. Exposés 543–560. IV, 317 pages. 1981.

Vol. 843: Functional Analysis, Holomorphy, and Approximation Theory. Proceedings. Edited by S. Machado. VI, 636 pages. 1981.

Vol. 844: Groupe de Brauer. Proceedings. Edited by M. Kervaire and M. Ojanguren. VII, 274 pages. 1981.

Vol. 845: A. Tannenbaum, Invariance and System Theory: Algebraic and Geometric Aspects. X, 161 pages. 1981.

Vol. 846: Ordinary and Partial Differential Equations, Proceedings. Edited by W. N. Everitt and B. D. Sleeman. XIV, 384 pages. 1981.

Vol. 847: U. Koschorke, Vector Fields and Other Vector Bundle Morphisms – A Singularity Approach. IV, 304 pages. 1981.

Vol. 848: Algebra, Carbondale 1980. Proceedings. Ed. by R. K. Amayo. VI, 298 pages. 1981.

Vol. 849: P. Major, Multiple Wiener-Itô Integrals. VII, 127 pages. 1981.

Vol. 850: Séminaire de Probabilités XV. 1979/80. Avec table générale des exposés de 1966/67 à 1978/79. Edited by J. Azéma and M. Yor. IV, 704 pages. 1981.

Vol. 851: Stochastic Integrals. Proceedings, 1980. Edited by D. Williams. IX, 540 pages. 1981.

Vol. 852: L. Schwartz, Geometry and Probability in Banach Spaces. X, 101 pages. 1981.

Vol. 853: N. Boboc, G. Bucur, A. Cornea, Order and Convexity in Potential Theory: H-Cones. IV, 286 pages. 1981.

Vol. 854: Algebraic K-Theory. Evanston 1980. Proceedings. Edited by E. M. Friedlander and M. R. Stein. V, 517 pages. 1981.

Vol. 855: Semigroups. Proceedings 1978. Edited by H. Jürgensen, M. Petrich and H. J. Weinert. V, 221 pages. 1981.

Vol. 856: R. Lascar, Propagation des Singularités des Solutions d'Equations Pseudo-Différentielles à Caractéristiques de Multiplicités Variables. VIII, 237 pages. 1981.

Vol. 857: M. Miyanishi. Non-complete Algebraic Surfaces. XVIII, 244 pages. 1981.

Vol. 858: E. A. Coddington, H. S. V. de Snoo: Regular Boundary Value Problems Associated with Pairs of Ordinary Differential Expressions. V, 225 pages. 1981.

Vol. 859: Logic Year 1979–80. Proceedings. Edited by M. Lerman, J. Schmerl and R. Soare. VIII, 326 pages. 1981.

Vol. 860: Probability in Banach Spaces III. Proceedings, 1980. Edited by A. Beck. VI, 329 pages. 1981.

Vol. 861: Analytical Methods in Probability Theory. Proceedings 1980. Edited by D. Dugué, E. Lukacs, V. K. Rohatgi. X, 183 pages. 1981.

Vol. 862: Algebraic Geometry. Proceedings 1980. Edited by A. Libgober and P. Wagreich. V, 281 pages. 1981.

Vol. 863: Processus Aléatoires à Deux Indices. Proceedings, 1980. Edited by H. Korezlioglu, G. Mazziotto and J. Szpirglas. V, 274 pages. 1981.

Vol. 864: Complex Analysis and Spectral Theory. Proceedings, 1979/80. Edited by V. P. Havin and N. K. Nikol'skii, VI, 480 pages. 1981.

Vol. 865: R. W. Bruggeman, Fourier Coefficients of Automorphic Forms. III, 201 pages. 1981.

Vol. 866: J.-M. Bismut, Mécanique Aléatoire. XVI, 563 pages. 1981.

Vol. 867: Séminaire d'Algèbre Paul Dubreil et Marie-Paule Malliavin. Proceedings, 1980. Edited by M.-P. Malliavin. V, 476 pages. 1981.

Vol. 868: Surfaces Algébriques. Proceedings 1976–78. Edited by J. Giraud, L. Illusie et M. Raynaud. V, 314 pages. 1981.

Vol. 869: A. V. Zelevinsky, Representations of Finite Classical Groups. IV, 184 pages. 1981.

Vol. 870: Shape Theory and Geometric Topology. Proceedings, 1981. Edited by S. Mardešić and J. Segal. V, 265 pages. 1981.

Vol. 871: Continuous Lattices. Proceedings, 1979. Edited by B. Banaschewski and R.-E. Hoffmann. X, 413 pages. 1981.

Vol. 872: Set Theory and Model Theory. Proceedings, 1979. Edited by R. B. Jensen and A. Prestel. V, 174 pages. 1981.

Vol. 873: Constructive Mathematics, Proceedings, 1980. Edited by F. Richman. VII, 347 pages. 1981.

Vol. 874: Abelian Group Theory. Proceedings, 1981. Edited by R. Göbel and E. Walker. XXI, 447 pages. 1981.

Vol. 875: H. Zieschang, Finite Groups of Mapping Classes of Surfaces. VIII, 340 pages. 1981.

Vol. 876: J. P. Bickel, N. El Karoui and M. Yor. Ecole d'Eté de Probabilités de Saint-Flour IX – 1979. Edited by P. L. Hennequin. XI, 280 pages. 1981.

Vol. 877: J. Erven, B.-J. Falkowski, Low Order Cohomology and Applications. VI, 126 pages. 1981.

Vol. 878: Numerical Solution of Nonlinear Equations. Proceedings, 1980. Edited by E. L. Allgower, K. Glashoff, and H.-O. Peitgen. XIV, 440 pages. 1981.

Vol. 879: V. V. Sazonov, Normal Approximation – Some Recent Advances. VII, 105 pages. 1981.

Vol. 880: Non Commutative Harmonic Analysis and Lie Groups. Proceedings, 1980. Edited by J. Carmona and M. Vergne. IV, 553 pages. 1981.

Vol. 881: R. Lutz, M. Goze, Nonstandard Analysis. XIV, 261 pages. 1981.

Vol. 882: Integral Representations and Applications. Proceedings, 1980. Edited by K. Roggenkamp. XII, 479 pages. 1981.

Vol. 883: Cylindric Set Algebras. By L. Henkin, J. D. Monk, A. Tarski, H. Andréka, and I. Németi. VII, 323 pages. 1981.

Vol. 884: Combinatorial Mathematics VIII. Proceedings, 1980. Edited by K. L. McAvaney. XIII, 359 pages. 1981.

Vol. 885: Combinatorics and Graph Theory. Edited by S. B. Rao. Proceedings, 1980. VII, 500 pages. 1981.

Vol. 886: Fixed Point Theory. Proceedings, 1980. Edited by E. Fadell and G. Fournier. XII, 511 pages. 1981.

Vol. 887: F. van Oystaeyen, A. Verschoren, Non-commutative Algebraic Geometry, VI, 404 pages. 1981.

Vol. 888: Padé Approximation and its Applications. Proceedings, 1980. Edited by M. G. de Bruin and H. van Rossum. VI, 383 pages. 1981.

Vol. 889: J. Bourgain, New Classes of $\mathcal{L}^p$-Spaces. V, 143 pages. 1981.

Vol. 890: Model Theory and Arithmetic. Proceedings, 1979/80. Edited by C. Berline, K. McAloon, and J.-P. Ressayre. VI, 306 pages. 1981.

Vol. 891: Logic Symposia, Hakone, 1979, 1980. Proceedings, 1979, 1980. Edited by G. H. Müller, G. Takeuti, and T. Tugué. XI, 394 pages. 1981.

Vol. 892: H. Cajar, Billingsley Dimension in Probability Spaces. III, 106 pages. 1981.

Vol. 893: Geometries and Groups. Proceedings. Edited by M. Aigner and D. Jungnickel. X, 250 pages. 1981.

Vol. 894: Geometry Symposium. Utrecht 1980, Proceedings. Edited by E. Looijenga, D. Siersma, and F. Takens. V, 153 pages. 1981.

Vol. 895: J.A. Hillman, Alexander Ideals of Links. V, 178 pages. 1981.

Vol. 896: B. Angéniol, Familles de Cycles Algébriques – Schéma de Chow. VI, 140 pages. 1981.

Vol. 897: W. Buchholz, S. Feferman, W. Pohlers, W. Sieg, Iterated Inductive Definitions and Subsystems of Analysis: Recent Proof-Theoretical Studies. V, 383 pages. 1981.

Vol. 898: Dynamical Systems and Turbulence, Warwick, 1980. Proceedings. Edited by D. Rand and L.-S. Young. VI, 390 pages. 1981.

Vol. 899: Analytic Number Theory. Proceedings, 1980. Edited by M.I. Knopp. X, 478 pages. 1981.

Vol. 900: P. Deligne, J. S. Milne, A. Ogus, and K.-Y. Shih, Hodge Cycles, Motives, and Shimura Varieties. V, 414 pages. 1982.

Vol. 901: Séminaire Bourbaki vol. 1980/81 Exposés 561–578. III, 299 pages. 1981.

Vol. 902: F. Dumortier, P.R. Rodrigues, and R. Roussarie, Germs of Diffeomorphisms in the Plane. IV, 197 pages. 1981.

Vol. 903: Representations of Algebras. Proceedings, 1980. Edited by M. Auslander and E. Lluis. XV, 371 pages. 1981.

Vol. 904: K. Donner, Extension of Positive Operators and Korovkin Theorems. XII, 182 pages. 1982.

Vol. 905: Differential Geometric Methods in Mathematical Physics. Proceedings, 1980. Edited by H.-D. Doebner, S.J. Andersson, and H.R. Petry. VI, 309 pages. 1982.

Vol. 906: Séminaire de Théorie du Potentiel, Paris, No. 6. Proceedings. Edité par F. Hirsch et G. Mokobodzki. IV, 328 pages. 1982.

Vol. 907: P. Schenzel, Dualisierende Komplexe in der lokalen Algebra und Buchsbaum-Ringe. VII, 161 Seiten. 1982.

Vol. 908: Harmonic Analysis. Proceedings, 1981. Edited by F. Ricci and G. Weiss. V, 325 pages. 1982.

Vol. 909: Numerical Analysis. Proceedings, 1981. Edited by J.P. Hennart. VII, 247 pages. 1982.

Vol. 910: S.S. Abhyankar, Weighted Expansions for Canonical Desingularization. VII, 236 pages. 1982.

Vol. 911: O.G. Jørsboe, L. Mejlbro, The Carleson-Hunt Theorem on Fourier Series. IV, 123 pages. 1982.

Vol. 912: Numerical Analysis. Proceedings, 1981. Edited by G. A. Watson. XIII, 245 pages. 1982.

Vol. 913: O. Tammi, Extremum Problems for Bounded Univalent Functions II. VI, 168 pages. 1982.

Vol. 914: M. L. Warshauer, The Witt Group of Degree k Maps and Asymmetric Inner Product Spaces. IV, 269 pages. 1982.

Vol. 915: Categorical Aspects of Topology and Analysis. Proceedings, 1981. Edited by B. Banaschewski. XI, 385 pages. 1982.

Vol. 916: K.-U. Grusa, Zweidimensionale, interpolierende Lg-Splines und ihre Anwendungen. VIII, 238 Seiten. 1982.

Vol. 917: Brauer Groups in Ring Theory and Algebraic Geometry. Proceedings, 1981. Edited by F. van Oystaeyen and A. Verschoren. VIII, 300 pages. 1982.

Vol. 918: Z. Semadeni, Schauder Bases in Banach Spaces of Continuous Functions. V, 136 pages. 1982.

Vol. 919: Séminaire Pierre Lelong – Henri Skoda (Analyse) Années 1980/81 et Colloque de Wimereux, Mai 1981. Proceedings. Edité par P. Lelong et H. Skoda. VII, 383 pages. 1982.

Vol. 920: Séminaire de Probabilités XVI, 1980/81. Proceedings. Edité par J. Azéma et M. Yor. V, 622 pages. 1982.

Vol. 921: Séminaire de Probabilités XVI, 1980/81. Supplément: Géométrie Différentielle Stochastique. Proceedings. Edité par J. Azéma et M. Yor. III, 285 pages. 1982.

Vol. 922: B. Dacorogna, Weak Continuity and Weak Lower Semicontinuity of Non-Linear Functionals. V, 120 pages. 1982.

Vol. 923: Functional Analysis in Markov Processes. Proceedings, 1981. Edited by M. Fukushima. V, 307 pages. 1982.

Vol. 924: Séminaire d'Algèbre Paul Dubreil et Marie-Paule Malliavin. Proceedings, 1981. Edité par M.-P. Malliavin. V, 461 pages. 1982.

Vol. 925: The Riemann Problem, Complete Integrability and Arithmetic Applications. Proceedings, 1979-1980. Edited by D. Chudnovsky and G. Chudnovsky. VI, 373 pages. 1982.

Vol. 926: Geometric Techniques in Gauge Theories. Proceedings, 1981. Edited by R. Martini and E.M.de Jager. IX, 219 pages. 1982.